Nanocrystals

Alejandro J. Paredes
Editor

Nanocrystals

A Universal Tool for the Formulation of Poorly Soluble Drugs

Springer

Editor
Alejandro J. Paredes
School of Pharmacy
Queen's University Belfast
Belfast, UK

ISBN 978-3-032-10237-9 ISBN 978-3-032-10238-6 (eBook)
https://doi.org/10.1007/978-3-032-10238-6

This Springer imprint is published by the registered company Springer Nature Switzerland AG
The registered company address is: Gewerbestrasse 11, 6330 Cham, Switzerland

Preface

Nanocrystals have emerged as one of the most versatile and clinically relevant platforms for the delivery of poorly soluble drugs. In recent decades, continuous advances in materials science, formulation strategies, and manufacturing technologies have transformed nanocrystals into an established drug delivery technology with multiple products on the market. Their unique ability to improve drug dissolution, enhance bioavailability, and enable controlled and targeted delivery has positioned nanocrystals as a key formulation approach in modern pharmaceutical sciences.

This book brings together key aspects of nanocrystal research, development, and application in a comprehensive and structured manner. Starting from the fundamentals of nanocrystal technology, it guides the reader through the state of the art in manufacturing approaches, formulation development, and therapeutic applications across multiple routes of administration. Each chapter addresses both opportunities and challenges, providing insights into oral, transdermal, parenteral, mucosal, ocular, pulmonary, and central nervous system delivery, as well as emerging fields such as microneedle-mediated administration and cosmeceuticals. Importantly, several chapters also address the role of nanocrystals as highly loaded nanovectors, discussing their interactions with biological surfaces, an area of research still in its infancy.

The goal of this book is to serve as both an introduction for new researchers and a detailed reference for experienced scientists working in drug delivery, formulation development, and pharmaceutical technology. By covering the breadth of applications, the book highlights the adaptability of nanocrystals and their growing role in addressing complex therapeutic needs.

I am deeply grateful to the contributors for sharing their expertise and for crafting chapters that reflect both the scientific rigor and the translational impact of nanocrystal research. Their efforts ensure that this book will serve as a valuable resource for academics, students, and professionals in the pharmaceutical sciences.

It is my hope that this volume will inspire continued innovation in the field and foster further integration of nanocrystal-based formulations into clinical practice, ultimately benefiting patients worldwide.

Belfast, UK Alejandro J. Paredes

Contents

About the Editor

Alejandro J. Paredes is a Pharmacist (2011) and holds a PhD (2016), from the National University of Córdoba, Argentina, and a Postgraduate Certificate in Higher Education and Teaching from Queen's University Belfast (2022). He is a Senior Lecturer in the School of Pharmacy at Queen's, where he leads a research group focused on controlled and targeted drug delivery using nanocrystals, and teaches across the BSc and MPharm programs. He is a named inventor on two patents, has authored numerous peer-reviewed articles, and is the editor of two textbooks. He has secured prestigious research funding from the EPSRC, the Academy of Medical Sciences, the Royal Society, the Northern Ireland Department for the Economy, and Applied Microbiology International. In 2024, he received the Emerging Scientist Award from the Academy of Pharmaceutical Sciences (UK) and a Springboard Award from the Academy of Medical Sciences (UK). He is a Fellow of the Higher Education Academy (UK) and of CICOPS (Italy), a Trustee of the British Society for Nanomedicine, and an active member of several scientific societies, contributing in various capacities. He also serves on the editorial boards of *Materials Today Bio*, *AAPS PharmSciTech*, and *Drug Delivery and Translational Research.*

Nanocrystals: Fundamentals of the Technology and its Application in Drug Delivery

Lucia Lopez-Vidal, Elise Catlin, and Alejandro J. Paredes

Abstract Nanocrystals (NCs) have emerged as an effective strategy for enhancing the bioavailability of poorly water-soluble drugs. By reducing particle size, NCs increase the surface area of the drug, thereby improving its dissolution rate and absorption. This approach addresses key challenges associated with the poor aqueous solubility of many therapeutic agents, leading to improved drug delivery and therapeutic efficacy. NCs offer several advantages, including high drug loading capacity, enhanced physical stability, and the potential for sustained or targeted release, making them a valuable tool in pharmaceutical formulations. This chapter discusses key strategies for stabilizing NCs, including the use of stabilizers that prevent aggregation and improve physical stability. It details commonly employed stabilizing agents, such as ionic surfactants and polymers, along with their mechanisms of action. Additionally, characterization techniques such as dynamic light scattering (DLS), laser diffraction (LD), and zeta potential analysis are discussed in terms of their role in assessing particle size, size distribution, and surface charge of NCs. Finally, the chapter highlights how these methods support the optimization of drug formulations, enhancing their performance in terms of dissolution, bioavailability, and overall therapeutic effectiveness.

Keywords Nanocrystals · Poorly water-soluble drugs · Drug delivery · Stabilizing agents

1 Introduction

Since the 1990s, advances in molecular pharmacology, polymer materials, drug delivery systems, and other areas, alongside emerging new technologies, have significantly transformed drug dosage forms and pharmaceutical preparations, paving the way in the era of Drug Delivery Systems (DDS) [1]. A key focus of these

L. Lopez-Vidal · E. Catlin · A. J. Paredes (✉)
School of Pharmacy, Queen's University Belfast, Belfast, UK
e-mail: l.lopezvidal@qub.ac.uk; ecatlin01@qub.ac.uk; a.paredes@qub.ac.uk

A. J. Paredes (ed.), *Nanocrystals*, https://doi.org/10.1007/978-3-032-10238-6_1

advances has been to address the challenges presented by poorly water-soluble drugs, a prevalent issue affecting both approved molecules and emerging active candidates [2, 3]. Despite the potent receptor-binding activity that these molecules may exhibit, their efficacy is often constrained by pharmacokinetic barriers. Specifically, if these drugs cannot dissolve in sufficient quantities or remain dissolved for an adequate duration within the gastrointestinal tract, they will be unable to permeate biological membranes required for therapeutic action. Consequently, numerous drugs experience poor dissolution and absorption, therefore to low active concentrations within target tissues.

The low solubility of an active pharmaceutical ingredient (API) is primarily linked to two factors, the hydrophobic nature of the molecule, which limits its ability to form hydrogen bonds with water, and a high crystalline lattice energy that hinders the separation of molecules needed for solution formation [4]. In recent years, the number of poorly soluble APIs has been on a steady rise, as it is now estimated that 70–90% of drugs in the development pipeline and 40% of currently marketed drugs exhibit poor water solubility and bioavailability [5]. Previously considered high-risk for development, such compounds have been given a new lease of life with an increased focus on solubility research. The industry's perspective has shifted from avoidance to acceptance. A costly process, drug development demands substantial financial investment, approaching $500 million, with prolonged timeframes of 10–20 years. As a result, addressing solubility limitations has become a prominent challenge and key interest for pharmaceutical and drug development sectors. Several strategies are employed to enhance the solubility of active compounds. At the molecular level, these include the use of co-solvents [6], salt formation, prodrug synthesis [7], and cyclodextrin complexation [8], all of which have shown positive effects on drug dissolution. At a particle level, various approaches can be applied, including co-crystallization [9], the use of metastable polymorphs [10], solid dispersions [11], co-amorphous systems, and micronization [12], among others. However, some of these strategies face limitations, mostly related to the intrinsic physicochemical properties of the API molecule. For example, a drug must possess a degree of lipophilicity in order to be soluble in the lipid phase, required for incorporation into lipid-based nanocarriers, while a specific molecular size is prerequisite for complexation within a cyclodextrin ring. Furthermore, nanocarriers often have limited drug loading capacities (5–30% w/w), limited long-term stability, and may require the addition of organic solvents [13]. In this context, the formulation of nanocrystals (NCs) has emerged as an attractive and flexible approach for the delivery of poorly water-soluble active compounds.

2 Nanocrystals

NCs were first developed in the early 1990s using wet milling techniques adapted from the pigment and photographic industries [14]. NCs can be defined as crystalline particles composed of pure drug with a size below 1μm. Unlike conventional

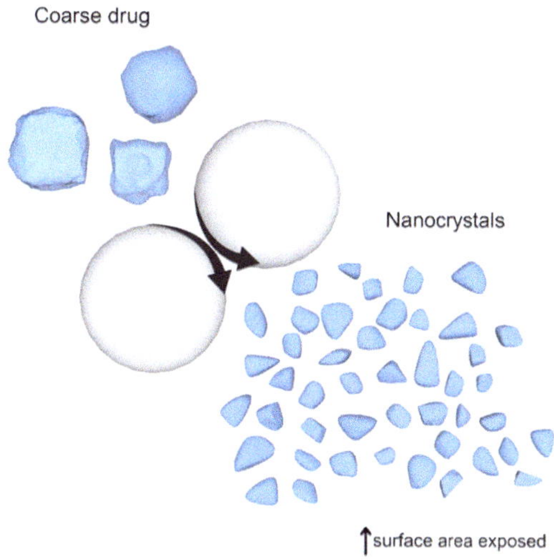

Fig. 1 Increase in surface area due to the reduction in particle size

polymer and lipid nanoparticles, which require carrier materials, NCs are surrounded by a thin layer of stabilizer, such as surfactants or polymeric stabilizers, to prevent agglomeration. When drug particles are in suspension, these systems are commonly referred to as nanosuspensions (NS), while the term NCs denotes solid, redispersible powders obtained through solvent removal techniques such as spray drying or lyophilization. Reducing particle size to the nanometer range increases the specific surface area exposed (Fig. 1), which grants NCs several distinctive properties.

– Increased Saturation Solubility

Saturation solubility (Ss) refers to the amount of drug that can dissolve until the saturation point in a medium at a given temperature. The Ss of solids in liquids is a thermodynamic property that can be predicted from the melting point or the water–oil partition coefficient of the compound [15]. However, when the particle size of a solid is reduced to the nanoscale, solubility ceases to be solely a thermodynamic property and instead becomes dependent on the size and shape of the solid, being reported as "apparent" saturation solubility [16]. Ostwald first presented a relationship for this size dependency in spherical particles, later corrected by Freundlich, now known as the Ostwald–Freundlich equation [17]. Equation 1 shows the modified Kelvin and Ostwald–Freundlich equations, where it can be seen that the increase in the curvature of nanoparticles results in increased dissolution pressure and, hence, drug solubility [18].

$$\ln \frac{S}{S0} = \frac{2\gamma Vm}{rRT} = \frac{2\gamma M}{r\rho RT} \tag{1}$$

where S is the drug solubility at temperature T, S_0 is the solubility of infinitely big particle material, R is the ideal gas constant (R ≈ 8314 J/mol.K), V_m is the molar volume, T is the temperature in kelvin, r is the particle radius, γ is the surface free energy, and M and ρ are the molecular weight and density of the compound, respectively.

– Increased Dissolution Rate

As demonstrated by the Noyes–Whitney equation (Eq. 2) [19], the dissolution rate of a drug is influenced, among other factors, by the exposed surface area to the solvent and the saturation solubility discussed above—two factors that are enhanced in NCs.

$$\frac{dC}{dt} = \frac{AD\,(Ss - C)}{h} \tag{2}$$

where dC/dt is the dissolution rate, A is the area available for dissolution, D is the compound's diffusion coefficient, C is the drug concentration in the medium, and h is the thickness of the diffusion layer.

– Increased Bioadhesion

The aforementioned increase in the exposed surface area of nano systems also enhances the contact area, promoting adhesion to mucus or other biological/cellular barriers, depending on the surface charge. This mucoadhesive effect has been shown to contribute to an increase in drug absorption [20].

In summary, an increased specific surface area of NCs leads to enhanced dissolution rate, saturation solubility, and bioadhesion. Collectively, these factors translate to improvements in the bioavailability of poorly soluble drugs. These systems also offer a number of advantages, such as ease of production and sterilization, high scalability, and versatility in administration via different routes [21].

2.1 *Advances in Nanocrystal Production and Application*

With respect to NCs in the drug delivery domain, research was initiated with the patent filed by Liversidge et al. (1992), which described dispersible particles of a crystalline pharmacological substance with a surface modifier adsorbed onto its surface, in an amount sufficient to maintain an effective average particle size of approximately 400 nm [14]. Subsequently, Müller et al. (1999) introduced and patented a drug delivery system, which is defined as comprising particles of at least one pure active compound, which may be defined as insoluble, poorly soluble, or moderately soluble in water, aqueous media, and/or organic solvents, and which possesses the following characteristics: is solid at room temperature, has an average diameter, determined by photon correlation spectroscopy, in the range of 10 nm to

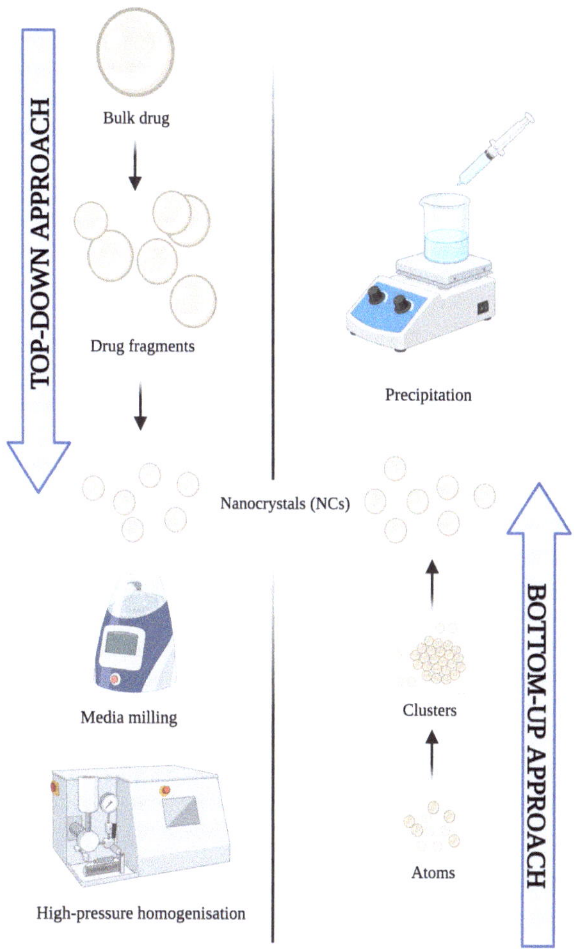

Fig. 2 A schematic diagram visually depicting the formation of NCs by the two different approaches with respective examples of each method in each class. (Diagram produced using BioRender.com 2023)

1000 nm and the proportion of particles larger than 5μm in the total population is less than 0.1%, according to the numerical distribution determined using a Coulter counter. The authors describe that the reduction in particle size was carried out without the need to first convert them to a molten state, employing cavitation, shear, and impact forces through the application of a high amount of energy. To achieve this, they prepared a drug suspension in an aqueous solution with the addition of a surfactant, which was then subjected to bead milling for several hours or days [22]. Over time, the technology of NCs production has evolved considerably, extending beyond the laboratory scale to include applications in industrial production. Currently, the techniques for obtaining NCs can be divided into two main categories: top-down and bottom-up techniques, as highlighted in Fig. 2. Bottom-up techniques involve dissolving the API in a medium where it is soluble (typically organic solvents) and subsequently precipitating it under controlled conditions to achieve smaller particle sizes with high monodispersity. However, scaling up these techniques to an industrial level possesses significant challenges as a result of the specific conditions essential to ensure the formation, controlled size, and stability of the NCs. Conversely, top-down techniques are high-energy processes in which micron-sized drug crystals are reduced to the nanoscale through mechanical attrition or high-pressure methods. Generally, the top-down approach is simpler to perform and more scalable, although the particle size obtained is usually larger compared to the bottom-up approach, and the particle size distribution, as determined by the polydispersity index, is also often broader [23]. Crucially, this topic will be addressed further in greater depth in Chap. 2.

NCs were originally developed as a strategy to enhance the oral absorption of poorly water-soluble drugs by increasing the saturation solubility and dissolution rate of APIs. Recently, a study demonstrated superior in vivo performance of NCs compared to solid dispersion formulations, another widely employed technique in pharmaceutical technology to optimize the unfavorable properties of APIs with low aqueous solubility [24]. However, the high drug load in NCs, which is significantly greater than that of other nanoparticles, combined with their ability to dissolve rapidly within the stomach's fluid volume (i.e., following oral administration), has shown promising potential for NCs beyond the scope of oral drug delivery. Importantly, examples in the literature show that NCs can circulate as intact particles within the body for varying periods, depending on the route of administration and physicochemical properties of the drug. This permits exploitation of NCs as targeted drug delivery systems through surface modification with the addition of functional ligands [25]. This innovative approach paves the way for new applications of these systems, as it enables more effective targeting of specific organs and, in the future, the potential development of drug NCs with immunotherapy agents to overcome multidrug resistance in cancer therapy [26]. Currently, a substantial number of scientific articles and numerous invention patents related to these technologies have been published, demonstrating their utility. Figure 3 illustrates the increasing

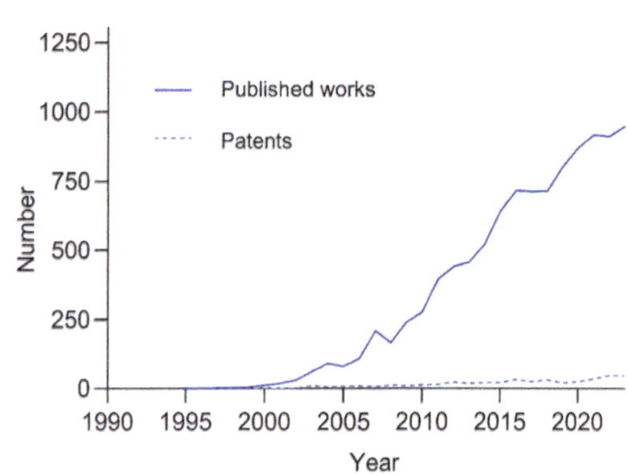

Fig. 3 Number of published articles and patents related to drug NCs found in the Scopus and Espacenet databases, respectively

trend in the number of publications (Search equation: "'Drug' AND ('nanocrystal' OR 'nanocrystals')" on Scopus) and patents on the subject (Search conducted in "title, abstract OR Claims" for terms "drug" OR "drugs" AND "nanocrystal" OR "nanocrystals" using the Espacenet search engine (Espacenet—Patent Search n.d.)). As reflected in the increasing number of publications and patents, interest in this technology has been steadily growing since the 2000s.

The rising interest of researchers in this technology, combined with the clear advantages offered by NCs, and the pressing need within pharmaceutical sciences to address poor water solubility challenges associated with the large number of active compounds, has led to the successful commercialization of various pharmaceutical NCs. The first drug containing NCs to reach the market was Gris-Peg®, approved in 1982 and marketed under the premise of "ultramicrosized" particles with crystalline properties, developed using a bottom-up approach. This drug is currently available, produced by Novartis, in 125 mg and 250 mg tablets, indicated for the treatment of ringworm infections. Following this, Cesamet® was commercialized, also developed with a bottom-up approach by the company Lilly, containing nabilone as an antiemetic for oral administration. In 2000, Rapamune® tablets produced by Wyeth, formulated with sirolimus NCs, were launched as immunosuppressants, offering a 21% improvement in bioavailability over the oral solution. The introduction of Emend® in 2003 by Merck & Co marked the arrival of an aprepitant NCs formulation, which demonstrated enhanced absorption, reduced food–drug interactions, and superior bioavailability compared to its micronized counterpart. Similarly, Tricor® (2004) by Abbot Laboratories and Triglide® (2005) by First Horizon Pharmaceutical presented substantial bioavailability improvements over coarse and micronized fenofibrate suspensions, with minimal dependence on food intake. The most recent products released to the market are Cabenuva™ and Invega Hafyera®. Cabenuva™ is a long-acting injectable formulation containing rilpivirine and cabotegravir, leveraging nanocrystal technology for the treatment of HIV-1 infection in individuals aged 12 years and older who weigh at least 35 kg (77 lbs). This injectable therapy helps maintain undetectable viral loads for up to 2 months, eliminating the need for daily pill intake. Invega Hafyera®, on the other hand, is a long-acting injectable antipsychotic that employs nanocrystal technology for the extended-release delivery of paliperidone palmitate.

At present, nearly 35 products containing NCs have been brought to market. Of those for which production methods are known, more than 90% are produced using top-down techniques, with media milling being the predominant method. Regarding routes of administration, 47% of the products are intended for oral delivery, making it the most prevalent route. In Table 1, the drugs containing NCs that have reached the market can be observed along with the producing company, the intended route of administration, dosage form, therapeutic indication, approval date, and current status.

Table 1 Marketed products containing NCs, along with the producing company, the intended route of administration, dosage form, therapeutic indication, approval date, and current status

Commercial name	Active	Company	Technology applied/approach	Route of administration/ dosage form	Indication of treatment	Date of approval	Current status (2024)
Gris-peg®	Griseofulvin	Novartis	Co-precipitation, BU	Oral/tablet	Ringworm infection	1982	On the market as 125 and 250 mg tablets with generic versions approved
Cesamet®	Nabilone	Lilly	Co-precipitation, BU	Oral/capsule	Antiemetic	1985	No longer on the market was discontinued
Neprelan®	Naproxen sodium	Wyeth	NanoCrystal® ElanNanosystems (MM), TD	Oral/tablet	Anti-inflammatory	1996	On the market as 375, 500, and 750 mg tablets with generic equivalent versions
Verelan® PM	Verapamil hydrochloride	Schwarz pharma	NanoCrystal® ElanNanosystems TD	Oral/capsules	Hypertension, angina, heart rhythm disorders	1998	On the market as 100, 200, and 300 mg tablets with generic versions available
Rapamune®	Sirolimus	Wyeth	NanoCrystal® ElanNanosystems (MM), TD	Oral/tablet	Immunosuppressant for renal transplantation	1999	No longer produced after 2023, due to sufficient production of generic forms
Megace ES®	Megestrol acetate	Par pharmaceutical Inc	NanoCrystal® ElanNanosystems (MM), TD	Oral/ Nanosuspension	Treatment of anorexia, cachexia and weight loss in patients diagnosed with acquired immunodeficiency syndrome	2001	Available as a 625 mg/ 5 ml concentrated formula with a generic version available

Ritalin LA®	Methyl-phenidate hydrochloride	Novartis	NanoCrystal® ElanNanosystems (MM), TD	Oral/tablet	Muscle relaxant	2002	On the market as 10, 20, 30, and 40 mg tablets. With 60 mg tablets now discontinued
Herbesser®	Diltiazem hydrochloride	Mitsubishi Tanabe pharmaceuticals	MM, TD	Oral/capsule	Angina pectoris and hypertension	2002	On the market as 90, 100, and 200 mg capsules
Zanaflex™	Tizanidine hydrochloride	Acorda	NanoCrystal® ElanNanosystems (MM), TD	Oral/capsule and tablets	Muscle relaxant	2002	On the market as 2, 4, and 6 mg capsules and 4 mg tablets with generic forms available
Avinza®	Morphine sulphate	King Pharmaceuticals	NanoCrystal® ElanNanosystems (MM), TD	Oral/tablet	Treatment of moderate to severe pain	2002	Discontinued in the US with generic versions approved
Emend®	Aprepitant	Merck & Co.	NanoCrystal® ElanNanosystems (MM), TD	Oral/capsule and suspension	Antiemetic	2003	On the market as 80 and 125 mg capsules and 125 mg powder for suspension. Generic versions are on the market
Vitoss®	Calcium phosphate	Stryker	Not available	Injection/foam packs and strips	Bone substitute permits cell adhesion and growth	2003	On the market sold block and morsel forms
OsSatura®	Hydroxyapatite	IsoTis Orthobiologics	Not available	Injection/paste	Bone substitute permits cell adhesion and growth	2003	Available as 5, 10, 15, and 30 cc
Tricor®	Fenofibrate	Abbot Laboratories	NanoCrystal® ElanNanosystems (MM), TD	Oral/tablet	Hypercholesterolemia	2004	All strengths (48, 54, 145, and 160 mg) of tablets have been discontinued

(continued)

Table 1 (continued)

Commercial name	Active	Company	Technology applied/approach	Route of administration/ dosage form	Indication of treatment	Date of approval	Current status (2024)
Ostim®	Hydroxyapatite	Heraseus Kulzer	Not available	Injection/paste	Bone substitute permits cell adhesion and growth	2004	Available as 0.2-, 1-, and 2-ml syringes
NanOss®	Hydroxyapatite	RTI surgical	Not available	Injection/paste	Bone substitute permits cell adhesion and growth	2005	On the market as 1, 2, 5, 10, and 20 cc
Ilevro	Nepafenac	Novartis	MM, TD	Ophthalmic/ nanosuspension	NSAID pain and inflammation after cataract	2005	On the market as a 0.3% formulation
Triglide®	Fenofibrate	First horizon pharmaceutical	IDD-P® Skyepharma production, HPH, TD	Oral/tablet	Hypercholesterolemia	2005	Available as a 160 mg tablet with no approved generic versions
Focalin XR®	Dexmethylphenidate hydrochloride	Novartis	NanoCrystal® ElanNanosystems (MM), TD	Oral/capsule	Attention deficit and hyperactivity disorder	2005	On the market as 5, 10, 15, 20, 25, 30, 35, and 40 mg capsules
EquivaBone®	Hydroxyapatite	Zimmer Biomet	Not available	Injection/paste	Bone substitute	2009	On the market as 2.5, 5, and 10 cc
Invega Sustenna®	Paliperidone palmitate	Janssen Pharmaceuticals	HPH, TD	Parental/ nanosuspension	Schizophrenia	2009	On the market as 39, 78, 117, 156, and 234 mg formulations given as a monthly injection
Ilevro®	Nepafenac	Alcon/Novartis	HPH, TD	Ophthalmic/eye drops (0.3% suspension)	Pain and inflammation associated with cataract surgery	2012	On the market; no generic version currently available

Ryanodex®	Dantrolene sodium	Eagle pharmaceuticals	Not available	Intravenous/ freeze-dried powder for injection	Malignant hypothermia	2014	Available as 250 mg vials with no generic versions available
Invega Trinza®	Paliperidone palimate	Janssen Pharmaceuticals	HPH, TD	Parental/ nanosuspension	Schizophrenia	2015	On the market as 273, 410, 546, and 819 mg formulations administered every 3 months as an injection
Aristada®	Aripiprazole lauroxi	Alkermes	HPH, TD	Long-acting injectable	Schizophrenia	2015	On the market in multiple strengths (441 mg to 1064 mg) with ongoing use
Acticoat™	Silver	Smith & Nephew	SILCRYST™ patented technology, Nucryst pharmaceuticals, TD	Absorbent dressing	Inflammation and infection	2016	On the market as 5 × 5, 10 × 10, 10 × 20 and 20 × 40 cm dressings
Zorvolex®	Diclofenac	Iroko pharmaceuticals	SoluMatrix fine particle technology™	Oral/capsule	NSAID management of mild to moderate acute pain	2016	Discontinued in the US
Yonsa®	Abiraterone acetate	Sun pharmaceutical	Dry milling, TD	Oral tablets	Metastatic castration-resistant prostate cancer (mCRPC)	2018	On the market in 125 mg tablets
Inveltys®	Loteprednol etabonate	Kala pharmaceuticals	MM, TD	Oftalmic/eye drop suspension	Post-operative ocular inflammation and pain	2018	On the market; twice-daily dosing, suspension 1% (10 mg/mL)
Anjeso™	Meloxicam	Baudax bio Inc	NanoCrystal® Alkermes pharma Ireland limited, (MM), TD	Intravenous/ nanosuspension	NSAID manage moderate to severe pain	2020	Discontinued within the US

(continued)

Table 1 (continued)

Commercial name	Active	Company	Technology applied/approach	Route of administration/ dosage form	Indication of treatment	Date of approval	Current status (2024)
Eysuvis®	Loteprednol etabonate	Kala pharmaceuticals	AMPPLIFY®, TD	Ophthalmic/ suspension	Short-term treatment dry eye	2020	On the market in 0.25%
Cabenuva™	Cabotegravir and rilpivirine	Janssen Pharmaceuticals	Not available	Intramuscular/ nanosuspension	HIV-1 infection	2021	On the market administered either monthly (400 mg cabotegravir/ 600 mg rilpivirine dose) or every 2 months (400 mg cabotegravir/600 mg rilpivirine)
Invega Hafyera®	Paliperidone palmitate	Janssen Pharmaceuticals	HPH, TD	Parental/ nanosuspension	Schizophrenia	2021	On the market as 1092 and 1560 mg injections administered bi-yearly
Ztalmy®	Ganoxolone	Marinus pharmaceuticals	Not available	Oral suspension	Seizures associated with cyclin-dependent kinase-like 5 (CDKL5) deficiency disorder (CDD)	2022	Approved and marketed in the US and EU

Table adapted from Catlin et al. [54]
BU bottom up, *HPH* high pressure homogenization, *MM* media milling, *TD* top down

2.2 Nanocrystals Stabilization

The reduction in particle size that occurs during the production of NCs results in an increase in Gibbs free energy, generating thermodynamic instability within the system. Particles will tend to increase in size to reduce the additional surface energy [27]. A nanosuspension is a thermodynamically unstable heterogeneous aqueous dispersion, and aggregation among crystals is one of the primary causes of its low stability. Particles in suspension exhibit Brownian motion, which may lead to collisions and fusion due to inter-particle attraction and van der Waals forces, resulting in an increase in particle size and polydispersity index [28]. This represents one of the major challenges in the production of NCs, as changes in particle size directly affect the drug's solubility and dissolution rate, as previously discussed. Another potential cause of instability is known as "Ostwald ripening" (crystal growth). The broader the particle size distribution (PDI), the more likely this phenomenon is to occur, where crystals of varying sizes undergo size change due to differences in their solubility. Smaller crystals exhibit higher surface energy and saturation solubility than larger ones, creating a concentration gradient that drives interaction between smaller and larger crystals, leading to material transfer with larger particles growing at the expense of smaller ones [29]. Finally, another cause of instability is sedimentation. According to Stokes' law, drug nanocrystals tend to sediment in the product medium, such as water or other liquids. In a suspension, larger particles naturally settle under the influence of gravity. In nanosuspensions, sedimentation behavior can be categorized into two types: flocculation and deflocculation. For NCs, it is acceptable for sedimentation to occur as long as it does so slowly and the particles can be easily redispersed [29, 30]. These physical instabilities represent one of the major challenges in NCs production, creating the need for stabilizing agents to obtain NCs as stable systems. Stabilizing agents can act by effectively reducing surface energy, thereby preventing aggregation or the phenomenon of Ostwald ripening [31, 32].

– Stabilizing Agents

The stabilizing agents commonly used for the preparation of NCs can be divided into three major groups based on their stabilization mechanisms: steric, electrostatic, and a combination of steric-electrostatic. In general, ionic surfactants stabilize nanoparticles due to electrostatic repulsion between each particle. In this process, when the stabilizing agent adheres to the drug's surface, a double electric layer is formed, creating a charge around the drug. When two nanoparticles in suspension collide, the same surface charge surrounding each particle repels each other, keeping them separated and unable to come close in proximity, thus preventing agglomeration and particle size increase. On the other hand, polymers and nonionic stabilizers act by creating a spatial barrier between particles through the adsorption of hydrophobic molecules on the surface of drug NCs. In this case, the polymers adsorb onto the NCs' surface, with their long chains extending outward, thereby limiting the movement of nanoparticles in suspension and preventing agglomeration [29]. The action mechanisms of the two types of stabilizers can be seen in Fig. 4.

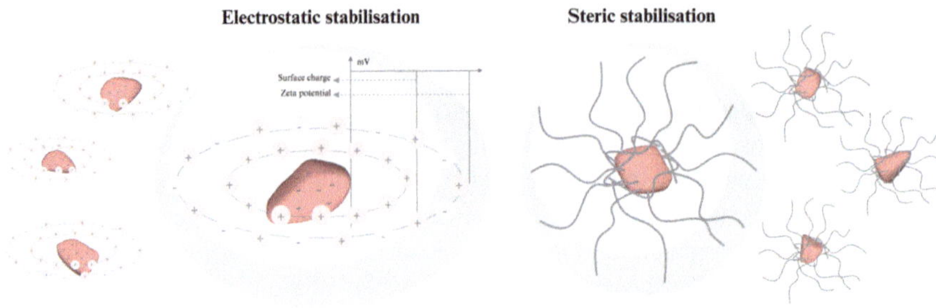

Fig. 4 Representation of the mechanisms of action of various types of stabilizers. (Figure adapted from Li et al. [29])

When selecting a stabilizing agent, several factors must be considered. Utilization of ionic surfactants, which provide stability via electrostatic repulsion, may be hindered by the presence of electrolytes or under highly acidic conditions of the medium, inhibiting the system's ability to maintain stability [29]. This becomes particularly relevant for orally administered drugs, as they will be exposed to the acidic gastric environment where electrostatic repulsion can be disrupted by the influence of electrolytes in the gastrointestinal system. In such cases, steric stabilization has proven more effective in preventing agglomeration, as it remains unaffected by charged ions [33]. However, selecting a steric stabilizing agent is a complex process, as successful system stabilization depends on multiple factors. Log P and enthalpy are known to correlate directly with the likelihood of achieving stable nanosuspension via steric stabilization. Furthermore, drugs with higher enthalpy values are generally better suited for nanosuspension formulation, regardless of the stabilizing agent employed [34]. Poloxamers, for instance, are FDA-approved as food additives and pharmaceutical ingredients, considered nontoxic, and are among the most commonly employed stabilizers in the preparation of NCs. Frequently used poloxamers include Pluronics® such as poloxamer 407 (Pluronic® F127) and poloxamer 188 (Pluronic® F68), both of which are nonionic linear triblock copolymeric stabilizers. Additionally, inverted poloxamers like Pluronic® Reverse 17R4 exist, although linear structures are generally deemed more effective as stabilizing agents [35]. However, poloxamers are not universally effective for all drugs. Their stabilizing ability relies on physisorption; if hydrophobic/hydrophilic interactions are weak, poor surface adsorption will occur, leading to insufficient and ineffective stabilization, leaving the resultant NCs prone to aggregation [36]. Another widely used stabilizing agent is hydroxypropyl methylcellulose (HPMC), a low-toxicity, semi-synthetic, nonionic cellulose derivative approved for pharmaceutical formulations [37]. Vitamin E (Tocopherol Polyethylene Glycol Succinate—TPGS), due to its amphiphilic nature characterized by a lipophilic tail and a hydrophilic polar head, is also frequently utilized as a stabilizing agent. While primarily employed in oral formulations, TPGS is also being investigated for parenteral and topical applications, as well as other routes of administration

Table 2 Different stabilizers used in the production of nanocrystals

Category		Stabilizers	Mechanism
Surfactants	Ionic	SLS	Electrostatic
		Centrimonium chloride	Electrostatic
		Dowfax 2A1	Electrostatic
		DOSS	Electrostatic
	Nonionic	Poloxamer 188	Steric
		Poloxamer 407	Steric
		Pluronic® reverse 17R4	Steric
		TPGS	Steric
		Tween 80	Steric
		Plantacare 2000	Steric
		Saponins	Steric
	Amphoteric	Lecithin	Steric
		Lipoid S75	Steric
Polymers	Synthetic	PVP	Steric
		PVA	Steric
		Soluplus®	Steric
	Semisynthetic	HPMC	Steric
		HPC	Steric
		HEC	Steric
		CMC	Electrostatic
		CMC-Na	Electrostatic
	Natural	Sodium alginate	Steric
		Chitosan	Electrostatic/steric

Table adapted from Li et al. [29] and updated
SLS sodium lauryl sulfate, *DOSS* dioctyl sodium sulfosuccinate, *TPGS* tocopherol polyethylene glycol succinate, *PVP* polyvinylpyrrolidone, *PVA* polyvinyl alcohol, *HPMC* hydroxypropyl methylcellulose, *HPC* hydroxypropyl cellulose, *HEC* hydroxyethyl cellulose, *CMC* carboxymethyl cellulose, *CMC-Na* sodium carboxymethyl cellulose

[38]. In NC stabilization, TPGS has been extensively reported as an effective agent, with its use well-documented in the literature [29]. Table 2 presents the stabilizers commonly reported in the literature for obtaining NCs, along with their classification.

2.3 Nanocrystals Characterization

Nanotechnology has emerged as one of the most relevant approaches in pharmaceutical sciences, as such the need for appropriate techniques to analyze the aforementioned systems has significantly increased. Currently, various techniques are commonly employed for the characterization of NCs in both physical and chemical terms. One key parameter when discussing NCs is particle size. The definition of the term "nano" varies depending on the discipline; in some fields of science, particles are considered "nano" if their size is below 100 nm. However, in pharmaceutical sciences, the term "nano" is generally applied to particles smaller than 1000 nm [39]. Several techniques are available for determining nanometric particle size, including dynamic light scattering (DLS), laser diffraction (LD), and microscopy.

Optical microscopy offers the advantage of providing images of nanoparticles; however, it lacks statistical significance. LD is a robust technique capable of analyzing particles of varying sizes within a single measurement. In this regard, LD provides a volume distribution and has a typical measurement range of approximately 0.05–80μm, with a maximum of 2000μm, depending on the equipment used. The standard characterization parameters in LD measured the distribution at 50%, 90%, and 99%, represented by D50, D90, and D99, respectively (i.e., D50 indicates that 50% of the particle volume is below the specified size) [40]. However, using this technique requires knowledge of specific optical parameters of the analyzed particles, which are not always known and can be challenging to determine. The Fraunhofer approximation is an alternative, but it is only valid for particles at least five times larger than the wavelength of light, thereby excluding those smaller than 3.5μm [41]. DLS is arguably the most widely used technique for measuring particle size in nanometric systems. Also referred to as photon correlation spectroscopy, this technique offers the advantage of enabling simple measurements with rapid and accurate results, but the main limitation of this technique is its inability to analyze particles larger than 6μm. As we know, the size and distribution of size within the sample are important characteristics of NCs, as they influence other properties such as physical stability, saturation solubility, dissolution rate, and even clinical efficacy. The smaller the particle size, the higher the surface energy of the particles, which promotes aggregation. The DLS technique provides not only the mean particle diameter but also the width of the particle size distribution, referred to as the "polydispersity index" (PDI), another key parameter in the development of NCs. The PDI ranges from 0 to 1; the lower the value, the more homogeneous the particle size distribution in the sample [40]. Additionally, the surface charge of nanocrystals is a key parameter that provides information regarding the physical stability of nanosuspensions. The greater the charge (in absolute value), the higher the electrostatic repulsion between the particles, and thus, the greater the physical stability. The surface charge of the particles is ideally quantified in terms of "zeta potential" which is measured through the electrophoretic mobility of the particles in an electric field. In general, particles inherently possess a surface charge due to the dissociation of surface functional groups, known as the Nernst potential [42]. The degree of dissociation of these functional groups depends on the pH of the suspension, meaning that zeta potential is a pH-dependent characteristic. According to the literature, a zeta potential of at least −30 mV is desired [43]. However, the stability of NCs in suspension will depend on the surface charge, provided the repulsion between particles is electrostatic rather than steric. To put it simply, a zeta potential close to neutrality does not necessarily indicate physical instability of the system if steric stabilizers have been used to obtain the NCs.

The solid-state properties (polymorphic crystalline form, solvated form (especially hydrate), and degree of crystallinity) influence the apparent solubility and, therefore, the dissolution rate. In turn, different conditions and procedures used in the manufacture of nanocrystals can impact the resulting solid-state form. This is why other characterization techniques commonly used for NCs include X-ray diffraction (XRD), a widely applied method to determine the crystalline state of a sample and identify potential polymorphic changes. This involves comparing the X-ray diffraction pattern of crystalline substances with that of a pure sample. Each crystalline substance generates a unique pattern, while mixtures of substances produce a combined pattern [44]. Fourier-transform infrared spectroscopy (FT-IR) is used to assess the chemical properties of the drug and its potential interactions with excipients by comparing the spectrum of the NCs with those of the pure components and their physical mixture. This technique is capable of detecting changes in the overall composition of biomolecules by identifying modifications in functional groups. It identifies structural differences in molecular binding between entities, which can reveal details about the existence of their interactions [45]. Thermal analysis techniques are also extensively utilized, as they provide valuable insights. Differential scanning calorimetry (DSC) allows for the comparison of the thermal behavior of solid NCs with that of the pure components to quantify amorphous or crystalline phases. Key parameters of amorphous or crystalline materials assessed using conventional DSC include the glass transition endotherm, crystallization exotherm, and melting endotherm [46]. DSC studies are useful for verifying the crystallinity state of the drug and its interaction with excipients after nanocrystal production. Certain top-down techniques, such as high-pressure homogenization or wet milling, can result in particles with an amorphous fraction, thereby improving saturation solubility. In this regard, DSC is performed on the pure drug, the physical mixture of the drug and excipients (stabilizers), and the final formulation, which may be in a dry form. Thermogravimetric analysis (TGA), on the other hand, measures mass changes as a function of temperature, offering information on processes such as vaporization and decomposition, as well as enabling compositional analysis [47]. Raman spectroscopy, based on the inelastic scattering of monochromatic light from a laser source, is a powerful, nondestructive analytical technique. It is used to obtain valuable information on the structural configuration, chemical composition, defects, doping mechanisms, phase transitions, crystallinity, and surface interactions of nanomaterials [48]. Raman spectra of pure samples are compared with those of the NCs produced to derive these insights.

2.4 Drug Delivery Using Nanocrystals

To date, NCs have been extensively employed for the delivery of hydrophobic drugs through various routes. The formulation of drug NCs has emerged as one of the preferred strategies for enabling the administration of high doses of hydrophobic drugs via otherwise challenging delivery pathways. The opportunities presented by the administration of NCs through different routes will be briefly outlined below and examined in greater depth throughout the entire book. The oral route of administration is the most commonly chosen for pharmacotherapy, as evidenced by the products that have successfully reached the market. Its popularity is attributed to the convenience it offers patients, the safety associated with this route, and its cost-effectiveness. However, when low-solubility drugs are administered orally, their absorption is often poor and erratic, influenced by a multitude of factors, such as concurrent intake with food, the fat content of said food, and gastrointestinal motility. To address these challenges, the formulation of NCs has been increasingly employed to enhance the bioavailability of such drugs [21]. On the other hand, injections are frequently utilized in clinical practice when a rapid onset of action, independence from the digestive system, or avoidance of hepatic metabolism is required. Common routes of injectable administration include intravenous, subcutaneous, intradermal, and intramuscular. Injectable NC formulations offer a significant improvement due to their nanometric size, high drug-loading capacity, and the need for small administration volumes [49]. At the same time, the reduction in the use of co-solvents also minimizes the dose volume and the potential for toxic or allergic reactions. Alternatively, the largest organ of the body, the skin, has been exploited as a drug delivery route. Despite the physiological barrier function, protecting the body from external sources, various technologies have been developed to enhance permeation through it [50]. In this regard, there are two main drug absorption pathways through the skin, transdermal and transappendageal, and in both, NCs have shown improvements in drug delivery [21]. One of the most interesting systems for drug delivery through the skin is microneedles (MNs), which provide efficient delivery of drug transdermally, bypassing the formidable barrier of the *stratum corneum*, in a non-invasive manner as opposed to traditional methods. The integration of MNs and NCs has been explored recently with promising results, as MNs provide a direct pathway for NCs, which exhibit increased saturation solubility, allowing a greater amount of drug molecules to be available for absorption into the dermal microcirculation located in the viable epidermis [51]. Regarding ocular administration, the low aqueous solubility of drugs poses a significant challenge for this route due to the absence of a biological release medium for drug dissolution and the complex physiological barrier of the eye. In this context, NCs present a promising solution, offering the capability to overcome the ocular physiological barrier and enhance drug bioavailability through this route [52]. Regarding intranasal administration, NCs delivered via this route facilitate drug transport across the mucosal barrier by increasing the contact area and residence time with the mucosa [21]. Finally, with respect to pulmonary administration of drugs, the particle size of said drug and the inhalation process significantly influence regional lung deposition, making NCs a promising carrier strategy to enhance drug delivery to the alveolar regions and improve pharmacotherapy [53]. This book will address the NC technology, with chapters organised in line with administration routes.

3 Conclusion

The strategy of producing drug NCs has demonstrated significant potential in improving the undesirable properties of poorly water-soluble drugs. Originally conceived as a method to enhance the bioavailability of such drugs, the formulation of NCs has generated considerable interest from both academia and industry, as evidenced by the growing number of publications and patents year on year. NCs have made a significant impact on pharmaceutical sciences, particularly in the delivery of poorly soluble drugs across various administration routes. Such advantages arise due to their small size and high surface area and distinctive physicochemical properties, enabling efficient delivery of low-solubility compounds, a high drug loading capacity, enhanced stability, and potential for both sustained and targeted drug release. As a result, NCs have emerged as a groundbreaking tool in drug delivery, offering a versatile platform for the transportation and subsequent delivery of therapeutic agents.

References

1. H. Park, A. Otte, K. Park, Evolution of drug delivery systems: From 1950 to 2020 and beyond. J. Control. Release **342**, 53–65 (2022). https://doi.org/10.1016/J.JCONREL.2021.12.030
2. S. Kalepu, V. Nekkanti, Insoluble drug delivery strategies: Review of recent advances and business prospects. Acta Pharm. Sin. B **5**, 442–453 (2015). https://doi.org/10.1016/J.APSB.2015.07.003
3. C.A. Lipinski, Drug-like properties and the causes of poor solubility and poor permeability. J. Pharmacol. Toxicol. Methods **44**, 235–249 (2000). https://doi.org/10.1016/S1056-8719(00)00107-6
4. J.E. Kipp, The role of solid nanoparticle technology in the parenteral delivery of poorly water-soluble drugs. Int. J. Pharm. **284**, 109–122 (2004). https://doi.org/10.1016/J.IJPHARM.2004.07.019
5. J.M. Ting, W.W. Porter, J.M. Mecca, F.S. Bates, T.M. Reineke, Advances in polymer design for enhancing oral drug solubility and delivery. Bioconjug. Chem. **29**, 939–952 (2018). https://doi.org/10.1021/ACS.BIOCONJCHEM.7B00646/ASSET/IMAGES/LARGE/BC-2017-00646K_0006.JPEG
6. K. Kawakami, N. Oda, K. Miyoshi, T. Funaki, Y. Ida, Solubilization behavior of a poorly soluble drug under combined use of surfactants and cosolvents. Eur. J. Pharm. Sci. **28**, 7–14 (2006). https://doi.org/10.1016/J.EJPS.2005.11.012
7. D.H. Jornada, G.F. Dos Santos Fernandes, D.E. Chiba, T.R.F. De Melo, J.L. Dos Santos, M.C. Chung, The prodrug approach: A successful tool for improving drug solubility. Molecules **21** (2015). https://doi.org/10.3390/MOLECULES21010042
8. B. dos Santos Lima, S. Shanmugam, et al., Inclusion complex with cyclodextrins enhances the bioavailability of flavonoid compounds: a systematic review. Phytochem. Rev. **18**, 1337–1359 (2019). https://doi.org/10.1007/S11101-019-09650-Y/TABLES/1
9. K.V. Drozd, A.N. Manin, D.E. Boycov, G.L. Perlovich, Simultaneous improvement of dissolution behavior and Oral bioavailability of antifungal Miconazole via Cocrystal and salt formation. Pharmaceutics **14** (2022). https://doi.org/10.3390/PHARMACEUTICS14051107

10. E. Spoletti, V. Verma, C. Cappuccino, M. Lusi, Solid solution polymorphs afford two highly soluble co-drug forms of tolbutamide and chlorpropamide. Chem. Commun. (Camb.) **59**, 14321–14324 (2023). https://doi.org/10.1039/D3CC04725C
11. P. Pandi, R. Bulusu, N. Kommineni, W. Khan, M. Singh, Amorphous solid dispersions: An update for preparation, characterization, mechanism on bioavailability, stability, regulatory considerations and marketed products. Int. J. Pharm. **586**, 119560 (2020). https://doi.org/10.1016/J.IJPHARM.2020.119560
12. K.R. Vandana, Y.P. Raju, V.H. Chowdary, M. Sushma, N.V. Kumar, An overview on in situ micronization technique – an emerging novel concept in advanced drug delivery. Saudi Pharm. J. **22**, 283 (2013). https://doi.org/10.1016/J.JSPS.2013.05.004
13. Y. Ding, T. Zhao, J. Fang, J. Song, H. Dong, J. Liu, S. Li, M. Zhao, Recent developments in the use of nanocrystals to improve bioavailability of APIs. Wiley Interdiscip. Rev. Nanomed. Nanobiotechnol. **16** (2024). https://doi.org/10.1002/WNAN.1958
14. GG Liversidge, KC Cundy, JF Bishop, DA Czekai, U.S. Patent No. 5,145,684. Washington, DC: U.S. Patent and Trademark Office (1992)
15. A. Mihranyan, M. Strømme, Solubility of fractal nanoparticles. Surf. Sci. **601**, 315–319 (2007). https://doi.org/10.1016/J.SUSC.2006.09.037
16. G. Kaptay, On the size and shape dependence of the solubility of nano-particles in solutions. Int. J. Pharm. **430**, 253–257 (2012). https://doi.org/10.1016/J.IJPHARM.2012.03.038
17. R. Defay, I. Prigogine, A. Bellemans, D.H. Everett, Surface tension and adsorption (1966)
18. D.R. Ely, R. Edwin García, M. Thommes, Ostwald–Freundlich diffusion-limited dissolution kinetics of nanoparticles. Powder Technol. **257**, 120–123 (2014). https://doi.org/10.1016/J.POWTEC.2014.01.095
19. A.A. Noyes, W.R. Whitney, The rate of solution of solid substances in their own solutions. J. Am. Chem. Soc. **19**, 930–934 (1897). https://doi.org/10.1021/JA02086A003/ASSET/JA02086A003.FP.PNG_V03
20. G. Kaur, J. Grewal, K. Jyoti, U.K. Jain, R. Chandra, J. Madan, Oral controlled and sustained drug delivery systems: Concepts, advances, preclinical, and clinical status, in *Drug Targeting and Stimuli Sensitive Drug Delivery Systems*, (2018), pp. 567–626. https://doi.org/10.1016/B978-0-12-813689-8.00015-X
21. M.B. McGuckin, J. Wang, R. Ghanma, N. Qin, S.D. Palma, R.F. Donnelly, A.J. Paredes, Nanocrystals as a master key to deliver hydrophobic drugs via multiple administration routes. J. Control. Release **345**, 334–353 (2022). https://doi.org/10.1016/J.JCONREL.2022.03.012
22. RH Müller, R Becker, B Kruss, K. Peters, Pharmaceutical nanosuspensions for medicament administration as systems with increased saturation solubility and rate of solution (1999)
23. A. Tuomela, J. Saarinen, C.J. Strachan, J. Hirvonen, L. Peltonen, Production, applications and in vivo fate of drug nanocrystals. J Drug Deliv Sci Technol **34**, 21–31 (2016). https://doi.org/10.1016/J.JDDST.2016.02.006
24. L. Lopez-Vidal, M. Tinti, M.E. Melian, L. Canton, M. Lorenzutti, L. Schofs, M.L. Formica, A.J. Paredes, S.S. Bruni, N. Litterio, R. Faccio, S.D. Palma, J.P. Real, In vivo pharmacokinetic study and PBPK modeling: Comparison between 3D-printed nanocrystals and solid dispersions. Int. J. Pharm. **669**, 125063 (2025). https://doi.org/10.1016/J.IJPHARM.2024.125063
25. M.G. Fuster, J. Wang, O. Fandiño, G. Víllora, A.J. Paredes, Folic acid-decorated nanocrystals as highly loaded Trojan horses to target cancer cells. Mol. Pharm. **21**, 2781–2794 (2024). https://doi.org/10.1021/ACS.MOLPHARMACEUT.3C01186/ASSET/IMAGES/LARGE/MP3C01186_0010.JPEG
26. P. Lhaglham, L. Jiramonai, Y. Jia, B. Huang, Y. Huang, X. Gao, J. Zhang, X.J. Liang, M. Zhu, Drug nanocrystals: Surface engineering and its applications in targeted delivery. IScience **27**, 111185 (2024). https://doi.org/10.1016/J.ISCI.2024.111185
27. T.L. Chang, H. Zhan, D. Liang, J.F. Liang, Nanocrystal technology for drug formulation and delivery. Front. Chem. Sci. Eng. **9**, 1–14 (2015). https://doi.org/10.1007/S11705-015-1509-3/METRICS

28. F. Le Berre, G. Chauveteau, E. Pefferkorn, Perikinetic and Orthokinetic aggregation of hydrated colloids. J. Colloid Interface Sci. **199**, 1–12 (1998). https://doi.org/10.1006/JCIS.1997.5307
29. J. Li, Z. Wang, H. Zhang, J. Gao, A. Zheng, Progress in the development of stabilization strategies for nanocrystal preparations. Drug Deliv. **28**, 19–36 (2021). https://doi.org/10.1080/10717544.2020.1856224
30. N.A. Martínez, F. Fernández-Álvarez, Á.V. Delgado, M.L. Badillo-García, J. Raba, S.E. Cerutti, J.L. Arias, First steps in the formulation of praziquantel nanosuspensions for pharmaceutical applications. Pharm. Dev. Technol. **25**, 892–898 (2020). https://doi.org/10.1080/10837450.2020.1756320
31. B. Derjaguin, L. Landau, Theory of the stability of strongly charged lyophobic sols and of the adhesion of strongly charged particles in solutions of electrolytes. Prog. Surf. Sci. **43**, 30–59 (1993). https://doi.org/10.1016/0079-6816(93)90013-L
32. L. Wu, J. Zhang, W. Watanabe, Physical and chemical stability of drug nanoparticles. Adv. Drug Deliv. Rev. **63**, 456–469 (2011). https://doi.org/10.1016/J.ADDR.2011.02.001
33. H. Rachmawati, A. Rahma, L. Al Shaal, R.H. Müller, C.M. Keck, Destabilization mechanism of ionic surfactant on curcumin nanocrystal against electrolytes. Sci. Pharm. **84**, 685–693 (2016). https://doi.org/10.3390/SCIPHARM84040685
34. M. George, I. Ghosh, Identifying the correlation between drug/stabilizer properties and critical quality attributes (CQAs) of nanosuspension formulation prepared by wet media milling technology. Eur. J. Pharm. Sci. **48**, 142–152 (2013). https://doi.org/10.1016/J.EJPS.2012.10.004
35. P. Liu, T. Viitala, A. Kartal-Hodzic, H. Liang, T. Laaksonen, J. Hirvonen, L. Peltonen, Interaction studies between indomethacin nanocrystals and PEO/PPO copolymer stabilizers. Pharm. Res. **32**, 628–639 (2015). https://doi.org/10.1007/S11095-014-1491-3
36. A. Tuomela, J. Hirvonen, L. Peltonen, Stabilizing agents for drug nanocrystals: Effect on bioavailability. Pharmaceutics **8**, 16 (2016). https://doi.org/10.3390/PHARMACEUTICS8020016
37. Inactive Ingredient Search for Approved Drug Products (n.d.)
38. M.A. Farooq, N.L. Trevaskis, TPGS decorated liposomes as multifunctional Nano-delivery systems. Pharm. Res. **40**, 245 (2022). https://doi.org/10.1007/S11095-022-03424-6
39. J. Kreuter, Nanoparticulate systems for brain delivery of drugs. Adv. Drug Deliv. Rev. **47**, 65–81 (2001). https://doi.org/10.1016/S0169-409X(00)00122-8
40. M.M. Chogale, V.N. Ghodake, V.B. Patravale, Performance parameters and characterizations of nanocrystals: A brief review. Pharmaceutics **8**, 26 (2016). https://doi.org/10.3390/PHARMACEUTICS8030026
41. C.M. Keck, Particle size analysis of nanocrystals: Improved analysis method. Int. J. Pharm. **390**, 3–12 (2010). https://doi.org/10.1016/J.IJPHARM.2009.08.042
42. E. Pardhi, R. Vasave, V. Srivastava, R. Yadav, N.K. Mehra, Nanocrystal technologies in biomedical science: From the bench to the clinic. Drug Discov. Today **29**, 103913 (2024). https://doi.org/10.1016/J.DRUDIS.2024.103913
43. R.H. Muller, C.M. Keck, Challenges and solutions for the delivery of biotech drugs – a review of drug nanocrystal technology and lipid nanoparticles. J. Biotechnol. **113**, 151–170 (2004). https://doi.org/10.1016/J.JBIOTEC.2004.06.007
44. M.A. Jahangir, S.S. Imam, A. Muheem, A. Chettupalli, F.A. Al-Abbasi, M.S. Nadeem, I. Kazmi, M. Afzal, S. Alshehri, Nanocrystals: Characterization overview, applications in drug delivery, and their toxicity concerns. J. Pharm. Innov. **17**, 237–248 (2022). https://doi.org/10.1007/S12247-020-09499-1/FIGURES/3
45. M.M. Eid, Characterization of nanoparticles by FTIR and FTIR-microscopy, in *Handbook of Consumer Nanoproducts*, (2022), pp. 645–673. https://doi.org/10.1007/978-981-16-8698-6_89
46. P. Gill, T.T. Moghadam, B. Ranjbar, Differential scanning calorimetry techniques: Applications in biology and nanoscience. J. Biomol. Tech. **21**, 167 (2010)

47. S. Loganathan, R.B. Valapa, R.K. Mishra, G. Pugazhenthi, S. Thomas, Thermogravimetric analysis for characterization of nanomaterials, in *Thermal and Rheological Measurement Techniques for Nanomaterials Characterization*, vol. 3, (2017), pp. 67–108. https://doi.org/10.1016/B978-0-323-46139-9.00004-9
48. G. Gouadec, P. Colomban, Raman spectroscopy of nanomaterials: How spectra relate to disorder, particle size and mechanical properties. Prog. Cryst. Growth Charact. Mater. **53**, 1–56 (2007). https://doi.org/10.1016/J.PCRYSGROW.2007.01.001
49. Y. Lu, Y. Li, W. Wu, Injected nanocrystals for targeted drug delivery. Acta Pharm. Sin. B **6**, 106–113 (2016). https://doi.org/10.1016/J.APSB.2015.11.005
50. A.Z. Alkilani, M.T.C. McCrudden, R.F. Donnelly, Transdermal drug delivery: Innovative pharmaceutical developments based on disruption of the barrier properties of the stratum corneum. Pharmaceutics **7**, 438–470 (2015). https://doi.org/10.3390/PHARMACEUTICS7040438
51. T.M. Tuan-Mahmood, M.T.C. McCrudden, B.M. Torrisi, E. McAlister, M.J. Garland, T.R.R. Singh, R.F. Donnelly, Microneedles for intradermal and transdermal drug delivery. Eur. J. Pharm. Sci. **50**, 623–637 (2013). https://doi.org/10.1016/J.EJPS.2013.05.005
52. O.P. Sharma, V. Patel, T. Mehta, Nanocrystal for ocular drug delivery: Hope or hype. Drug Deliv. Transl. Res. **6**, 399–413 (2016). https://doi.org/10.1007/S13346-016-0292-0/TABLES/3
53. A. Kumar, F. Chen, A. Mozhi, X. Zhang, Y. Zhao, X. Xue, Y. Hao, X. Zhang, P.C. Wang, X.J. Liang, Innovative pharmaceutical development based on unique properties of nanoscale delivery formulation. Nanoscale **5**, 8307 (2013). https://doi.org/10.1039/C3NR01525D
54. E.J. Catlin, L. Lopez-Vidal, R.F. Donnelly, A.J. Paredes, A dynamic duo comes of age: Nanocrystals and microneedles for hydrophobic drug delivery. Expert. Opin. Drug. Deliv. **22**, 1509–1526 (2025). https://doi.org/10.1080/17425247.2025.2531059

Manufacturing Techniques for Solid Drug Nanoparticles

Peuli Ghosh, Marta Bergillos Ruiz, Ajay Kumar, Peter Davern, Benjamin K. Hodnett, Ake Rasmuson, Samira Kaissi, and Sarah P. Hudson

Abstract Poor aqueous solubility remains a critical barrier in the development of many active pharmaceutical ingredients (APIs), with nearly 70% of drug candidates exhibiting limited dissolution rates and reduced bioavailability. Solid drug nanoparticle-based drug delivery has emerged as a transformative strategy to address these challenges, leveraging size reduction to enhance dissolution rates and solubility as described by the Noyes–Whitney and Ostwald–Freundlich principles. This chapter provides a comprehensive overview of nanoparticle production strategies, focusing on both top–down (wet milling, high-pressure homogenization) and bottom-up (crystallization and precipitation) approaches. While top–down techniques have dominated commercial applications, they are constrained by energy-intensive processing and potential thermal degradation. Conversely, bottom-up methods offer advantages in cost efficiency and scalability, but present challenges in particle size control, amorphous content, and long-term stability. Strategies for overcoming these limitations, including the use of stabilizers and advanced isolation methods such as spray drying, freeze drying, and supercritical fluid technologies, are critically discussed. By highlighting recent advancements in manufacturing and stabilization, this chapter emphasizes the pivotal role of drug nanoparticles in shaping next-generation pharmaceutical formulations and underscores the need for continued innovation to achieve effective, stable, and scalable delivery systems.

Keywords Poor aqueous solubility · Drug nanoparticles · Nanoparticle drug delivery · Top–down processing · Bottom-up processing · Nanoparticle isolation · Spray drying · Freeze drying · Formulation stabilizers and formulation stability · Carrier particles · Pharmaceutical nanotechnology

---

P. Ghosh · M. B. Ruiz · A. Kumar · P. Davern · B. K. Hodnett · A. Rasmuson · S. Kaissi · S. P. Hudson (✉)
Department of Chemical Sciences, SSPC, Research Ireland's National Centre for Pharmaceuticals, Bernal Institute, University of Limerick, Limerick, Ireland
e-mail: peuli.ghosh@ul.ie; Marta.BergillosRuiz@ul.ie; Peter.davern@ul.ie; Kieran.Hodnett@ul.ie; ake.rasmuson@ul.ie; Sarah.hudson@ul.ie

A. J. Paredes (ed.), *Nanocrystals*, https://doi.org/10.1007/978-3-032-10238-6_2

1 Introduction

Approximately 70% of drugs in development exhibit poor water solubility and slow dissolution rates [1, 2]. Poorly water-soluble drugs present a challenge in drug formulation as low solubility leads to inconsistent dissolution rates, thus lowering the bioavailability of the drugs inside the human body. This low aqueous solubility can also result in either accumulation of the drug in tissue or elimination through first-pass metabolism in the intestine or liver because of their high hydrophobicity and low solubility [3].

Nanoparticle formation [4–7] is a classical approach to enhance dissolution rates and solubilities of poorly water-soluble drugs. A decrease in particle size leads to an increase in the specific surface area, which is directly proportional to the dissolution rate, as described by the Noyes-Whitney equation [8]. It is also possible to achieve an increase in solubility when API crystals are reduced to the nanoscale as explained by the Ostwald–Freundlich equation where reducing the radius of the particles exponentially increases the apparent solubility of the drug [8, 9]. However, this effect is not significant until the particle size is below ~200 nm.

Nanosizing of solid drug crystals can be performed by either (i) a particle size reduction (top–down) approach or (ii) a crystallization or precipitation (bottom-up) approach, Fig. 1. The particle size reduction approach can be performed by, for example, wet milling [10, 11] and high-pressure homogenization [12] where mechanical energy is provided to larger particles to generate smaller counterparts.

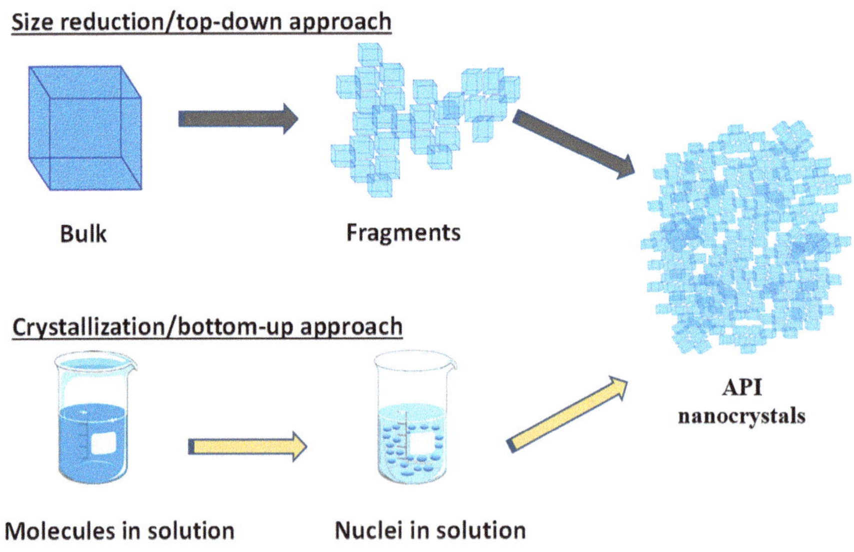

Fig. 1 Schematic diagram of size reduction and crystallization/precipitation methods for producing drug nanoparticles

To date, approximately 19 solid API nanoparticle oral dosage forms are available commercially, and all have been prepared by the top–down approach [13].

In top–down or, particle size reduction approaches, mechanical energy is used to reduce the size of large particles to nanometre ranges using various techniques such as wet ball milling (Nanocrystal® technology [14)] and high-pressure homogenization (Dissocubes®, Nanopure® technologies [14]). However, this approach has limitations such as long processing times, high energy requirements, and risk of thermal or mechanical degradation [7, 15].

Bottom-up or crystallization (or precipitation) approaches begin with API molecules present in solution followed by the (nucleation and) growth of API crystals to form particles of the desired size in a controlled manner via molecular association. This method can be performed by crystallization or precipitation of API particles through liquid antisolvent precipitation, sono-precipitation or reaction precipitation, supercritical spray drying, or evaporation methods [7, 16, 17]. Compared to mechanical size reduction, crystallization or precipitation methods are fast, cost-efficient, and have low energy requirements [18, 19]. However, these processes also face limitations, namely (i) controlling the particle size distribution, particularly in the nanometer range, (ii) a high degree of amorphous content, and (iii) challenges in retaining the nano-dimensions of the particles during subsequent isolation to dryness. There is a tendency toward high degrees of amorphous content when high levels of supersaturation are used during the bottom-up approach. Frequently, solid drug nanoparticles can appear amorphous during X-ray diffraction analysis due to their size as well as lack of order. Thus, in this chapter, the term drug "nanoparticles" refers to both nanocrystals and amorphous nanoparticles.

Regardless of the way they are produced, nanosuspensions tend to agglomerate or undergo Ostwald ripening to reduce their high surface energies [20]. Additives can be used to stabilize nanosuspensions but are system specific [21]. To increase the shelf life of nanoparticles, it is often necessary to isolate them from suspension into dried powder forms.

2 Size Reduction Approach

2.1 Dry Milling

Conventional size reduction methods include the dry–impact milling technique, which is a straightforward method that can be tested on most APIs, regardless of their solubility in aqueous or nonaqueous solvents [22]. The main properties determining the final size of the crystals after milling are crystal hardness and the presence of defects in the starting crystals. These properties determine if the starting drug crystals can be milled to the nano-scale [23, 24]. In general, to get particle sizes to the nanometre size range requires long processing times and high energy input, leading to very high-energy consumption [25]. The heat dissipated from the high energy input can lead to a change in the API's crystalline state (such as polymorphic transformation or amorphization) [26]. Also, milling has limited control over the particle morphology and surface properties.

2.2 Wet Milling

Wet milling was first developed by Liversidge and co-workers [27]. It involves the continuous stirring and milling of a drug slurry within an aqueous or nonaqueous liquid-based specialized media. The active component of the mill is the milling chamber, which is charged with milling aids, Fig. 2. This equipment can operate either in a batch or a re-circulation mode. Commonly used milling aids include ceramic, stainless steel, polystyrene, or glass balls typically coated to prevent sample contamination due to erosion [28, 29]. During the process, a crude slurry composed of the drug particles, water, and additives is introduced into the milling chamber. There, the slurry is transformed into a nanoparticle dispersion via the combination of high-energy shear and impaction force between the milling pearls and solid drug particles, resulting in the formation of nanoparticles (~ 400 nm) [30–32]. Several parameters, e.g., milling time, speed, and temperature, need to be controlled to achieve optimal results [33]. Before milling initiation, the addition of drug-specific additives to the slurry is essential to minimize the energy available for the formation

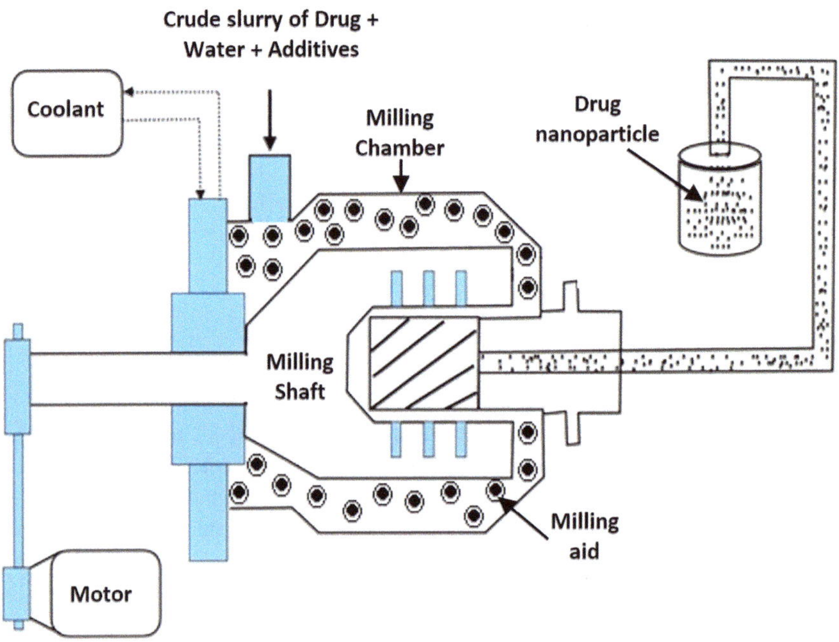

Fig. 2 Schematic representation of the wet milling process [22, 27]

of new particle surfaces, to inhibit crystal growth, and to prevent agglomeration. Scaling up wet milling technology involves challenges related to process parameters, mathematical modelling, environmental impact, and product quality [34]. Key factors in scale-up include maintaining similarity in Reynolds number, Euler number, and specific power consumption, with traditional approaches focusing on rotor tip speed for consistent particle sizes [35]. Common options of milling equipment are ball mills, planetary mills, vibration mills, and attritors. For early drug research, planetary bead-milling enables wet milling of milligram-scale quantities, achieving nanometre-sized particles efficiently [36]. For larger batch sizes or more viscous formulations, mills with external stirring devices are ideal because the rotating stirrer generates strong shear forces [34].

The main drawbacks of this technique are traces of solvent and un-milled drug material remaining after size reduction, as well as the need for preliminary downsizing of the starting material. Some examples include the preparation of niclosamide nanoparticles (235 nm) by wet milling by Ye et al. [37] for parenteral administration, while Patel et al. [38] utilized wet milling to produce nanosuspensions of efavirenz (320 nm) with improved oral bioavailability. Zhai and colleagues improved the wet milling process of caffeine nanoparticles (250–660 nm) for dermal delivery by using Carbopol® 981 as an additive in a medium of ethanol–propylene glycol (3:7), yielding nanoparticles that were stable for up to 60 days at 4–25 °C [39]. Kakran et al. [40] reported the preparation by wet milling of quercetin nanoparticles (276 nm) with superior saturation solubility as compared with coarse quercetin particles.

2.3 *High-Pressure Homogenization (HPH)*

High-pressure homogenization (HPH) is also widely used to produce drug nanoparticles via the "top–down approach" and was first developed by Muller et al. in 1994 [41]. There are two types of HPH techniques, namely, microfluidization and piston-gap HPH; the latter is the more widely used technique to produce drug nanoparticles [42].

Different technologies that use piston-gap HPH include the DissoCubes® platform (SkyePharma AG, Muttenz, Switzerland) [43] and NanoPure® (PharmaSol, Berlin, Germany) [14]. The latter technology uses nonaqueous media to produce water-free nanoparticles. In the piston gap HPH method, a micronized drug powder is dispersed in a solvent and subsequently forced by a piston through the tiny homogenization gap at pressures ranging up to 4000 bar, but typically 1500–2000 bar [14, 44]. The width of the homogenization gap ranges from 5 to 20 μm depending on the viscosity of the suspension and the applied pressure, Fig. 3. According to Bernoulli's law, a high streaming velocity of the suspension increases the dynamic pressure inside the piston gap, which is neutralized by a reduction of the static pressure below the vapour pressure of the solvent [14]. This low pressure leads to boiling of the solvent and the formation of gas bubbles inside

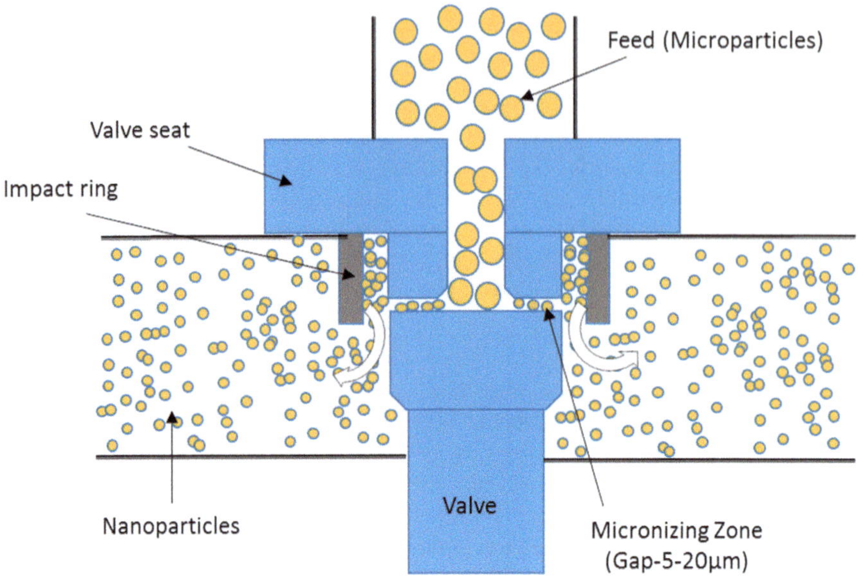

Fig. 3 Schematic diagram of high-pressure homogenization process. Drug microparticles are forced through a tiny gap in the micronizing zone, which creates high turbulence and shear along with compression, acceleration, pressure drop, and impact leading to the formation of nanoparticles [22]

the solution. When the solution leaves the piston gap, the static pressure on the solution rises to atmospheric pressure, forcing the gas bubbles to collapse immediately [14]. This rupture of the gas bubbles generates shock waves, which contribute to size reduction of the drug particles.

Insoluble Drug Delivery–Particles (IDDP™) technology (SkyePharma Canada Inc.) uses microfluidization techniques to achieve the production of submicron particles of poorly soluble drugs [14]. Microfluidization uses jet stream homogenizers such as the Microfluidizer® from Microfluidics Inc. to generate the nanoparticles via the collision of two fluid streams under pressure up to 1700 bar. The collision happens within two different types of chambers ("Y" and "Z" types). In the "Y" chamber, two streams of the suspension collide at one point, which induces the breakage of the particles; in the "Z" chamber, the stream changes direction, which produces collisions and shear forces that reduce the particle size [28].

From HPH techniques, drug particle sizes of 50–400 nm can be obtained [42]. Several variables affect the size of the final product, such as hardness of the starting drug particles, residence time within the piston gap, or the velocity of material transfer during the HPH process [45, 46]. Concentrated drug suspensions can be used to increase the number of particle interactions resulting in greater particle size reduction, but this is restricted to a suspension loading upper limit of 10% w/w due to (i) difficulties in stabilizing highly concentrated suspensions, and

(ii) the large amount of energy required to process highly viscous slurries [47, 48]. HPH is well-suited for the development of parenteral and oral formulations of poorly water-soluble drugs because there is no risk of contamination from milling aids. HPH also eliminates potential microbial contaminants [45, 46, 49, 50]. There are several disadvantages of using HPH, including long processing times when multiple homogenization cycles are required to reach the desired particle size distribution. Furthermore, a larger minimum batch size is required as compared with milling because of the higher void volume relative to most piston gap homogenizers. Also, the drug must be pre-milled before HPH to prevent blockage of apertures in the piston gap [48, 50]. A comprehensive comparison between the advantages and disadvantages of different top–down approaches is listed in Table 1.

Additives are also critical to preserve the particle size distribution after generation of the nanoparticles by these top–down methods, during long-term storage (in suspension) or during subsequent isolation and drying steps. This is discussed in more detail in Sect. 3.1.5.

3 Crystallization/Precipitation (Bottom-Up) Approach

Bottom-up approaches to produce nanoparticles often involve two steps. The first is the generation of a supersaturated solution followed by the formation of nuclei, which subsequently grow. The second step is the isolation of the nanoparticles into dry powder by removing the solvent. Crystallization/precipitation platform technologies include liquid antisolvent crystallization [54], supercritical fluid crystallization [25, 32, 55, 56], and sono-crystallization [57, 58] and common nanoparticle separation technologies include spray drying [17, 59, 60] and freeze drying [61, 62]. In liquid antisolvent precipitation, the formation of the nanosuspension and the subsequent drying step are distinct processes, while in others, like spray drying and several supercritical fluid crystallization, they happen simultaneously.

Simplified rates of nucleation (N) and growth (G), defined by Nyvlt in 1968 (Eqs. 1 and 2), are expressed as a power function of the supersaturation (S). Typical exponent values for in the range of 5–15 for nucleation and 1–2 for growth [63, 64]. Thus, there is a stronger dependence of nucleation rate on supersaturation (S) as compared with the growth rate, resulting in the formation of a large number of small nuclei at high *S*.

$$N = k'S^n \tag{1}$$

$$G = k''S^g \tag{2}$$

where N = nucleation rate, k' = nucleation rate constant, n = kinetic order of nucleation, S = supersaturation, G = growth rate, k'' = growth rate constant, g = kinetic order of growth.

Table 1 Advantages and disadvantages of commercial top–down size reduction methods for drug nanoparticle production

Top–down method	Advantages	Disadvantages	Commercial names (manufacturers)
Dry impact milling	*Applicable to most drugs*—Can be used regardless of solubility in aqueous or nonaqueous solvents *Straightforward process*—Simple and widely tested method for size reduction	*High energy consumption*—Requires long processing times and significant energy input *Heat generation*—Excessive heat can cause polymorphic transformation or amorphization *Limited control over particle properties*—Limited control over particle morphology and surface characteristics	Retsch™ MM 200 Mixer Mill [51]
Wet milling	*Effective nanoparticle formation*—Produces nanoparticles (~400 nm) with improved solubility and bioavailability *Versatile medium compatibility*—Can be performed in aqueous or non-aqueous media *Controlled process parameters*—Milling time, speed, and temperature can be optimized for better results *Scalability options*—Suitable for small-scale (planetary bead milling) and large-scale (stirred mills) production	*Time-consuming process*—Longer runs required to reduce particle size in nanometer range *Contaminated product*—Traces of solvents and some unmilled drug may remain post-milling *Requires preliminary downsizing*—Initial reduction of drug particle size is necessary before milling *Scale-up challenges*—Requires careful control of process parameters and modelling for large-scale production	PE075 laboratory batch mill (Netzsch technology) [52]
High-pressure homogenization	*Produces ultra-fine nanoparticles* —Achieves particle sizes as small as 50 nm *Contamination-free process*—No milling aids, reducing the risk of contamination *Versatile for drug formulations*—Suitable for both oral and parenteral drug delivery *Effective size reduction*—Utilizes high-pressure shock waves for efficient particle breakdown	*Long processing time*—Requires multiple homogenization cycles for optimal size reduction *High energy consumption*—Processing highly viscous suspensions demands significant energy *Large minimum batch size*—Requires more material compared with milling *Pre-milling requirement*—API must be pre-milled to micrometer range to prevent blockage in the piston gap	Dissocubes® (SkyePharma) [43], Microfluidizer® (microfluidics International Corp) [53]

As a result of fast nucleation, ongoing growth is restricted due to a reduction in the amount of solute available. Thus, by increasing the nucleation rate to an extent where the growth becomes negligible, drug nanoparticles can be obtained. Unfortunately, this can also result in the generation of a metastable, or sometimes amorphous solid form.

3.1 *Liquid Antisolvent Crystallization/Precipitation*

Liquid antisolvent crystallization/precipitation (LAS) requires low energy consumption, conventional process parameters (temperature and pressure), feasible scale-up to a continuous process, and has no requirement for expensive equipment [65] [66]. Using this technique, drug nanoparticles (<100 nm) [67] can be produced due to the high supersaturation levels that can be induced. Additives used during LAS to stabilize the nanoparticles can help to control the resulting particle habit and polymorphic form [4].

The basic requirement of the LAS process is sufficient solubility of the drug in a solvent, which is miscible with an antisolvent that has low drug solubility. Additionally, solvent toxicity, boiling point, and solvent–additive interactions should be taken into consideration [25]. In this method, the drug solution is quickly added to the antisolvent under rapid stirring. Rapid mixing of the antisolvent and solvent leads to homogeneous supersaturation. The lower the drug's solubility in the resulting antisolvent/solvent mixture, the higher the level of supersaturation reached, resulting in high nucleation rates to produce small or nanosized particles as discussed earlier [68].

Often, high supersaturations and fast nucleation rates result in amorphous nanoparticles; in these situations, the term precipitation is more accurate to use than crystallization. The properties of the resulting drug particles can be altered by controlling the parameters employed, such as the mixing rate, temperature, drug concentration, antisolvent to solvent ratio, and the inclusion of stabilizing additives.

3.1.1 Mixing Rate

The LAS process can be divided into two steps: the mixing of the drug solution and antisolvent streams to generate a supersaturated solution followed by crystallization/ precipitation of the particles [20]. The mixing rate controls the particle size and particle size distribution in the final suspension. Mixing is needed to generate a uniform supersaturation because addition of the API solution to the antisolvent generates a momentary localized increase in supersaturation. To obtain drug nanoparticles with a narrow size distribution, a uniform supersaturation is necessary. The total mixing process is controlled by two main timescales, namely, the mixing time (t_{mix}) and the nucleation time ($t_{nucleation}$) [20, 69]. The efficiency of mixing can be described by the dimensionless Damköhler number (Da) which is the ratio of t_{mix}

and $t_{nucleation}$ [20, 69–71]. When Da > 1, the mixing is slower than nucleation and thus controls the process. As the mixing process is slow, attainment of homogeneous supersaturation is also slow [69], resulting in broad particle size distributions. In contrast, when Da < 1, mixing time is less than nucleation time, and $t_{nucleation}$ controls the overall process. As mixing is fast, a homogeneous supersaturation is achieved, resulting in a narrow particle size distribution. Hence, it is necessary to maintain a short t_{mix} so a Da < 1 can be achieved. Increasing the stirring speed increases mixing, which enhances the mass transfer and diffusion rate between multiphases [25, 72]. Thus, a homogeneous supersaturation can be achieved in a shorter time by increasing the stirring speeds, resulting in narrower particle size distributions and smaller particle sizes.

Zhang et al. reported that during liquid antisolvent crystallization of megestrol acetate, increasing the stirring speed from 500 to 2500 rpm intensified the mixing, which resulted in a decrease in the mean particle size from 263 to 208 nm and a narrower particle size distribution [73]. Similar trends were observed for curcumin [74], beclomethasone dipropionate [75] and griseofulvin [69]. In small-scale laboratory equipment, agitation is often achieved by magnetic stirring, but this is not sufficient at a larger scale. Special mixing devices are reported in the literature for liquid antisolvent crystallization to generate intense mixing and molecular diffusion at a large scale. Such devices include the static mixer [76], impinging jet mixer [77], coaxial turbulent jet mixer [78], and a T-mixer [70]. These devices facilitate the reduction in mixing time to milliseconds or microseconds. The mixing properties of fluidic devices have also been investigated as a basis for selecting appropriate devices and associated operating parameters to carry out processes that are kinetically fast and mixing-controlled, such as anti-solvent crystallization/precipitation and nanoparticle synthesis [79].

3.1.2 Temperature

The temperature of the antisolvent-solvent mixture influences the drug nanoparticle size generated during liquid antisolvent precipitation. With a decrease in temperature, the drug solubility tends to decrease in the solvent-antisolvent mixture, which leads to an increase in supersaturation. Higher supersaturation increases the nucleation rate resulting in the generation of smaller particles. Lower solubility also lessens the Ostwald ripening process [25, 69]. Reducing the temperature also increases the viscosity of the system. At high viscosity, a reduction in particle mobility is observed, which lowers the collision rate between particles and thus the extent of particle agglomeration. Zhang et al. observed a reduction in solubility and an increase in supersaturation of megestrol acetate during LAS when the temperature was decreased from 30 to 3 °C [73]. This subsequently resulted in a particle size reduction from 2 μm to 240 nm. Similar trends of reduction in particle size with lowering temperature during LAS were also observed for beclomethasone dipropionate [75] and griseofulvin [69].

3.1.3 Solvent/Antisolvent Ratio

The solvent/antisolvent (AS/S) ratio influences particle size and particle size distribution generated during liquid antisolvent crystallization. An increase in the AS/S ratio tends to lead to an initial rapid decrease in the solubility of the drug, up to an optimal ratio where further increases have a negligible impact, resulting in a higher supersaturation, which further enhances the nucleation rate and particle size reduction. It should be noted that a very high AS/S ratio can affect mixing times which can negatively impact the particle size distribution [69, 80]. Zhang et al. reported that the solubility of megestrol acetate (MA) drastically decreased when the AS/S ratio changed from 5 to 8 and then slowly decreased before the AS/S ratio reached 25 [73]. At an AS/S ratio of 5, MA had a higher solubility in the acetone-water mixture, which led to a lower supersaturation and thus slower nucleation rate and larger particle size. As the AS/S ratio increased, supersaturation also increased up to a ratio of 8, which resulted in a decrease in particle size [73]. Similar results were observed for beclomethasone dipropionate [75], simvastatin [81], griseofulvin [69], betulin [82], arteimisinin [83], where with the increase in AS/S ratio up to an optimal ratio, smaller particle sizes were obtained, after which particle size remains constant irrespective of further increase in the AS/S ratio.

3.1.4 Drug Concentration

The concentration of drug during a LAS crystallization process also impacts the resulting particle size and size distribution. With an increase in drug concentration, the supersaturation of the drug increases at a fixed solvent-antisolvent mixture. The increase in supersaturation leads to a high nucleation rate and thus decrease in particle size, yet very high drug concentrations, due to formation of numerous nuclei per unit volume in the AS/S mixture, which can result in increased frequency of collisions and subsequent aggregation. Dalvi et al. studied the effect of griseofulvin concentrations within a range of 5–40 mg/mL and observed that the particle size initially decreased with an increase in concentration up to 10 mg/mL but any further increase resulted in an increase in particle size [69]. Similar results were observed for glibenclamide [84] and L-nimodipine [85].

3.1.5 Additives

Employment of additives during the antisolvent crystallization method is a common strategy to help generate small particle sizes with a narrow size distribution. Additives are also critical to preserve the particle size distribution during long-term storage (in suspension) or during subsequent isolation and drying steps. During the LAS process, the additives are required to promote nucleation and inhibit or at least slow down the growth of particles. It is difficult to probe the action of additives during liquid antisolvent crystallization for the generation of drug nanosuspensions due to the exceptionally high supersaturations typically employed, which result in very fast nucleation and growth rates. Thus frequently, the effect of additives on conventional nucleation and crystal growth experiments at moderate supersaturation is studied and the results are linked to LAS at higher supersaturations [86]. The presence of additives can impact the nucleation kinetics by changing (i) the drug solubility, (ii) the nucleus–solution interfacial energy, (iii) the pre-exponential factor, or by (iv) switching the nucleation process from homogeneous to heterogeneous. Theoretical modelling by molecular dynamic simulations of interactions between additive, solute, and solvent has been performed to probe the influence of additives on nucleation [87]. It has been suggested that surfactants are most effective at increasing nucleation rates by reducing the nucleus–solution interfacial energy upon absorption onto the interface [87]. Surface active oligomeric additives were shown to increase nucleation rates by providing a nucleation center acting as a heterogeneous surface for nucleation [88]. It has also been shown that polyvinylpyrrolidone increased the nucleation rate of naproxen crystals by acting as a heterogeneous nucleation center in the crystallization process [89]. Notably, effect of additives can be varied by changing the additive concentration [90]. In the induction time experiment for vanillin, a reduction in induction time was observed at low additive concentration, whereas it had the opposite effect at high additive concentration [90]. The surfactant sodium docusate was found to enhance the nucleation rate of a mefenamic acid nanosuspension when it was present during nucleation and hydroxypropylmethyl cellulose (HPMC), added 5 s after nucleation, helped to prevent polymorphic transformation [4].

The size of drug nanoparticles prepared by both wet media milling and liquid antisolvent crystallization/precipitation and the respective additives used are compared in Table 2. The morphology of mefenamic acid nanoparticles prepared with liquid antisolvent precipitation showed more mature crystal habitat with distinctive edges compared to media-milled nanoparticles, Fig. 4.

3.2 Spray Drying

During spray drying, a drug solution (or nanosuspension) is pumped through a nozzle and atomized into droplets into a hot airflow or vacuum chamber. As a result, the solvent evaporates, and a dry powder forms [96–98]. During this process, porous or hollow low-density material can form due to the rapid evaporation of solvent content [99, 100]. Rotary wheel (or disk) atomizers, pressure nozzles, and two-fluid nozzles are the three major types of atomizers used in practice. During atomization, the atomizer turns the drug solution into small liquid droplets. In general, high atomizer pressure leads to smaller droplets and as a result, smaller particle sizes are obtained. There are three different air current systems with respect to spray drying [60, 97, 101]. If a heated drying medium is flowing in the same direction as the atomized feed, then it is termed as co-current flow. In this air current system,

Table 2 A comparative list of drug nanoparticles formed by both media milling and antisolvent crystallization technique

API	Wet milling	Liquid antisolvent crystallization/precipitation	Refs.
Fenofibrate	Particle size—285 nm Additive—SDS and HPMC-E3 Recirculation mode, 2 hour	Particle size—210 ± 8 nm Additive—HPMC	[91, 92]
Glyburide	Particle size—163 nm Additive—Kolliphor P188 120 min grinding	Particle size—298 ± 3.0 nm Additive—HPMC 15cps and PVPK-30	[19, 51]
Ibuprofen	Particle size—603 nm Additive—Pharmacoat 60 min grinding	Particle size—92 ± 2.5 nm Additive—HPMC and Pluronic F127	[19, 93]
Naproxen	Particle size—32 nm Additive—$ZrOCl_2$, electrostatic stabilization 6 hours grinding	Particle size—119.6 ± 12.3 nm Additive—Poloxamer-188	[52, 94]
Mefenamic acid	Particle size—172 ± 15 nm Additive—HPMC (SSL grade) 5 hours grinding	Particle size—312.0 ± 0.5 nm Additive—DOSS and HPMC	[4, 95]

Fig. 4 SEM images of mefenamic acid nanoparticles prepared by (**a**) media milling [95]; (**b**) antisolvent crystallization [4]

contact between the substance and hot air is minimal because of the rapid cooling of the air further below the chamber. It is mainly used for thermosensitive substances requiring minimum heat contact. Most food products (e.g., milk powder, coffee, flavors) are produced using a co-current spray drying system, due to its suitability for thermosensitive compounds A similar rationale applies in pharmaceutical applications for heat-sensitive drugs.

If the direction of the heated drying medium and atomized feed is opposite, then the spray drying process is termed counter-current flow. It is mainly used for non-heat-sensitive materials. This flow combines heat treatment and agglomeration effects, which result in increased powder flowability and median particle size. The combination of co-current and counter-current flow in a spray dryer is termed

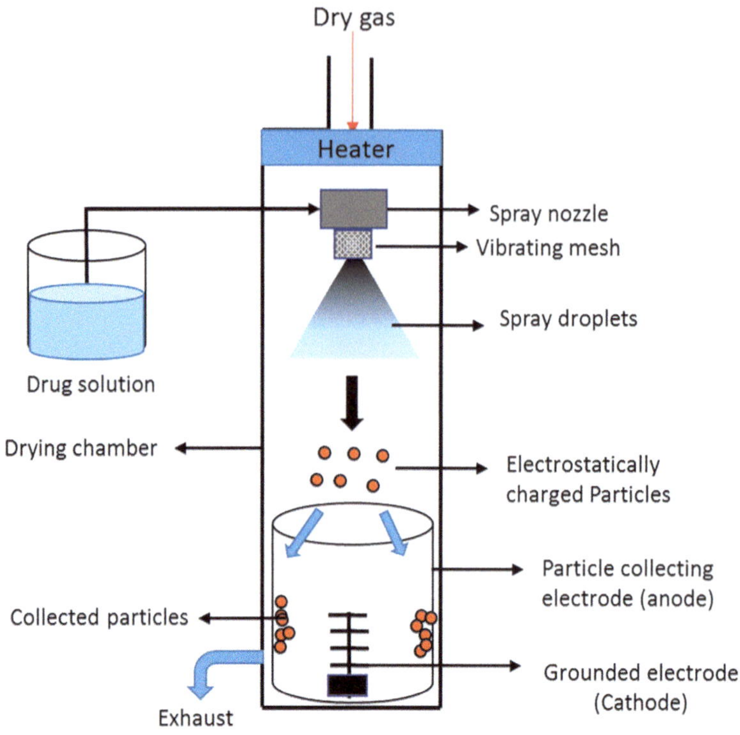

Fig. 5 Schematic diagram of nano spray dryer B-90 [104]

mixed-flow spray drying. For spray drying processes, faster drying due to higher temperature is suitable as slow drying rates may produce stickier particles, which lead to wall deposition and low yield [102]. Thus, spray drying is not appropriate for thermosensitive compounds as it can result in the formation of different polymorphic forms due to high temperature exposure. Also, different process parameters such as feed rate and inlet temperature can be controlled to achieve desired yield, particle morphology, and moisture content [102, 103].

Until recently, spray drying has been limited to the production of micron-sized particles. However, particle sizes in the range of 300–5000 nm can be produced by the new generation of spray dryers (for example, the Buchi B-90 from BUCHI Labortechnik AG, Flawil, Switzerland, Fig. 5) [105, 106]. These spray dryers require only milligram quantities of drug. Drug nanoparticles can be obtained directly from solution in high yields (70–90%). The novelty of this spray dryer lies in the spray head (vibrating mesh spray technology), heating system architecture, and electrostatic particle collector [105]. Baba et al. prepared nanoparticles of calpain inhibitor with an average size of 300 nm using the nanospray dryer, used in the treatment of diseases such as Alzheimer's or Parkinson's disease [104]. In another study, Harsha et al. produced nanoparticles of vildaglipcin for the treatment of diabetes with a particle size of 445 nm [107]. The yields obtained with spray drying can be limited (50–70%), but the nano spray dryer B-90 has increased these yields to 70–90% [59].

3.3 Supercritical Fluid Antisolvent Technique (SAS)

Supercritical fluid (SCF) technologies as alternatives to using organic solvents have been studied for the generation of drug nanoparticles (Fig. 6). SCFs exist at a temperature and pressure above the critical point of a phase that combines the properties of gases and liquids as it can effuse through solids like a gas and dissolve materials like a liquid [109, 110]. It is very easy to tune SCF solvent properties by very small changes in pressure and temperature close to the critical point [111]. Carbon dioxide (CO_2) is the most commonly used supercritical fluid for pharmaceutical applications because it forms at a very practical critical temperature (304.2 K, 31.1 °C) and pressure (7.38 MPa; 1070 psi) [112]. It is also nontoxic, non-flammable, inexpensive, and can be recycled.

In the supercritical fluid antisolvent (SAS) technique, supercritical fluid is used as the antisolvent, and depending on the experimental set up, the SCF can interact with the drug solution in different ways [113]. For example, in the *gaseous antisolvent* (GAS) method, SCF is passed through the drug solution in the form of gas, whereas in the *aerosol solvent extraction system* (ASES) or the *precipitation from compressed antisolvent* (PCA) method, the drug solution is sprayed in the form of fine mist into the chamber of compressed SCF [110, 114]. In both methods, dried nanoparticles can then be collected for downstream processing.

Higher supersaturation levels are generated in the PCA process compared with the GAS process because of the higher surface area of the aerosolized drug solution,

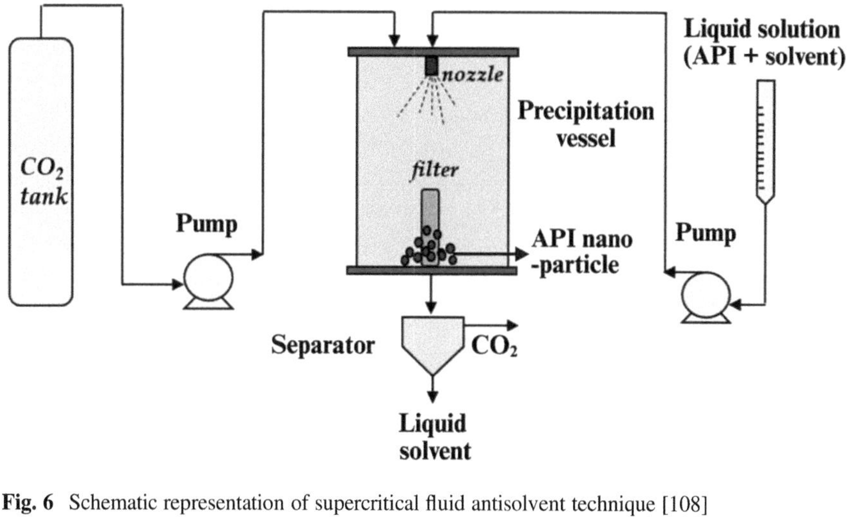

Fig. 6 Schematic representation of supercritical fluid antisolvent technique [108]

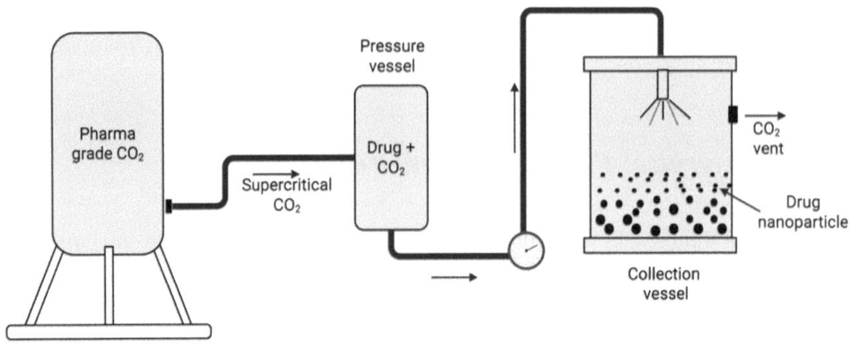

Fig. 7 Schematic representation of Nanoform's controlled drug solubility in supercritical CO_2 technology. (Created with Biorender®)

which results in a faster rate of mass transfer into the antisolvent [115]. Based on this principle, York and co-workers developed the solution enhanced dispersion method by SCF (SEDS) [116]. In this method, conditions are adjusted such that CO_2 remains in a supercritical state and a two-channelled coaxial nozzle is used. This allows a high-velocity stream of supercritical CO_2 to disperse and combine with a continuously flowing stream of drug solution; as a result, rapid nucleation and subsequent particle formation occur due to high rates of mass transfer between the SCF and the drug-containing solvent. Recent studies by Padrela et al. [117] demonstrated polymorphic control over carbamazepine using the GAS method through additive selection. In contrast, solvent crystallization through the evaporation method provided mixtures of different polymorphs when using the same set of additives. Thus, the SAS technique can also be used to tune the polymorphic outcome in the final drug powder.

However, these processes have several disadvantages, such as the need for pressure pumps to generate high pressures, temperatures, and the need for specially designed nozzles, which increase the cost of the technology [109, 110].

A recent commercially available method—Controlled Expansion of Supercritical Solutions (CESS®) technology generates nanoparticles directly from solution is offered by Nanoform, a rapidly growing Finnish company working with numerous multinational biopharmaceutical industries. In this bottom-up method (Fig. 7), the drug solubility in supercritical CO_2 is controlled at different temperatures and pressures. Particle size, morphology, and polymorphic properties can be controlled in this process.

3.4 Scaling Up of Precipitation Approaches

Antisolvent crystallization/precipitation for the production of drug nanoparticles with a narrow particle size distribution is difficult to scale up to industrial scale via batch mode as it requires a homogenous high level of supersaturation [118, 119]. While attaining a homogenous supersaturation distribution in a stirred vessel is facile at the lab-scale owing to rapid recirculation and higher shear [120], scaling up the stirred vessel could lead to generation of supersaturation gradients affected by factors such as the rate of stirring, impeller design [121], and the location of feed pipes [120].

In recent years, different methods have been used to intensify the mixing within tubular reactors, such as the development and application of continuous oscillatory baffled crystallizers [122, 123] and static mixers [124, 125] to help achieve near ideal plug flow-like behavior. Flow through curved pipes, like helical coils and coiled flow inverters, is another technique adopted for the intensification of mixing, where the occurrence of Dean vortices due to the pipe curvature brings about thorough mixing in the radial direction, while operating in the laminar regime reduces the dispersion coefficient [126, 127].

3.4.1 Continuous Crystallization/Precipitation Modes

Apart from the challenges in crystallizer design for scale up, there are also batch-to-batch product variability and process inefficiencies to consider [128]. In contrast, operating the process in continuous mode has many advantages like high productivity and better control over critical quality criteria [128, 129]. Plug-flow-like conditions can also ensure that a narrow particle size distribution is obtained [128]. Currently, there are limited data available in the literature on continuous antisolvent crystallization to form drug nanoparticles. In the reports available, drug nanoparticles are mainly isolated from suspension to dried powder by coupling antisolvent precipitation with a spray drying process [130–132].

Mixing setups such as impinging jet mixers [133, 134] or microfluidic antisolvent crystallizers [135, 136] provide rapid and homogeneous mixing and can be used in a continuous manner. Continuous antisolvent crystallization/precipitation can also be performed using hollow fibre membrane modules for better control over mixing to generate drug nanoparticles [137–139].

Continuous crystallizers in the pharmaceutical industry can be broadly classified into mixed suspension mixed product removal (MSMPR) and tubular crystallizers [140]. The most eminent difference between the two is the residence time distribution (RTD) of the crystallizers. The single-stage MSMPR displays a broad RTD (similar to that of continuous stirred tank reactor (CSTR)) while a narrower RTD (characteristic of plug flow reactors) is displayed by tubular crystallizers or a cascade of MSMPR (similar to CTSR-in-series) [141, 142].

Yuancai et al. developed a continuous antisolvent crystallization method of a poorly water-soluble drug, spironolactone, using a static mixer compromising 6–18 mixing elements to achieve rapid and homogeneous turbulent mixing and generate nanometre range (~500 nm) nanoparticles [143]. Several studies have been reported to generate micron/nano size drug particles combining cooling and antisolvent crystallization using continuous mixed suspension mixed product removal systems [144–147].

4 Isolation and Drying of Drug Nanoparticles

It is necessary to isolate drug nanoparticles from suspension into dried powder forms to increase their shelf life or for downstream processing into the final drug product. Isolation of nanosuspensions to their respective dried forms is challenging as they are prone to aggregate during separation from suspension to lower their high surface energy [5] or they do not separate easily. If filtration is utilized, then filter paper pores can become clogged by very small particles, resulting in unreasonably long filtering times for larger quantities (and agglomeration and particle growth during this process). Spray drying [148], described earlier in Sect. 3.1.3, and freeze drying [149] are separation techniques for nanosuspensions widely used in the pharmaceutical industry. Carrier particle-mediated isolation [150–154] has recently been reported in the literature as an alternative technology to isolate nanoparticles from suspension by simple and inexpensive filtration without the use of any harsh process conditions such as high temperatures or pressures.

4.1 Freeze Drying

Freeze-drying, also known as lyophilization or cryodesiccation, is an isolation technique that can be employed to remove solvents from materials. During the freeze-drying process, the nanosuspension is first frozen at normal pressure and then gradually heated under vacuum to produce a dried porous material [61]. While freezing can prevent diffusion and crystal growth, nanoparticle aggregation and fusion due to the capillary stresses caused during the process can occur [155]. Thus, maintaining re-dispersibility of the drug nanoparticles is one of the main challenges faced during freeze drying. This problem can be solved by incorporating a good dispersant into the nanosuspension prior to freeze-drying that can rapidly dissolve in contact with water or physiological fluid and prevents agglomeration [61, 62]. Normally, sugars and polymers are used as dispersants, and they can also act as cryoprotectants during the freezing process [61, 156]. Freeze-drying is generally a time and cost-inefficient process, and certain organic solvents present in the nanosuspension can cause problems with the process [156].

4.2 Carrier Particle-Mediated Isolation

An alternative, efficient and cost-effective method to isolate drug nanoparticles from suspension was recently reported through a carrier particle-mediated filtration process, Fig. 8. This technique involves the attachment of drug nanoparticles to the surface of insoluble carrier particles, which can subsequently be filtered to dryness. The faster dissolution rate or enhanced solubility profile of the drug nanoparticles will be retained if they attach without aggregating on the surface of the carrier particle. Khan et al. were the first to use carrier microparticles for nanocrystal

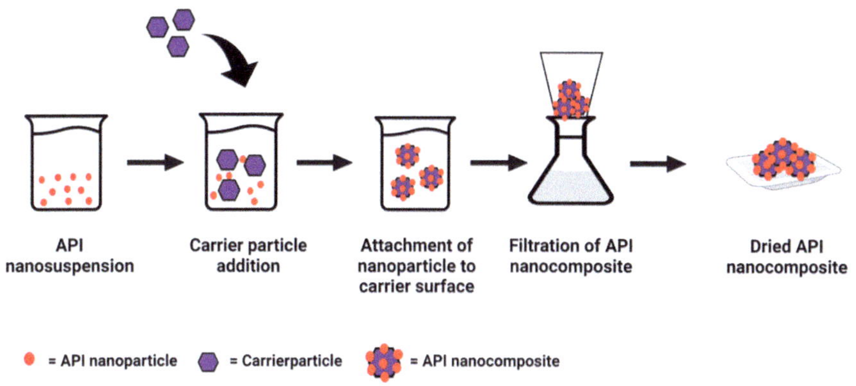

Fig. 8 Schematic representation of nanoparticle isolation and drying through a carrier particle-mediated filtration process. (Created with Biorender®)

Table 3 A list of carrier particles that have been used in nanocomposite formulations from the literature

Carrier particle	Drugs	Refs.
Dibasic calcium phosphate[a]	Ibuprofen, Glibenclamide	[157]
Montmorillonite[a]	Dalcetrapib, valsartan, clozapine, carbamazepine, curcumin	[152–154]
Protamine functionalized montmorillonite (PA-MMT)[b]	Fenofibrate, Mefenamic acid, valsartan, clozapine, carbamazepine, curcumin	[150, 152, 153]
Mesoporous silicate SBA-15[b]	Valsartan	[151]
Mesoporous organosilane ESE[b]	Valsartan	[151]

[a]From supplier (Sigma-Aldrich)
[b]Developed in-house

recovery from suspension [157]. In this process, nanosuspension of two drugs—ibuprofen and glibenclamide were prepared by either wet milling or an antisolvent crystallization process followed by the addition of insoluble dibasic calcium phosphate as a carrier particle. The nanoparticles attached onto the surface of large insoluble carrier particles and were separated easily by filtration using 0.4 μm filter paper. Dissolution studies of the resultant dry powders showed rapid dissolution rates comparable to the original aqueous nanosuspensions [157]. Tierney et al. used functionalized montmorillonite clay (MMT) as a carrier particle added during or, shortly after the formation of nanoparticles to isolate fenofibrate and mefenamic acid nanoparticles. A fast-dissolving dried nanocomposite montmorillonite (MMT) powder with up to 9.1 w/w% for fenofibrate and 4.8 w/w% for mefenamic acid was produced. Kumar et al. further developed this technology in a semi-continuous process to generate gram-scale dry fast-dissolving nanocomposite powders [158]. In this process, valsartan nanocomposite powders with MMT were generated on a nominal reactor scale of 60 mL at a loading of up to 30 w/w% valsartan to MMT. The application of carrier particles to separate drug nanoparticles from suspension into the dry state has high industrial appeal due to the simple infrastructure requirement. Further examples of this technology with other drugs are described in Table 3.

5 Additives to Stabilize Drug Nanoparticles

Whether drug nanoparticles are produced by size reduction or crystallization/precipitation methods, additives are usually required to stabilize the resulting nanosuspension or dry nanoparticles. The additive should prevent Ostwald ripening, agglomeration, and polymorphic transformation (unless desired). Generally regarded as safe (GRAS) surfactants and polymers are frequently used as additives for this purpose. The interaction of drug nanoparticles through attractive and steric forces, named colloidal interactions, directly affects nanoparticle stability. Therefore, manipulating these colloidal interactions can result in stable nanoparticles [26, 159].

To achieve steric stabilization, nonionic or amphiphilic polymers (e.g., polyvinyl alcohol, polyvinylpyrrolidone, polyethylene glycol, pluronic F-127 and F128, hydroxyl-propyl methylcellulose) are commonly used [159]. The anchor hydrophobic segments of these additives are adsorbed onto the drug particles, extending their hydrophilic tail segments into a hydrophilic solvent (or vice versa depending on the lipophilicity of the solvent being used). When one particle approaches another, for the stabilizing segments to interact with each other, the solvent molecules that have solvated the tail segments need to be displaced, Fig. 9. When the stabilizing segment is well solvated, particles are prevented from agglomeration as this desolvation or displacement of solvent molecules is energetically unfavourable [26].

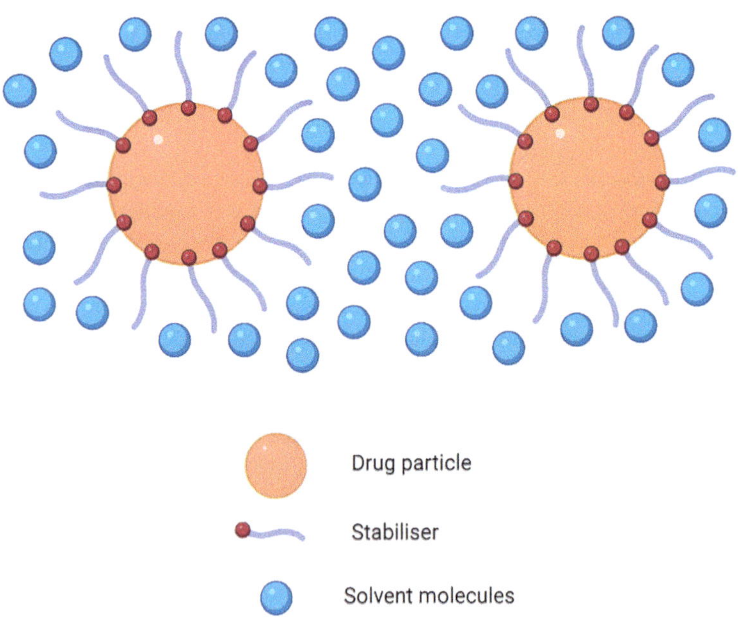

Fig. 9 Diagram of the steric stabilization of nanoparticles in suspension. (Created with BioRender®)

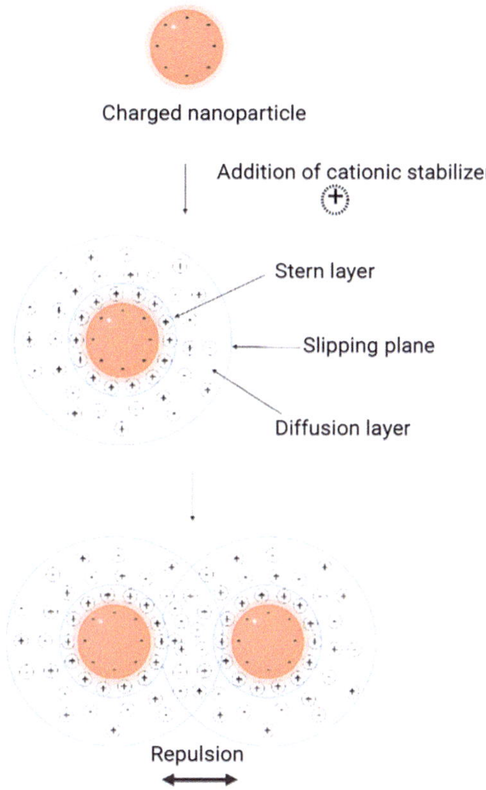

Fig. 10 DVLO theory for electrostatic stabilization. (Created with Biorender®)

To achieve electrostatic stabilization, ionic surfactants (e.g., sodium dodecyl sulphate or sodium lauryl sulphate, sodium docusate, sodium deoxycholate, lecithin) and charged polymers (e.g., chitosan, sodium alginate) are generally used to stabilize the drug nanoparticles via electrostatic repulsion [159]. The stabilization provided by electrostatic repulsion is described by DLVO theory [160, 161] (named after Derjaguin, Landau, Verwey, and Overbeek), Fig. 10. As the particle moves, the ions that surround the particles move with it up to a boundary called the slipping plane. The potential at this slipping plane is called the zeta potential, and can be used to predict the suspension stability. Normally, drug nanoparticles with a zeta potential more positive than +30 mV or more negative than −30 mV are stable [162], as the high repulsion between the particles prevents their agglomeration. The concentration of surfactants also plays an important role in the electrostatic stabilization of drug nanoparticles. When surfactants are used as additives, care must be taken to stay below the surfactant critical micellar concentration to ensure sufficient adsorption to the particle surface. At concentrations above the critical micelle concentration (CMC), micelles will form leaving fewer surfactant molecules available for particle stabilization [69].

Table 4 A nonexhaustive list of additives that have been used in nanoparticle formulations from the literature

Additive	Drugs	Refs.
Polyvinyl alcohol	Curcumin, nitrendipine, cinnarizine, griseofulvin, indomethacin, itraconazole, loviride, mebendazole, naproxen, phenylbutazone, phenytoin	[163–165]
Polyvinylpyrrolidone K30	Curcumin, glibizide, paclitaxel, rosuvastatin	[163, 166–168]
Hydroxypropyl methylcellulose	Glibizide, fenofibrate, naproxen, carbamazepine	[166, 169–171]
Tween 20	Etodolac, paclitaxel, folic acid	[167, 172, 173]
Tween 80	Brinzolamide, etodolac, celecoxib, mefenamic acid	[172, 174–176]
Poloxamer 407 (Pluronic® F127)	Valsartan, naproxen, brinzolamide, nimodipine	[152, 174, 177, 178]
Poloxamer 188 (Pluronic® F68)	Brinzolamide, budesonide, indomethacin	[174, 179, 180]
Sodium lauryl sulfate	Glibizide, naproxen, mefenamic acid	[166, 176, 177]
Vitamin E tocopherol polyethylene glycol succinate	Curcumin, naproxen, loviridine, itraconazole	[163, 165, 177]
Combination of Hydroxypropyl methylcellulose and sodium docusate	Fenofibrate, dalcetrapib	[181]
Combination of Poloxamer hydroxypropylcellulose and sodium lauryl sulphate	Azodicarbonamide, fenofibrate, griseofulvin, ibuprofen, phenylbutazone	[182]

Although there is limited data in the literature on the impact of using multiple additives for the stabilization of drug nanosuspensions, examples can be found in Table 4, suggesting this could provide enhanced stabilization.

The presence of additives can also affect the solid-state or polymorphic transformations of the nanoparticles. As mentioned before, additives interact with drug particles either by adsorption on the particle surfaces or by modifying the solution environment. This interaction can stabilize metastable or amorphous forms that would normally transition to a more thermodynamically stable state by dissolution and crystallization [183]. This transition is known as solution-mediated polymorphic transformation, and is induced by the difference in solubility between different polymorphs [184].

6 Conclusion

The emergence of drug nanoparticles as a transformative solution to address the challenges of poor water solubility in active pharmaceutical ingredients has greatly enhanced drug bioavailability and therapeutic efficacy. Size reduction approaches, such as wet milling and high-pressure homogenization, have been widely employed for drug nanoparticle production but come with limitations like energy intensity and long processing times. Crystallization/precipitation methods present an attractive alternative due to their cost-effectiveness, scalability, and low energy requirements. However, challenges in stabilizing particle size, polymorphic form, and maintaining long-term stability highlight the complexities associated with nanosuspension and nanoparticle handling and isolation.

Advances in drug nanoparticle isolation techniques such as spray drying, freeze drying, and emerging supercritical fluid technologies and carrier particle-mediated methods have significantly improved the ability to preserve nanoparticle properties while ensuring scalability and minimizing agglomeration risks. Thus, drug nanoparticles are set to play a pivotal role in developing next-generation pharmaceutical formulations. This chapter underscores the importance of continuous innovation in manufacturing techniques to meet the increasing demand for effective, scalable, and stable drug nanoparticle systems.

Looking ahead, the successful translation of these advances into scalable manufacturing processes and commercially available formulations will be pivotal. Bridging laboratory innovation with industrial production not only accelerates the availability of more effective therapies but also reshapes the pharmaceutical landscape by lowering development barriers, improving patient access, and enhancing healthcare outcomes worldwide.

References

1. T. Schikarski, H. Trzenschiok, M. Avila, W. Peukert, Impact of solvent properties on the precipitation of active pharmaceutical ingredients. Powder Technol. **415** (2023). https://doi.org/10.1016/j.powtec.2022.118032
2. A.B. Anane-Adjei, E. Jacobs, S.C. Nash, S. Askin, R. Soundararajan, M. Kyobula, J. Booth, A. Campbell, Amorphous solid dispersions: Utilization and challenges in preclinical drug development within AstraZeneca. Int. J. Pharm. (2022). https://doi.org/10.1016/j.ijpharm.2021.121387
3. A.O. Oladipo, S.L. Lebelo, T.A.M. Msagati, Nanocarrier design–function relationship: The prodigious role of properties in regulating biocompatibility for drug delivery applications. Chem. Biol. Interact. **25** (2023). https://doi.org/10.1016/j.cbi.2023.110466
4. K. Bodnar, S.P. Hudson, Å.C. Rasmuson, Stepwise use of additives for improved control over Formation & Stability of Mefenamic acid nanocrystals produced by Antisolvent precipitation. Cryst. Growth Des. **17**(2), 454–466 (2017). https://doi.org/10.1021/acs.cgd.6b01256
5. T.B. Tierney, Y. Guo, S. Beloshapkin, Å.C. Rasmuson, S.P. Hudson, Investigation of the particle growth of fenofibrate following antisolvent precipitation and freeze–drying. Cryst. Growth Des. **15**(11), 5213–5222 (2015)

6. T.B. Tierney, Å.C. Rasmuson, S.P. Hudson, Size and shape control of Micron-sized salicylic acid crystals during Antisolvent crystallization. Org. Process. Res. Dev. **21**(11), 1732–1740 (2017). https://doi.org/10.1021/acs.oprd.7b00181
7. V. Verma, K.M. Ryan, L. Padrela, Production and isolation of pharmaceutical drug nanoparticles. Int. J. Pharm. (2021). https://doi.org/10.1016/j.ijpharm.2021.120708
8. J.E. Kipp, The role of solid nanoparticle technology in the parenteral delivery of poorly water-soluble drugs. Int. J. Pharm. **13**, 109–122 (2004). https://doi.org/10.1016/j.ijpharm.2004.07.019
9. J. Brouwers, M.E. Brewster, P. Augustijns, Supersaturating drug delivery systems: The answer to solubility-limited oral bioavailability? J. Pharm. Sci., 2549–2572 (2009). https://doi.org/10.1002/jps.21650
10. F. Hládek, S. Chvíla, O. Navrátil, M. Balouch, F. Štěpánek, Systematic investigation of wet-milling kinetics and colloidal stability of pharmaceutical nanocrystals. Cryst. Growth Des. **22**(12), 6928–6940 (2022). https://doi.org/10.1021/acs.cgd.2c00546
11. M. Malamatari, K.M.G. Taylor, S. Malamataris, D. Douroumis, K. Kachrimanis, Pharmaceutical nanocrystals: Production by wet milling and applications. Drug Discov. Today, 534–547. https://doi.org/10.1016/j.drudis.2018.01.016
12. K.S. Yadav, K. Kale, High pressure homogenizer in pharmaceuticals: Understanding its critical processing parameters and applications. J. Pharm. Innov. **1**, 690–701 (2020). https://doi.org/10.1007/s12247-019-09413-4
13. A.A. Halwani, Development of pharmaceutical nanomedicines: From the bench to the market. Pharmaceutics (2022). https://doi.org/10.3390/pharmaceutics14010106
14. J.U.A.H. Junghanns, R.H. Müller, Nanocrystal technology, drug delivery and clinical applications. Int. J. Nanomedicine **3**(3), 295–309 (2008)
15. J.T. Joshi, A review on micronization techniques. J. Pharm. Sci. Technol. **3**(7), 651–681 (2011)
16. R. Kumar, S.V. Dalvi, P.F. Siril, Nanoparticle-based drugs and formulations: Current status and emerging applications. ACS Appl. Nano Mater. **26**, 4944–4961 (2020). https://doi.org/10.1021/acsanm.0c00606
17. A. O'Sullivan, K.M. Ryan, L. Padrela, Amorphization versus cocrystallization of celecoxib-tramadol hydrocholoride using CO2-assisted nano-spray drying. J CO_2 Util **73** (2023). https://doi.org/10.1016/j.jcou.2023.102529
18. K. Joshi, A. Chandra, K. Jain, S. Talegaonkar, Nanocrystalization: An emerging technology to enhance the bioavailability of poorly soluble drugs. Pharm Nanotechnol **7**(4), 259–278 (2019). https://doi.org/10.2174/2211738507666190405182524
19. S. Khan, M. de Matas, J. Zhang, J. Anwar, Nanocrystal preparation: Low-energy precipitation method revisited. Cryst. Growth Des. **13**(7), 2766–2777 (2013). https://doi.org/10.1021/cg4000473
20. M.E. Matteucci, M.A. Hotze, K.P. Johnston, R.O. Williams, Drug nanoparticles by Antisolvent precipitation: Mixing energy versus surfactant stabilization. Langmuir **22**(21), 8951–8959 (2006). https://doi.org/10.1021/la061122t
21. S.K. Poornachary, V.D. Chia, Y. Yani, G. Han, P.S. Chow, R.B.H. Tan, Anisotropic crystal growth inhibition by polymeric additives: Impact on modulation of naproxen crystal shape and size. Cryst. Growth Des. **17**(9), 4844–4854 (2017). https://doi.org/10.1021/acs.cgd.7b00802
22. Z.H. Loh, A.K. Samanta, P.W. Sia Heng, Overview of milling techniques for improving the solubility of poorly water-soluble drugs. Asian J. Pharm. Sci. **10**(4), 255–274 (2014). https://doi.org/10.1016/j.ajps.2014.12.006
23. R.H. Müller, K. Peters, Nanosuspensions for the formulation of poorly soluble drugs. I. Preparation by a size-reduction technique. Int. J. Pharm. **160**(2), 229–237 (1998). https://doi.org/10.1016/S0378-5173(97)00311-6
24. E. Merisko-Liversidge, G.G. Liversidge, Nanosizing for oral and parenteral drug delivery: A perspective on formulating poorly-water soluble compounds using wet media milling

technology. Adv. Drug Deliv. Rev. **63**(6), 427–440 (2011). https://doi.org/10.1016/j.addr.2010.12.007
25. B. Sinha, R.H. Müller, J.P. Möschwitzer, Bottom-up approaches for preparing drug nanocrystals: Formulations and factors affecting particle size. Int. J. Pharm. **453**(1), 126–141 (2013). https://doi.org/10.1016/j.ijpharm.2013.01.019
26. L. Wu, J. Zhang, W. Watanabe, Physical and chemical stability of drug nanoparticles. Adv. Drug Deliv. Rev. **63**(6), 456–469 (2011). https://doi.org/10.1016/j.addr.2011.02.001
27. E. Merisko-Liversidge, G.G. Liversidge, E.R. Cooper, Nanosizing: A formulation approach for poorly-water-soluble compounds. Eur. J. Pharm. Sci. **18**(2), 113–120 (2003). https://doi.org/10.1016/S0928-0987(02)00251-8
28. R. Al-Kassas, M. Bansal, J. Shaw, Nanosizing techniques for improving bioavailability of drugs. J. Control. Release, 202–212 (2017). https://doi.org/10.1016/j.jconrel.2017.06.003
29. V.B. Junyaprasert, B. Morakul, Nanocrystals for enhancement of oral bioavailability of poorly water-soluble drugs. Asian J. Pharm. Sci., 13–23 (2015). https://doi.org/10.1016/j.ajps.2014.08.005
30. M. Descamps, J.F. Willart, E. Dudognon, V. Caron, Transformation of pharmaceutical compounds upon milling and Comilling: The role of Tg. J. Pharm. Sci. **96**(5), 1398–1407 (2007). https://doi.org/10.1002/jps.20939
31. B. Van Eerdenbrugh, G. Van den Mooter, P. Augustijns, Top-down production of drug nanocrystals: Nanosuspension stabilization, miniaturization and transformation into solid products. Int. J. Pharm. **364**(1), 64–75 (2008). https://doi.org/10.1016/j.ijpharm.2008.07.023
32. N. Rasenack, B.W. Müller, Micron-size drug particles: Common and novel micronization techniques. Pharm. Dev. Technol. **9**(1), 1–13 (2004). https://doi.org/10.1081/PDT-120027417
33. M.M. Can, S. Ozcan, A. Ceylan, T. Firat, Effect of milling time on the synthesis of magnetite nanoparticles by wet milling. Mater. Sci. Eng. B **172**(1), 72–75 (2010). https://doi.org/10.1016/j.mseb.2010.04.019
34. R. Kumar, A.K. Thakur, P. Chaudhari, R.K. Arya, K.N. Gupta, D. Thapliyal, A. Bedar, R.S. Krishna, K.C. Pitchaiah, Nanoparticle preparation of pharmaceutical compounds via wet milling: Current status and future prospects. Powder Technol. (2024). https://doi.org/10.1016/j.powtec.2024.119430
35. A. Harter, L. Schenck, I. Lee, A. Cote, High-shear rotor-stator wet milling for drug substances: Expanding capability with improved scalability. Org. Process. Res. Dev. **17**(10), 1335–1344 (2013). https://doi.org/10.1021/op4001143
36. I. Ghosh, D. Schenck, S. Bose, C. Ruegger, Optimization of formulation and process parameters for the production of nanosuspension by wet media milling technique: Effect of vitamin e TPGS and nanocrystal particle size on oral absorption. Eur. J. Pharm. Sci. **47**(4), 718–728 (2012). https://doi.org/10.1016/j.ejps.2012.08.011
37. Y. Ye, X. Zhang, T. Zhang, H. Wang, B. Wu, Design and evaluation of injectable Niclosamide nanocrystals prepared by wet media milling technique. Drug Dev. Ind. Pharm. **41**(9), 1416–1424 (2015). https://doi.org/10.3109/03639045.2014.954585
38. G.V. Patel, V.B. Patel, A. Pathak, S.J. Rajput, Nanosuspension of Efavirenz for improved oral bioavailability: Formulation optimization, in vitro, in situ and in vivo evaluation. Drug Dev. Ind. Pharm. **40**(1), 80–91 (2014). https://doi.org/10.3109/03639045.2012.746362
39. X. Zhai, J. Lademann, C.M. Keck, R.H. Müller, Nanocrystals of medium soluble actives – novel concept for improved dermal delivery and production strategy. Int. J. Pharm. **470**(1–2), 141–150 (2014). https://doi.org/10.1016/j.ijpharm.2014.04.060
40. M. Kakran, R. Shegokar, N.G. Sahoo, L. Al Shaal, L. Li, R.H. Müller, Fabrication of quercetin nanocrystals: Comparison of different methods. Eur. J. Pharm. Biopharm. **80**(1), 113–121 (2012). https://doi.org/10.1016/j.ejpb.2011.08.006
41. R Muller, Pharmaceutical nanosuspensions for medicament administration as system of increased saturation solubility and rate of solution (1999)

42. C.M. Keck, R.H. Müller, Drug nanocrystals of poorly soluble drugs produced by high pressure homogenisation. Eur. J. Pharm. Biopharm., 3–16 (2006). https://doi.org/10.1016/j.ejpb.2005.05.009
43. RH Muller, R Becker, B Kruss, K Peters, 54 pharmaceutical nanosuspensions for medcament administrationas systems with increased saturation solubility and rate of solution 75 Inventors: 30 Foreign Application Priority Data; PCT Pub, 1999; Vol. 858
44. J.K. Patel, Y.V. Pathak, *Emerging Technologies for Nanoparticle Manufacturing* (Springer, Cham, 2021), pp. 109–128
45. J. Möschwitzer, G. Achleitner, H. Pomper, R.H. Müller, Development of an intravenously injectable chemically stable aqueous omeprazole formulation using nanosuspension technology. Eur. J. Pharm. Biopharm. **58**(3), 615–619 (2004). https://doi.org/10.1016/j.ejpb.2004.03.022
46. W. Sun, S. Mao, Y. Shi, L.C. Li, L. Fang, Nanonization of Itraconazole by high pressure homogenization: Stabilizer optimization and effect of particle size on oral absorption. J. Pharm. Sci. **100**(8), 3365–3373 (2011). https://doi.org/10.1002/jps.22587
47. B.E. Rabinow, Nanosuspensions in drug delivery. Nat. Rev. Drug Discov. **3**(9), 785–796 (2004). https://doi.org/10.1038/nrd1494
48. Y. Gao, Z. Li, M. Sun, H. Li, C. Guo, J. Cui, A. Li, F. Cao, Y. Xi, H. Lou, G. Zhai, Preparation, characterization, pharmacokinetics, and tissue distribution of curcumin nanosuspension with TPGS as stabilizer. Drug Dev. Ind. Pharm. **36**(10), 1225–1234 (2010). https://doi.org/10.3109/03639041003695139
49. N. Schöler, K. Krause, O. Kayser, R.H. Müller, K. Borner, H. Hahn, O. Liesenfeld, Atovaquone nanosuspensions show excellent therapeutic effect in a new murine model of reactivated toxoplasmosis. Antimicrob. Agents Chemother. **45**(6), 1771–1779 (2001). https://doi.org/10.1128/AAC.45.6.1771-1779.2001
50. N.G. Sahoo, M. Kakran, L.A. Shaal, L. Li, R.H. Müller, M. Pal, L.P. Tan, Preparation and characterization of quercetin nanocrystals. J. Pharm. Sci. **100**(6), 2379–2390 (2011). https://doi.org/10.1002/jps.22446
51. P.A. Mura, M. Cirri, A. Rossetti, D.A. Allemandi, A.J. Paredes, S.D. Palma, Preparation of glyburide nanocrystals with improved dissolution properties by dry-ball- and wet-bead-milling: Systematic comparison by experimental design of the performance of the two methods. J. Drug Deliv. Sci. Technol. **91** (2024). https://doi.org/10.1016/j.jddst.2023.105222
52. S. Maar, C. Damm, W. Peukert, Wet nanomilling of naproxen using a novel stabilization mechanism via zirconium complexation. Adv. Powder Technol. **33**(9) (2022). https://doi.org/10.1016/j.apt.2022.103723
53. S.C. Feijoo, W.W. Hayes, C.E. Watson, J.H. Martin, Effects of microfluidizer® technology on bacillus Licheniformis spores in ice cream mix. J. Dairy Sci. **80**(9), 2184–2187 (1997). https://doi.org/10.3168/jds.S0022-0302(97)76166-6
54. Y. Li, X. Zhao, Y. Zu, Y. Zhang, Y. Ge, C. Zhong, W. Wu, Preparation and characterization of micronized Ellagic acid using Antisolvent precipitation for oral delivery. Int. J. Pharm. **486**(1–2), 207–216 (2015). https://doi.org/10.1016/j.ijpharm.2015.03.071
55. B. Long, K.M. Ryan, L. Padrela, From batch to continuous – new opportunities for supercritical CO2 technology in pharmaceutical manufacturing. Eur. J. Pharm. Sci. **137**, 104971 (2019). https://doi.org/10.1016/j.ejps.2019.104971
56. L. Padrela, M.A. Rodrigues, A. Duarte, A.M.A. Dias, M.E.M. Braga, H.C. de Sousa, Supercritical carbon dioxide-based technologies for the production of drug nanoparticles/nanocrystals – a comprehensive review. Adv. Drug Deliv. Rev. **131**, 22–78 (2018). https://doi.org/10.1016/j.addr.2018.07.010
57. K.U. Pandey, S.K. Poornachary, S.V. Dalvi, Insights to the action of additives for stabilization of ultrafine particles of Fenofibrate in aqueous suspensions produced by Sonoprecipitation. Powder Technol. **363**, 310–325 (2020). https://doi.org/10.1016/j.powtec.2019.12.059

58. H.K. Chan, P.C.L. Kwok, Production methods for nanodrug particles using the bottom-up approach. Adv. Drug Deliv. Rev. **63**(6), 406–416 (2011). https://doi.org/10.1016/j.addr.2011.03.011
59. X. Li, N. Anton, C. Arpagaus, F. Belleteix, T.F. Vandamme, Nanoparticles by spray drying using innovative new technology: The Büchi Nano spray dryer B-90. J. Control. Release **147**(2), 304–310 (2010). https://doi.org/10.1016/j.jconrel.2010.07.113
60. F. Users, A.T.J. Hayward, J. Wiley, Spray Drying Handbook (Third Edition) by K. Masters, 687 Pages, George Godwin Limited (London) (1979); John Wiley & Sons (New York) (1979). AIChE J. **26**(4), 702–702 (1980). https://doi.org/10.1002/AIC.690260430
61. S. Jacob, A.B. Nair, J. Shah, Emerging role of nanosuspensions in drug delivery systems. Biomater Res **24**(1), 1–16 (2020). https://doi.org/10.1186/s40824-020-0184-8
62. P.F. Yue, G. Li, J.X. Dan, Z.F. Wu, C.H. Wang, W.F. Zhu, M. Yang, Study on formability of solid nanosuspensions during solidification: II novel roles of freezing stress and Cryoprotectant property. Int. J. Pharm. **475**(1–2), 35–48 (2014). https://doi.org/10.1016/j.ijpharm.2014.08.041
63. J. Nývlt, in *Kinetics of nucleation in solutions*, ed. by North-Holland Publishing Co, vol. 3, (1968)
64. A.S. Myerson, S.R. Anderson, R.C. Bennett, D. Green, P. Karpinski, *Handbook of Industrial Crystallization* (Butterworth-Heinemann, 2019)
65. J. Hu, W.K. Ng, Y. Dong, S. Shen, R.B.H. Tan, Continuous and scalable process for water-redispersible nanoformulation of poorly aqueous soluble APIs by Antisolvent precipitation and spray-drying. Int. J. Pharm. **404**(1–2), 198–204 (2011). https://doi.org/10.1016/j.ijpharm.2010.10.055
66. S. Bansal, M. Bansal, R. Kumria, Nanocrystals: Current strategies and trends. Int J Res Pharm Biomed Sci **4**(November), 10 (2012)
67. H. Trzenschiok, M. Distaso, W. Peukert, A new approach for the stabilization of amorphous drug nanoparticles during continuous Antisolvent precipitation. Chem. Eng. J. **361**, 428–438 (2019). https://doi.org/10.1016/j.cej.2018.12.075
68. S.L. Raghavan, K. Schuessel, A. Davis, J. Hadgraft, Formation and stabilisation of Triclosan colloidal suspensions using supersaturated systems. Int. J. Pharm. **261**(1–2), 153–158 (2003). https://doi.org/10.1016/S0378-5173(03)00299-0
69. S.V. Dalvi, R.N. Dave, Controlling particle size of a poorly water-soluble drug using ultrasound and stabilizers in Antisolvent precipitation. Ind. Eng. Chem. Res. **48**(16), 7581–7593 (2009). https://doi.org/10.1021/ie900248f
70. C. Beck, S.V. Dalvi, R.N. Dave, Controlled liquid Antisolvent precipitation using a rapid mixing device. Chem. Eng. Sci. **65**(21), 5669–5675 (2010). https://doi.org/10.1016/j.ces.2010.04.001
71. H. Rehage, M. Kind, The first Damköhler number and its importance for characterizing the influence of mixing on competitive chemical reactions. Chem. Eng. Sci. **229** (2021). https://doi.org/10.1016/j.ces.2020.116007
72. M. Torbacke, Å.C. Rasmuson, Influence of different scales of mixing in reaction crystallization. Chem. Eng. Sci. **56**(7), 2459–2473 (2001). https://doi.org/10.1016/S0009-2509(00)00452-8
73. Z.B. Zhang, Z.G. Shen, J.X. Wang, H. Zhao, J.F. Chen, J. Yun, Nanonization of Megestrol acetate by liquid precipitation. Ind. Eng. Chem. Res. **48**(18), 8493–8499 (2009). https://doi.org/10.1021/ie900944y
74. M. Kakran, N.G. Sahoo, I.L. Tan, L. Li, Preparation of nanoparticles of poorly water-soluble antioxidant curcumin by Antisolvent precipitation methods. J. Nanopart. Res. ***14***(3) (2012). https://doi.org/10.1007/s11051-012-0757-0
75. Z. Wang, J.F. Chen, Y. Le, Z.G. Shen, J. Yun, Preparation of ultrafine beclomethasone dipropionate drug powder by antisolvent precipitation. Ind. Eng. Chem. Res. **46**(14), 4839–4845 (2007)

76. E. Sah, H. Sah, Recent trends in preparation of poly(Lactide-co-Glycolide) nanoparticles by mixing polymeric organic solution with Antisolvent. J. Nanomater. **2015** (2015). https://doi.org/10.1155/2015/794601
77. H. Chiou, H.K. Chan, D. Heng, et al., A novel production method for inhalable cyclosporine a powders by confined liquid impinging jet precipitation. J. Aerosol Sci. **39**(6), 500–509 (2008). https://doi.org/10.1016/j.jaerosci.2008.02.002
78. J.M. Lim, A. Swami, L.M. Gilson, S. Chopra, S. Choi, J. Wu, R. Langer, R. Karnik, O.C. Farokhzad, Ultra-high throughput synthesis of nanoparticles with homogeneous size distribution using a coaxial turbulent jet mixer. ACS Nano **8**(6), 6056–6065 (2014). https://doi.org/10.1021/nn501371n
79. A.N. Joshi, A.V. Ganjare, S. Hudson, V.V. Ranade, Micro-mixing characteristics of a pinched-tube, a fluidic oscillator and a vortex-based cavitation device. Chem. Eng. J. **515** (2025). https://doi.org/10.1016/j.cej.2025.163657
80. R. Kumar, A.K. Thakur, N. Banerjee, A. Kumar, G.K. Gaurav, R.K. Arya, Liquid Antisolvent crystallization of pharmaceutical compounds: Current status and future perspectives. Drug Deliv. Transl. Res. **1**, 400–418 (2023). https://doi.org/10.1007/s13346-022-01219-1
81. T. Jiang, N. Han, B. Zhao, Y. Xie, S. Wang, Enhanced dissolution rate and oral bioavailability of simvastatin nanocrystal prepared by Sonoprecipitation. Drug Dev. Ind. Pharm. **38**(10), 1230–1239 (2012). https://doi.org/10.3109/03639045.2011.645830
82. X. Zhao, W. Wang, Y. Zu, Y. Zhang, Y. Li, W. Sun, C. Shan, Y. Ge, Preparation and characterization of Betulin nanoparticles for oral hypoglycemic drug by Antisolvent precipitation. Drug Deliv. **21**(6), 467–479 (2014). https://doi.org/10.3109/10717544.2014.881438
83. M. Kakran, N.G. Sahoo, L. Li, Z. Judeh, Particle size reduction of poorly water soluble Artemisinin via Antisolvent precipitation with a syringe pump. Powder Technol. **237**, 468–476 (2013). https://doi.org/10.1016/j.powtec.2012.12.029
84. R.D. Deshpande, D.V. Gowda, N.S.K.V. Vegesna, R. Vaghela, P.K. Kulkarni, The effect of Nanonization on poorly water soluble Glibenclamide using a liquid anti-solvent precipitation technique: Aqueous solubility, in vitro and in vivo study. RSC Adv. **5**(99), 81728–81738 (2015). https://doi.org/10.1039/c5ra12678a
85. Y. Zu, N. Li, X. Zhao, Y. Li, Y. Ge, W. Wang, K. Wang, Y. Liu, In vitro dissolution enhancement of micronized L-Nimodipine by Antisolvent re-crystallization from its crystal form H. Int. J. Pharm. **464**(1–2), 1–9 (2014). https://doi.org/10.1016/j.ijpharm.2014.01.020
86. C.C. Sun, W. Sun, S. Price, C. Hughes, J. Ter Horst, S. Veesler, K. Lewtas, A. Myerson, H. Pan, G. Coquerel, J. Van Den Ende, H. Meekes, M. Mazzotti, I. Rosbottom, F. Taulelle, S. Black, A. Mackenzie, S. Janbon, P. Vekilov, T. Threlfall, T. Turner, K. Back, H. Cuppen, D. Toroz, J. Sefcik, J. Lovelock, R. Hammond, N. Candoni, E. Simone, M. Ward, C.A. Bertran, T. Vetter, R. Sear, J. De Yoreo, R. Davey, J. Anwar, E. Santiso, D.T. Wu, K. Roberts, B. Peters, S. Schroeder, F. Jones, A. Rasmuson, H. Cölfen, J. Zeglinski, M. Salvalaglio, Solvent and additive interactions as determinants in the nucleation pathway: General discussion. Faraday Discuss. **179**(July), 383–420 (2015). https://doi.org/10.1039/c5fd90038g
87. J. Anwar, P.K. Boateng, R. Tamaki, S. Odedra, Mode of action and design rules for additives that modulate crystal nucleation. Angew. Chem. Int. Edition **48**(9), 1596–1600 (2009). https://doi.org/10.1002/anie.200804553
88. G.G. Poon, T. Lemke, C. Peter, V. Molinero, B. Peters, Soluble Oligomeric Nucleants: Simulations of chain length, binding strength, and volume fraction effects. J. Phys. Chem. Lett. **8**(23), 5815–5820 (2017). https://doi.org/10.1021/acs.jpclett.7b02651
89. S.K. Poornachary, G. Han, J.W. Kwek, P.S. Chow, R.B.H. Tan, Crystallizing micronized particles of a poorly water-soluble active pharmaceutical ingredient: Nucleation enhancement by polymeric additives. Cryst. Growth Des. **16**(2), 749–758 (2016). https://doi.org/10.1021/acs.cgd.5b01343

90. O. Pino-García, Å.C. Rasmuson, Influence of additives on nucleation of vanillin: Experiments and introductory molecular simulations. Cryst. Growth Des. **4**(5), 1025–1037 (2004). https://doi.org/10.1021/cg049955+
91. B.D. Kevadiya, M. Barvaliya, L. Zhang, A. Anovadiya, H. Brahmbhatt, P. Paul, C. Tripathi, Fenofibrate nanocrystals embedded in oral strip-films for bioavailability enhancement. Bioengineering **5**(1) (2018). https://doi.org/10.3390/bioengineering5010016
92. R. Kumar, P.F. Siril, Enhancing the solubility of Fenofibrate by nanocrystal formation and encapsulation. AAPS PharmSciTech **19**(1), 284–292 (2018). https://doi.org/10.1208/s12249-017-0840-z
93. A. Ouranidis, N. Gkampelis, E. Vardaka, A. Karagianni, D. Tsiptsios, I. Nikolakakis, K. Kachrimanis, Overcoming the solubility barrier of ibuprofen by the rational process design of a nanocrystal formulation. Pharmaceutics **12**(10), 1–32 (2020). https://doi.org/10.3390/pharmaceutics12100969
94. P. Sindhuri, Formulation characterization of statistically optimized naproxen nanocrystals with quality by design. Asian J. Pharm. **16**(1) (2022)
95. A. Ito, C. Konnerth, J. Schmidt, W. Peukert, Effect of polymer species and concentration on the production of Mefenamic acid nanoparticles by media milling. Eur. J. Pharm. Biopharm. **98**, 98–107 (2016). https://doi.org/10.1016/j.ejpb.2015.11.011
96. M. Malamatari, A. Charisi, S. Malamataris, K. Kachrimanis, I. Nikolakakis, Spray drying for the preparation of nanoparticle-based drug formulations as dry powders for inhalation. PRO **8**(7) (2020). https://doi.org/10.3390/pr8070788
97. U. Wais, A.W. Jackson, T. He, H. Zhang, Nanoformulation and encapsulation approaches for poorly water-soluble drug nanoparticles. Nanoscale **8**(4), 1746–1769 (2016). https://doi.org/10.1039/c5nr07161e
98. J. Broadhead, S.K. Edmond Rouan, C.T. Rhodes, The spray drying of pharmaceuticals. Drug Dev. Ind. Pharm. **18**(11–12), 1169–1206 (1992)
99. C.H. Voon, K.L. Foo, B.Y. Lim, S.C.B. Gopinath, Y. Al-Douri, Synthesis and preparation of metal oxide powders, in *Metal Oxide Powder Technologies*, (Elsevier, 2020), pp. 31–65
100. P.K. Mehrotra, *Powder Processing and Green Shaping*, vol 1 (Elsevier Ltd, 2014). https://doi.org/10.1016/B978-0-08-096527-7.00007-6
101. M.W. Woo, B. Bhandari, Spray drying for food powder production, in *Handbook of Food Powders: Processes and Properties*, (2013), pp. 29–56. https://doi.org/10.1533/9780857098672.1.29
102. A. Ziaee, A.B. Albadarin, L. Padrela, T. Femmer, E. O'Reilly, G. Walker, Spray drying of pharmaceuticals and biopharmaceuticals: Critical parameters and experimental process optimization approaches. Eur. J. Pharm. Sci. **127**. (2018), 300–318 (2019). https://doi.org/10.1016/j.ejps.2018.10.026
103. M. Malamatari, S. Somavarapu, K.M.G. Taylor, G. Buckton, Solidification of Nanosuspensions for the production of solid oral dosage forms and inhalable dry powders. Expert Opin. Drug Deliv. **13**(3), 435–450 (2016). https://doi.org/10.1517/17425247.2016.1142524
104. K. Baba, K. Nishida, Calpain inhibitor nanocrystals prepared using nano spray dryer B-90. Nanoscale Res. Lett. **7**(1), 436 (2012)
105. D. Heng, S.H. Lee, W.K. Ng, R.B. Tan, The nano spray dryer B-90. Expert Opin. Drug Deliv. **8**(7), 965–972 (2011). https://doi.org/10.1517/17425247.2011.588206
106. H.D. Williams, N.L. Trevaskis, S.A. Charman, R.M. Shanker, W.N. Charman, C.W. Pouton, C.J. Porter, Strategies to address low drug solubility in discovery and development. Pharmacol. Rev. **65**(1), 315–499 (2013)
107. S.N. Harsha, B.E. Aldhubiab, A.B. Nair, I.A. Alhaider, M. Attimarad, K.N. Venugopala, S. Srinivasan, N. Gangadhar, A.H. Asif, Nanoparticle formulation by Büchi B-90 Nano spray dryer for oral Mucoadhesion. Drug Des. Devel. Ther. **9**, 273 (2015). https://doi.org/10.2147/DDDT.S66654

108. P. Franco, I. De Marco, Nanoparticles and nanocrystals by supercritical CO2-assisted techniques for pharmaceutical applications: A review. Appl. Sci. (Switzerland) **11**(4), 1–27 (2021). https://doi.org/10.3390/app11041476
109. Palakodaty Srinivas, York Peter, Phase behavioral effects on particle formation processes using supercritical fluids. Pharm. Res. **16**(7), 976–985 (1999)
110. I. Pasquali, R. Bettini, F. Giordano, Solid-state chemistry and particle engineering with supercritical fluids in pharmaceutics. Eur. J. Pharm. Sci., 299–310 (2006). https://doi.org/10.1016/j.ejps.2005.11.007
111. E.M. Phillips, V.J. Stella, Rapid expansion from supercritical solutions: application to pharmaceutical processes. Int. J. Pharm. **94**(1–3), 1–10 (1993)
112. M. Bahrami, S. Ranjbarian, Production of micro- and nano-composite particles by supercritical carbon dioxide. J. Supercrit. Fluids, 263–283 (2007). https://doi.org/10.1016/j.supflu.2006.05.006
113. J. Jung, M. Perrut, Particle design using supercritical fluids: Literature and patent survey. J. Supercrit. Fluids **20**(3), 179–219 (2001)
114. G. Muhrer, U. Meier, F. Fusaro, S. Albano, M. Mazzotti, Use of compressed gas precipitation to enhance the dissolution behavior of a poorly water-soluble drug: Generation of drug microparticles and drug-polymer solid dispersions. Int. J. Pharm. **308**(1–2), 69–83 (2006). https://doi.org/10.1016/j.ijpharm.2005.10.026
115. F. Fusaro, M. Häuchen, M. Mazzotti, G. Muhrer, B. Subramaniam, Dense gas Antisolvent precipitation: A comparative investigation of the GAS and PCA techniques. Ind. Eng. Chem. Res. **44**(5), 1502–1509 (2005). https://doi.org/10.1021/ie049495h
116. R.K. Kankala, B.Q. Chen, C.G. Liu, H.X. Tang, S. Wang, et al., Solution-enhanced dispersion by supercritical fluids: An ecofriendly nanonization approach for processing biomaterials and pharmaceutical compounds. Int. J. Nanomedicine, 4227–4245 (2018). https://doi.org/10.2147/IJN.S166124
117. L. Padrela, J. Zeglinski, K.M. Ryan, Insight into the role of additives in controlling polymorphic outcome: A CO2-Antisolvent crystallization process of carbamazepine. Cryst. Growth Des. **17**(9), 4544–4553 (2017). https://doi.org/10.1021/acs.cgd.7b00163
118. S. Nandi, L. Padrela, L. Tajber, A. Collas, Development of Long-acting injectable suspensions by continuous Antisolvent crystallization: An integrated bottom-up process. Int. J. Pharm. **648** (2023). https://doi.org/10.1016/j.ijpharm.2023.123550
119. S. Jia, P. Yang, Z. Gao, Z. Li, C. Fang, J. Gong, Recent progress in Antisolvent crystallization. CrystEngComm **24**(17), 3122–3135 (2022). https://doi.org/10.1039/d2ce00059h
120. J.B. Joshi, N.K. Nere, C.V. Rane, B.N. Murthy, C.S. Mathpati, A.W. Patwardhan, V.V. Ranade, CFD simulation of stirred tanks: Comparison of turbulence models. Part I: Radial flow impellers. Can. J. Chem. Eng. **89**(1), 23–82 (2011). https://doi.org/10.1002/cjce.20446
121. A.R. Khopkar, V.V. Ranade, Stirred vessels: Computational modeling of multiphase flows and mixing, in *Chemical Engineering in the Pharmaceutical Industry: Active Pharmaceutical Ingredients*, (2019), pp. 261–317
122. A. Egedy, J.A. Oliva, B. Szilágyi, Z.K. Nagy, Experimental analysis and compartmental modeling of the residence time distribution in DN6 and DN15 continuous oscillatory baffled crystallizer (COBC) systems. Chem. Eng. Res. Des. **161**, 322–331 (2020). https://doi.org/10.1016/j.cherd.2020.07.011
123. S. Lawton, G. Steele, P. Shering, L. Zhao, I. Laird, X.W. Ni, Continuous crystallization of pharmaceuticals using a continuous oscillatory baffled crystallizer. Org. Process. Res. Dev. **13**(6), 1357–1363 (2009). https://doi.org/10.1021/op900237x
124. A.J. Alvarez, A.S. Myerson, Continuous plug flow crystallization of pharmaceutical compounds. Cryst. Growth Des. **10**(5), 2219–2228 (2010). https://doi.org/10.1021/cg901496s
125. J. Albertazzi, F. Florit, V. Busini, R. Rota, Mixing efficiency and residence time distributions of a side-injection tubular reactor equipped with static mixers. Ind. Eng. Chem. Res. **60**(29), 10595–10602 (2021). https://doi.org/10.1021/acs.iecr.1c00575

126. L. Hohmann, T. Greinert, O. Mierka, S. Turek, G. Schembecker, E. Bayraktar, K. Wohlgemuth, N. Kockmann, Analysis of crystal size dispersion effects in a continuous coiled tubular crystallizer: Experiments and modeling. Cryst. Growth Des. **18**(3), 1459–1473 (2018). https://doi.org/10.1021/acs.cgd.7b01383
127. L. Sharma, K.D.P. Nigam, S. Roy, Single phase mixing in coiled tubes and coiled flow inverters in different flow regimes. Chem. Eng. Sci. **160**, 227–235 (2017). https://doi.org/10.1016/j.ces.2016.11.034
128. D. Zhang, S. Xu, S. Du, J. Wang, J. Gong, Progress of pharmaceutical continuous crystallization. Engineering **3**(3), 354–364 (2017). https://doi.org/10.1016/J.ENG.2017.03.023
129. Z.K. Nagy, A. El Hagrasy, J. Litster (eds.), *Continuous Pharmaceutical Processing*, vol 42 (Springer, Cham, 2020)
130. S. Kumar, J. Shen, D.J. Burgess, Nano-amorphous spray dried powder to improve oral bioavailability of Itraconazole. J. Control. Release **192**, 95–102 (2014). https://doi.org/10.1016/j.jconrel.2014.06.059
131. Q. Ma, H. Sun, E. Che, X. Zheng, T. Jiang, C. Sun, S. Wang, Uniform nano-sized valsartan for dissolution and bioavailability enhancement: Influence of particle size and crystalline state. Int. J. Pharm. **441**(1–2), 75–81 (2013). https://doi.org/10.1016/j.ijpharm.2012.12.025
132. J. Hu, Y. Dong, W.K. Ng, G. Pastorin, Preparation of drug nanocrystals embedded in mannitol microcrystals via liquid Antisolvent precipitation followed by immediate (on-line) spray drying. Adv. Powder Technol. **29**(4), 957–963 (2018). https://doi.org/10.1016/j.apt.2018.01.013
133. K. Tacsi, Á. Joó, É. Pusztai, A. Domokos, Z.K. Nagy, G. Marosi, H. Pataki, Development of a triple impinging jet mixer for continuous Antisolvent crystallization of acetylsalicylic acid reaction mixture. Chem. Eng. Process. Process Intensif. **165** (2021). https://doi.org/10.1016/j.cep.2021.108446
134. M. Jiang, Y.E.D. Li, H.H. Tung, R.D. Braatz, Effect of jet velocity on crystal size distribution from Antisolvent and cooling crystallizations in a dual impinging jet mixer. Chem. Eng. Process. Process Intensif. **97**, 242–247 (2015). https://doi.org/10.1016/j.cep.2015.09.005
135. F. Liu, W. Luo, J. Qiu, Y. Guo, S. Zhao, B. Bao, Continuous Antisolvent crystallization of Dolutegravir sodium using microfluidics. Ind. Eng. Chem. Res. (2022). https://doi.org/10.1021/acs.iecr.2c00655
136. S. Nandi, L. Verstrepen, M. Hugo Silva, L. Padrela, L. Tajber, A. Collas, Continuous microfluidic Antisolvent crystallization as a bottom-up solution for the development of long-acting injectable formulations. Pharmaceutics ***16***(3) (2024). https://doi.org/10.3390/pharmaceutics16030376
137. R. Othman, G.T. Vladisavljevic, E. Simone, Z.K. Nagy, R.G. Holdich, Preparation of microcrystals of Piroxicam monohydrate by Antisolvent precipitation via microfabricated metallic membranes with ordered pore arrays. Cryst. Growth Des. **17**(12), 6692–6702 (2017). https://doi.org/10.1021/acs.cgd.7b01307
138. X. Jiang, Y. Shao, L. Sheng, P. Li, G. He, Membrane crystallization for process intensification and control: A review. Engineering **1**, 50–62 (2021). https://doi.org/10.1016/j.eng.2020.06.024
139. J.C.W. Fern, S. Ohsaki, S. Watano, R. Pfeffer, Continuous synthesis of nano-drug particles by Antisolvent crystallization using a porous hollow-fiber membrane module. Int. J. Pharm. **543**(1–2), 139–150 (2018). https://doi.org/10.1016/j.ijpharm.2018.03.025
140. J. Orehek, D. Teslić, B. Likozar, Continuous crystallization processes in pharmaceutical manufacturing: A review. Org. Process Res. Dev., 16–42 (2021). https://doi.org/10.1021/acs.oprd.0c00398
141. Y. Mo, K.F. Jensen, A miniature CSTR Cascade for continuous flow of reactions containing solids. React Chem Eng **1**(5), 501–507 (2016). https://doi.org/10.1039/c6re00132g
142. C. Hu, Reactor design and selection for effective continuous manufacturing of pharmaceuticals. J. Flow Chem. **11**(3), 243–263 (2021)

143. Y. Dong, W.K. Ng, J. Hu, S. Shen, R.B.H. Tan, A continuous and highly effective static mixing process for Antisolvent precipitation of nanoparticles of poorly water-soluble drugs. Int. J. Pharm. **386**(1–2), 256–261 (2010). https://doi.org/10.1016/j.ijpharm.2009.11.007
144. J.M. Schall, J.S. Mandur, R.D. Braatz, A.S. Myerson, Nucleation and growth kinetics for combined cooling and Antisolvent crystallization in a mixed-suspension, mixed-product removal system: Estimating solvent dependency. Cryst. Growth Des. **18**(3), 1560–1570 (2018). https://doi.org/10.1021/acs.cgd.7b01528
145. M. Lenka, D. Sarkar, Combined cooling and Antisolvent crystallization of L-asparagine monohydrate. Powder Technol. **334**, 106–116 (2018). https://doi.org/10.1016/j.powtec.2018.04.038
146. H. Zhang, J. Quon, A.J. Alvarez, J. Evans, A.S. Myerson, B. Trout, Development of continuous anti-solvent/cooling crystallization process using cascaded mixed suspension, mixed product removal crystallizers. Org. Process. Res. Dev. **16**(5), 915–924 (2012). https://doi.org/10.1021/op2002886
147. Y. Yang, Z.K. Nagy, Advanced control approaches for combined cooling/Antisolvent crystallization in continuous mixed suspension mixed product removal cascade crystallizers. Chem. Eng. Sci. **127**, 362–373 (2015). https://doi.org/10.1016/j.ces.2015.01.060
148. H.X. Zhang, J.X. Wang, Z.B. Zhang, Y. Le, Z.G. Shen, J.F. Chen, Micronization of atorvastatin calcium by Antisolvent precipitation process. Int. J. Pharm. **374**(1–2), 106–113 (2009). https://doi.org/10.1016/j.ijpharm.2009.02.015
149. A. Tuomela, J. Hirvonen, L. Peltonen, Stabilizing agents for drug nanocrystals: Effect on bioavailability. Pharmaceutics **8**(2) (2016). https://doi.org/10.3390/pharmaceutics8020016
150. T. Tierney, K. Bodnár, Å. Rasmuson, S. Hudson, Carrier particle design for stabilization and Isolation of drug nanoparticles. Int. J. Pharm. **518**(1–2), 111–118 (2017). https://doi.org/10.1016/j.ijpharm.2016.11.045
151. M. Bergillos-Ruiz, A. Kumar, B.K. Hodnett, P. Davern, Å. Rasmuson, S.P. Hudson, Impact of carrier particle surface properties on drug nanoparticle attachment. Int. J. Pharm., 123743 (2023). https://doi.org/10.1016/j.ijpharm.2023.123743
152. A. Kumar, P. Davern, B.K. Hodnett, S.P. Hudson, Carrier particle mediated stabilization and isolation of valsartan nanoparticles. Colloids Surf. B Biointerfaces **175**, 554–563 (2019). https://doi.org/10.1016/j.colsurfb.2018.12.021
153. A. Kumar, B.K. Hodnett, S. Hudson, P. Davern, Modification of the zeta potential of montmorillonite to achieve high active pharmaceutical ingredient nanoparticle loading and stabilization with optimum dissolution properties. Colloids Surf. B. **193** (2020). https://doi.org/10.1016/j.colsurfb.2020.111120
154. K. Bodnár, S.P. Hudson, Å.C. Rasmuson, Drug loading and dissolution properties of Dalcetrapib-montmorillonite nanocomposite microparticles. Org. Process. Res. Dev. **24**(6), 977–987 (2020). https://doi.org/10.1021/acs.oprd.9b00460
155. S. Kumar, R. Gokhale, D.J. Burgess, Sugars as bulking agents to prevent nano-crystal aggregation during spray or freeze-drying. Int. J. Pharm. **471**(1–2), 303–311 (2014). https://doi.org/10.1016/j.ijpharm.2014.05.060
156. W. Abdelwahed, G. Degobert, S. Stainmesse, H. Fessi, Freeze-drying of nanoparticles: formulation, process and storage considerations. Adv. Drug Deliv. Rev. **58**(15), 1688–1713 (2006)
157. S. Khan, M. de Matas, S. Plakkot, J. Anwar, Nanocrystal recovery by use of carrier particles. Cryst. Growth Des. **14**(3), 1003–1009 (2014)
158. A. Kumar, K.A. Ramisetty, S. Bordignon, B.K. Hodnett, P. Davern, S. Hudson, Preparation, stabilisation, isolation and tableting of valsartan nanoparticles using a semi-continuous carrier particle mediated process. Int. J. Pharm. **597**, 120199 (2021). https://doi.org/10.1016/j.ijpharm.2021.120199
159. A.A. Thorat, S.V. Dalvi, Liquid Antisolvent precipitation and stabilization of nanoparticles of poorly water soluble drugs in aqueous suspensions: Recent developments and future perspective. Chem. Eng. J. **1**, 1–34 (2012). https://doi.org/10.1016/j.cej.2011.12.044

160. B. Derjaguin, L. Landau, Theory of the stability of strongly charged lyophobic sols and of the adhesion of strongly charged particles in solutions of electrolytes. Prog. Surf. Sci. **43**(1–4), 30–59 (1993). https://doi.org/10.1016/0079-6816(93)90013-L
161. E.J.W. Verwey, Theory of the stability of lyophobic colloids. J. Phys. Colloid Chem. **51**(3), 631–636 (1947). https://doi.org/10.1021/J150453A001
162. P.R. Mishra, L. Shaal, et al., Production and characterization of Hesperetin Nanosuspensions for dermal delivery. Int. J. Pharm. **371**(1–2), 182–189 (2009). https://doi.org/10.1016/J.IJPHARM.2008.12.030
163. H. Rachmawati, L. Shaal, et al., Development of curcumin nanocrystal: Physical aspects. J. Pharm. Sci. **102**(1), 204–214 (2013). https://doi.org/10.1002/JPS.23335
164. D. Xia, P. Quan, H. Piao, H. Piao, S. Sun, Y. Yin, F. Cui, Preparation of stable nitrendipine nanosuspensions using the precipitation–ultrasonication method for enhancement of dissolution and oral bioavailability. Eur. J. Pharm. Sci. **40**(4), 325–334 (2010). https://doi.org/10.1016/J.EJPS.2010.04.006
165. B. Van Eerdenbrugh, J. Vermant, J.A. Martens, L. Froyen, J. Van Humbeeck, P. Augustijns, G. Van Den Mooter, A screening study of surface stabilization during the production of drug nanocrystals. J. Pharm. Sci. **98**(6), 2091–2103 (2009). https://doi.org/10.1002/JPS.21563
166. K.V. Mahesh, S.K. Singh, M. Gulati, A comparative study of top-down and bottom-up approaches for the preparation of Nanosuspensions of glipizide. Powder Technol. **256**, 436–449 (2014). https://doi.org/10.1016/J.POWTEC.2014.02.011
167. Y. Lu, Z.H. Wang, T. Li, H. McNally, K. Park, M. Sturek, Development and evaluation of transferrin-stabilized paclitaxel nanocrystal formulation. J. Control. Release **176**(1), 76–85 (2014). https://doi.org/10.1016/J.JCONREL.2013.12.018
168. A. Maged, A.A. Mahmoud, S. Salah, N.M. Abd-elmonsif, M.M. Ghorab, Spray-dried Rosuvastatin nanoparticles for promoting hair growth. AAPS PharmSciTech **21**(6), 1–12 (2020). https://doi.org/10.1208/S12249-020-01746-5/FIGURES/7
169. H. Yang, F. Teng, P. Wang, B. Tian, X. Lin, X. Hu, L. Zhang, K. Zhang, Y. Zhang, X. Tang, Investigation of a nanosuspension stabilized by Soluplus® to improve bioavailability. Int. J. Pharm. **477**(1–2), 88–95 (2014). https://doi.org/10.1016/J.IJPHARM.2014.10.025
170. B. Mishra, J. Sahoo, P.K. Dixit, Formulation and process optimization of naproxen nanosuspensions stabilized by Hydroxy propyl methyl cellulose. Carbohydr. Polym. **127**, 300–308 (2015). https://doi.org/10.1016/J.CARBPOL.2015.03.077
171. D. Douroumis, A. Fahr, Stable carbamazepine colloidal systems using the Cosolvent technique. Eur. J. Pharm. Sci. **30**(5), 367–374 (2007). https://doi.org/10.1016/J.EJPS.2006.12.003
172. S.A. Afifi, M.A. Hassan, A.S. Abdelhameed, K.A. Elkhodairy, Nanosuspension: An emerging trend for bioavailability enhancement of Etodolac. Int J Polym Sci **2015** (2015). https://doi.org/10.1155/2015/938594
173. J. Pardeike, D.M. Strohmeier, N. Schrödl, C. Voura, M. Gruber, J.G. Khinast, A. Zimmer, Nanosuspensions as advanced printing ink for accurate dosing of poorly soluble drugs in personalized medicines. Int. J. Pharm. **420**(1), 93–100 (2011). https://doi.org/10.1016/J.IJPHARM.2011.08.033
174. A. Tuomela, P. Liu, J. Puranen, S. Rönkkö, T. Laaksonen, G. Kalesnykas, O. Oksala, J. Ilkka, J. Laru, K. Järvinen, J. Hirvonen, L. Peltonen, Brinzolamide nanocrystal formulations for ophthalmic delivery: Reduction of elevated intraocular pressure in vivo. Int. J. Pharm. **467**(1–2), 34–41 (2014). https://doi.org/10.1016/J.IJPHARM.2014.03.048
175. A. Dolenc, J. Kristl, S. Baumgartner, O. Planinšek, Advantages of celecoxib nanosuspension formulation and transformation into tablets. Int. J. Pharm. **376**(1–2), 204–212 (2009). https://doi.org/10.1016/J.IJPHARM.2009.04.038
176. N. Perween, S. Alshehri, T.S. Easwari, V. Verma, M. Faiyazuddin, A. Alanazi, F. Shakeel, Investigating the feasibility of Mefenamic acid nanosuspension for pediatric delivery: Preparation, characterization, and role of excipients. Processes **9**(4), 574 (2021). https://doi.org/10.3390/PR9040574

177. M. George, I. Ghosh, Identifying the correlation between drug/stabilizer properties and critical quality attributes (CQAs) of nanosuspension formulation prepared by wet media milling technology. Eur. J. Pharm. Sci. **48**(1–2), 142–152 (2013). https://doi.org/10.1016/J.EJPS.2012.10.004
178. Q. Fu, J. Sun, D. Zhang, M. Li, Y. Wang, G. Ling, X. Liu, Y. Sun, X. Sui, C. Luo, L. Sun, X. Han, H. Lian, M. Zhu, S. Wang, Z. He, Nimodipine nanocrystals for oral bioavailability improvement: Preparation, characterization and pharmacokinetic studies. Colloids Surf. B Biointerfaces **109**, 161–166 (2013). https://doi.org/10.1016/J.COLSURFB.2013.01.066
179. J. Raula, A. Rahikkala, T. Halkola, J. Pessi, L. Peltonen, J. Hirvonen, K. Järvinen, T. Laaksonen, E.I. Kauppinen, Coated particle assemblies for the concomitant pulmonary Administration of Budesonide and Salbutamol Sulphate. Int. J. Pharm. **441**(1–2), 248–254 (2013). https://doi.org/10.1016/J.IJPHARM.2012.11.036
180. T. Laaksonen, P. Liu, A. Rahikkala, L. Peltonen, E.I. Kauppinen, J. Hirvonen, K. Järvinen, J. Raula, Intact nanoparticulate indomethacin in fast-dissolving carrier particles by combined wet milling and aerosol flow reactor methods. Pharm. Res. **28**(10), 2403–2411 (2011). https://doi.org/10.1007/S11095-011-0456-Z/FIGURES/7
181. P. Ghosh, A. Rasmuson, S.P. Hudson, Impact of additives on drug particles during liquid Antisolvent crystallization and subsequent freeze-drying. Org. Process. Res. Dev. **27**(11), 2020–2034 (2023). https://doi.org/10.1021/ACS.OPRD.3C00204/ASSET/IMAGES/LARGE/OP3C00204_0012.JPEG
182. E. Bilgili, M. Li, A. Afolabi, Is the combination of cellulosic polymers and anionic surfactants a good strategy for ensuring physical stability of BCS class II drug nanosuspensions? Pharm. Dev. Technol. **21**(4), 499–510 (2016). https://doi.org/10.3109/10837450.2015.1022788
183. M.A. O'Mahony, A. Maher, D.M. Croker, Å.C. Rasmuson, B.K. Hodnett, Examining solution and solid state composition for the solution-mediated polymorphic transformation of carbamazepine and Piracetam. Cryst. Growth Des. **12**(4), 1925–1932 (2012). https://doi.org/10.1021/CG201665Z/ASSET/IMAGES/LARGE/CG-2011-01665Z_0008.JPEG
184. J. Schöll, D. Bonalumi, L. Vicum, M. Mazzotti, M. Müller, In situ monitoring and modeling of the solvent-mediated polymorphic transformation of L-glutamic acid. Cryst. Growth Des. **6**(4), 881–891 (2006). https://doi.org/10.1021/cg0503402

Nanocrystals: Enhancing the Oral Bioavailability of Poorly Soluble Drugs

Lucia Lopez-Vidal, Pedro S. Londoño, Alan Rossetti, Juan Pablo Real, Alejandro J. Paredes, and Santiago D. Palma

Abstract Nanocrystals (NCs) have emerged as a highly effective approach to enhance the oral delivery of poorly water-soluble drugs, which represent a significant portion of new and existing therapeutics. Poor aqueous solubility limits drug dissolution and absorption, posing major challenges for oral bioavailability and clinical efficacy. By reducing drug particle size to the nanometre scale, NCs improve dissolution rate and saturation solubility, leading to more consistent pharmacokinetics and enhanced bioavailability. These improvements allow for lower doses, reduced dosing frequency, and increased patient compliance, key benefits for oral therapies. The field of oral NCs remains a focus of active research and commercial development, driven by the urgent need to address solubility issues affecting numerous drugs. Despite challenges in manufacturing scale-up and stability, advances in formulation science continue to expand the clinical applications of NCs. Overall, the NC technology provides a clinically validated and versatile platform to overcome solubility and absorption limitations, supporting the development of safer and more effective oral drug formulations. This chapter covers the oral administration route, its challenges, and how NCs overcome solubility and absorption issues, reviewing recent studies and marketed NC-based oral formulations.

Keywords Nanocrystals · Nanosuspensions · Oral route · Poorly soluble drugs · Bioavailability

---

L. Lopez-Vidal · P. S. Londoño · A. J. Paredes (✉)
School of Pharmacy, Queen's University Belfast, Belfast, UK
e-mail: l.lopezvidal@qub.ac.uk; plondonoruiz01@qub.ac.uk; a.paredes@qub.ac.uk

A. Rossetti · J. P. Real · S. D. Palma (✉)
Faculty of Chemical Sciences, Department of Pharmaceutical Sciences, National University of Córdoba (UNC), Cordoba, Argentina
e-mail: juan.real@unc.edu.ar; sdpalma@unc.edu.ar

A. J. Paredes (ed.), *Nanocrystals*, https://doi.org/10.1007/978-3-032-10238-6_3

1 Introduction

The oral route is the most widely used for the administration of drugs worldwide due to its fast onset of action, simplicity, convenience, and high patient acceptance; far surpassing the preference for parenteral or inhaled methods. The global market for oral solid dosage (OSD) pharmaceuticals demonstrates strong and sustained growth, with recent estimates valuing the market at approximately US$ 727.77 billion in 2023 and US$ 779–780 billion in 2024. Current projections indicate the market will exceed US$ 1.17–1.23 trillion by 2030–2032, reflecting a compound annual growth rate (CAGR) of around 6.9% to 7.1% over this period [1, 2]. Forecasts from other market research firms support a similar growth trajectory [3]. These data collectively underscore the sustained dominance and economic relevance of oral drug delivery systems, reaffirming their position as the preferred choice for both patients and pharmaceutical manufacturers. In addition, it allows treatments targeted to specific regions of the gastrointestinal (GI) tract, essential for pathologies, such as gastric cancers, infections, and gastroduodenal ulcers [4]. However, oral drug delivery faces significant challenges, particularly related to the poor water solubility and limited bioavailability of active pharmaceutical ingredients. Notably, 70–90% of drugs in development and approximately 40% of marketed drugs exhibit poor aqueous solubility, which compromises their therapeutic efficacy [5]. In addition, oral administration must overcome the chemical and biological instability of certain compounds and the physiological barriers of the digestive tract. Because of this, to optimise therapeutic outcomes, it is essential to understand the physicochemical properties of drug candidates, the biological barriers involved, and their pharmacokinetic and pharmacodynamic profiles, enabling the design of more efficient and effective delivery systems [6]. In this context, drugs characterised by low solubility and/or permeability pose a significant challenge, as these properties limit their oral bioavailability. A notable example is atorvastatin (ATV), a lipid-lowering statin marketed under the brand name Lipitor®, which generated global revenues of US$9.3 billion between 2002 and 2009. Despite its commercial success, ATV belongs to Class II of the Biopharmaceutical Classification System (BCS), characterised by low solubility and high permeability. This results in an absolute bioavailability of only 12% and systemic bioavailability of approximately 30%. Furthermore, its absorption can be reduced by up to 90% when taken with food, and chronic use of high doses has been associated with adverse effects such as hepatotoxicity and rhabdomyolysis. These limitations highlight the urgent need to develop innovative strategies that enhance both solubility and bioavailability of such compounds, aiming to optimise their therapeutic performance. In this regard, several studies have investigated the reformulation of ATV using nanotechnology, showing promising results [7, 8].

Given these challenges and the growing demand for effective oral drug delivery, nanocrystals (NCs) have emerged as a promising strategy to enhance the solubility, dissolution rate, and bioavailability of oral poorly water-soluble drugs. By reducing particle size to the nanometre scale, NCs improve drug absorption and therapeutic efficacy without the need for complex formulation. This chapter will first provide an overview of the oral route of administration, highlighting its advantages and limitations, before exploring how NCs technology addresses key barriers to successful oral drug delivery.

2 The Oral Route of Administration

Due to its ease of administration, painlessness, low cost, wide-ranging drug absorption and distribution, and high patient adherence, the oral route is often considered the first choice for pharmacotherapy [6]. However, despite its numerous advantages, oral drug administration continues to present challenges due to various anatomical, physiological, and biochemical factors. Understanding the physiology of the GI tract is crucial to comprehending how orally administered drugs are absorbed, as this complex and dynamic system is designed for both the digestion and absorption of nutrients and pharmacological compounds. Anatomically, the GI tract comprises the oral cavity, oesophagus, stomach, small intestine, and colon. Each of these regions exhibits distinct characteristics that influence drug stability, release, and absorption. For instance, the stomach provides a highly acidic environment (pH 1.0–2.5) that facilitates food breakdown but can also degrade acid-sensitive drugs, such as those containing peptide bonds or labile structures, making it one of the most aggressive barriers to oral absorption [9]. Moreover, the presence of pepsin and other proteolytic enzymes further contributes to the degradation of therapeutic molecules, particularly proteins and peptides. In contrast, the small intestine presents a more neutral pH (6.0–7.4) and a vast absorptive surface area, estimated at up to 200 m^2, thanks to the presence of villi and microvilli along the intestinal lumen. This, coupled with the availability of multiple transport pathways (passive, active, and carrier-mediated), makes it the primary site for the absorption of orally administered drugs. The intestinal mucosa also serves an immunological function, capable of recognising and transferring exogenous antigens to the immune system via M cells within the gut-associated lymphoid tissue (GALT) [10]. Nevertheless, the small intestine presents significant challenges. Factors such as rapid intestinal transit, enzymatic degradation by pancreatic enzymes, and the presence of bile salts can reduce the bioavailability of certain active pharmaceutical ingredients (APIs), especially those with poor solubility or high sensitivity in aqueous environments [11]. The intestinal mucous layer, a viscoelastic gel primarily composed of mucins, functions as a protective barrier over the intestinal epithelium. This water-rich, viscous layer features pore sizes ranging from 100 to 1500 nm, which can impede the passage of poorly water-soluble drugs, particularly those that are micronised. Following this layer, drugs must cross the unstirred water layer (UWL), an approximately 300 μm thick, hydrophilic zone adjacent to the epithelial membrane. The hydrophilicity and thickness of the UWL create a significant barrier to drug diffusion, especially for BCS class II drugs, for which high membrane permeability renders UWL penetration a critical rate-limiting step [12]. Once the drug has traversed the UWL, absorption may still be limited by efflux transporters such as P-glycoprotein (P-gp), also known as multidrug resistance protein (MDR1). Localised on the apical membrane of enterocytes, P-gp functions to expel drugs back into the intestinal lumen, thereby reducing their effective absorption [13]. Lipophilic drugs and various toxic compounds are common substrates of P-gp, which plays a protective role by limiting systemic exposure to potentially harmful substances. The action of P-gp is therefore a key determinant in the distribution and efficacy of many orally administered therapeutics [8, 14]. Gastric transit time is variable (ranging from 1 to 3 hours), whereas small intestinal transit tends to be consistent at approximately 3.5 hours. Peristaltic activity, modulated by the fed or fasted state, determines the speed of drug progression through the GI tract and thereby influences residence time and bioavailability. Drug absorption is most efficient in regions with a high density of membrane transporters. In summary, intestinal surface area, local pH, and transit time all play crucial roles in determining drug absorption and bioavailability, which in turn directly impacts clinical efficacy. The colon, by contrast, presents a higher pH (6.5–7.5), significantly longer residence time, and relatively low enzymatic activity compared to the upper GI tract. These features make it an appealing target for site-specific drug delivery systems, especially for the treatment of conditions such as ulcerative colitis, Crohn's disease, and colorectal cancer. In addition, the high interindividual variability in gastric emptying, colonic transit, and gut microbiota composition poses a major challenge to achieving effective and reproducible drug release in the colon. Among the key biochemical obstacles, pH variation across the GI tract, ranging from acidic in the stomach to mildly alkaline in the colon, not only impacts drug stability but also alters solubility and chemical form, thereby influencing absorption. Another significant challenge lies in the presence of digestive and metabolic enzymes, which may inactivate or transform a drug before it reaches systemic circulation. To address these limitations, oral formulation design generally pursues three main goals [15]: first, local drug delivery aimed at treating gastrointestinal diseases, where the drug needs to reach the mucosa but does not necessarily require systemic absorption; second, systemic drug delivery, in which the drug must cross the intestinal mucosa to enter systemic circulation; and thirdly, improving the dissolution rates of poorly water-soluble drugs, where enhancing solubility takes precedence over mucosal penetration. Drug absorption in the GI tract is governed by multiple factors, including available surface area, blood flow at the absorption site, the drug's physical state (e.g., solution, suspension, or solid dosage form), aqueous solubility, and the local concentration at the absorption site [15]. For absorption to occur, drugs must penetrate the epithelial barrier, which they may do via paracellular diffusion (between cells), transcellular diffusion (through cells), or active transport mechanisms, including receptor-mediated endocytosis and carrier-mediated uptake. When formulating drugs classified as BCS Class II or IV, dissolution is typically the rate-limiting step for absorption, rather than permeability.

In this context, it is hardly surprising that one of the major focuses of pharmaceutical sciences in recent years has been the search for strategies to improve the solubility and oral bioavailability of poorly soluble drugs. Among these strategies are prodrug formation [16], the use of cyclodextrins [17], salt formation [18], and the production of amorphous solid dispersions [19], among others. Within the realm of nanotechnology applied to this goal, the development of NCs stands out. By definition, NCs are nanoparticles with a high drug load, stabilised by an excipient, existing in a crystalline state and measuring less than 1 μm in size. Compared with other nanostructured drug delivery systems, NCs offer several advantages, including high drug loading, excellent long-term physical stability, ease of industrial scale-up, and the possibility of manufacturing them without the use of organic solvents, depending on the methodology employed [20–22]. Importantly, NCs present significant potential to enhance the oral bioavailability of drugs by increasing their dissolution rate and saturation solubility, thereby facilitating more efficient absorption across the GI tract. Their small particle size leads to a larger surface area, which accelerates dissolution and can help overcome the poor aqueous solubility that limits many drugs' effectiveness when administered orally.

3 Oral Absorption of Drug Nanocrystals

As previously mentioned, NCs have garnered considerable attention due to their ability to effectively enhance the oral bioavailability of poorly water-soluble drugs. However, initially, the underlying mechanism behind this improvement was not entirely clear, and two main hypotheses emerged. On the one hand, it has been proposed that particle size reduction and improved solubility play a decisive role in increasing bioavailability. On the other hand, some studies have suggested that the endocytosis of NCs is the most influential factor contributing to enhanced absorption [23]. Nevertheless, studies supporting the predominance of the endocytic route have largely relied on in vitro cell-based experiments, where the limited culture volume significantly constrains the assessment of nanoparticle dissolution behaviours. As a result, these studies may have overestimated the contribution of endocytosis to NC absorption. In contrast, in animal or in vivo models, formulations are typically subjected to a dissolution process prior to absorption, and this dissolution step plays a crucial role. In this sense, He et al. (2015) studied the uptake of Saquinavir (SQV) NCs in the intestinal tract [24]. For this, the NCs were labelled with ethyl rhodamine B, and the intestinal mucosal absorption was studied at different segments (duodenum, jejunum, and ileum) using confocal laser scanning microscope (CLSM) at 1 h after oral administration in rats by intragastric gavage. In this case, the authors compare the SQV NCs against coarse SQV crystals, and they observed a difference between the NCs and the coarse crystals, with a superior uptake from the NC formulation. The authors discuss that the efficient transport of SQV across Caco-2 monolayer was increased by the enhanced release of SQV from the NCs, and the rapid accumulation of the drug in cells was caused by the potential uptake of intact NCs via multiple endocytosis pathways. The authors also performed a pharmacokinetic study in male Wistar rats, and they observed a significant increase in SQV plasma concentration when they compared the NCs against the coarse crystal, both administered orally. The results showed that oral absorption of SQV was significantly improved by increased cellular uptake and transport resulting from enhanced drug dissolution and the cellular uptake of individual nanocrystals compared with the coarse crystals. Zhang et al. (2021) also investigated this phenomenon using a dual-fluorescence labelling strategy to identify the primary absorption route of orally administered NCs. Specifically, they employed coumarin 6 (C6), a poorly water-soluble molecule and aggregation-caused quenching (ACQ) fluorophore, in combination with MeTTMN, an aggregation-induced emission (AIE) fluorophore. Together, these molecules formed a FRET (Förster Resonance Energy Transfer) system, enabling fluorescence-based tracking of intact NC particles. Upon dissolution of the NCs, only the fluorescence of C6 remained detectable, thus allowing the researchers to distinguish whether absorption occurred predominantly as dissolved molecules or as intact nanoparticles via endocytosis. In this sense, the authors demonstrated in an in vivo model that C6/MeTTMN NCs (C6/M NCs) could be absorbed both in dissolved molecular form and as intact particles. However, dissolved C6 predominated, indicating that the primary mechanism by which NCs improved in vivo drug performance was the enhancement of dissolution. Nonetheless, the demonstrated absorption of intact NCs represents an intriguing strategy for the development of future formulations [25].

As noted earlier, enhanced dissolution is a key factor associated with improved oral absorption of drug formulations, particularly when dissolution is the rate-limiting step [26]. The oral performance of NC formulations is highly dependent on the state in which NCs are released in the GI tract. As summarised in Fig. 1, optimal bioavailability enhancement occurs when NCs are released as finely dispersed particles. While slight aggregation may not significantly impair the dissolution velocity, pronounced aggregation has been shown to markedly reduce it [27]. Aggregated NCs present a reduced specific surface area exposed to the

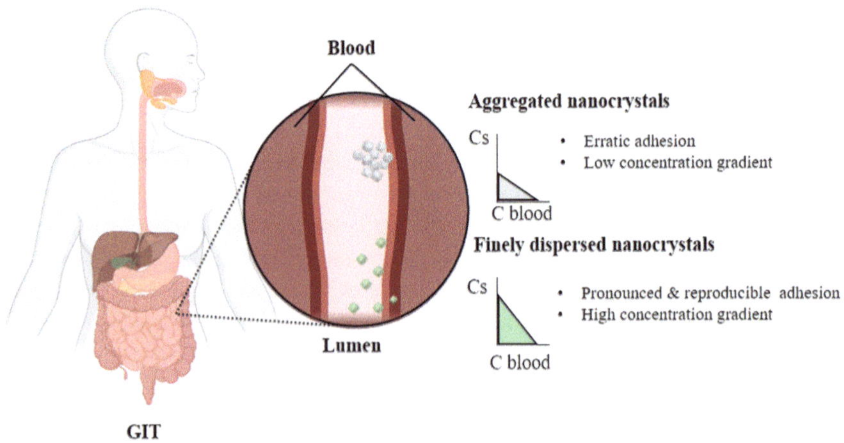

Fig. 1 Schematic representation of the oral administration of nanocrystals, finely dispersed vs. aggregated. (Figure adapted from Mauludin et al. [27])

dissolution medium, leading to a lower effective dissolution rate and, consequently, diminished drug absorption. Additionally, the absorption of intact nanoparticles can be compromised due to aggregation, further reducing the overall bioavailability of the drug. Furthermore, NCs administered in aggregated form may exhibit erratic absorption, with notable interindividual variability in both extent and rate of uptake. This has been attributed to incomplete or delayed dissolution, as well as physical factors such as sedimentation or altered interaction with the intestinal mucosa [28]. In this context, maintaining colloidal stability under physiological conditions, particularly in the acidic gastric environment, is critical. Poorly stabilised formulations may promote particle re-aggregation, thereby compromising the pharmacokinetic benefits typically associated with NCs [29, 30].

Nonetheless, a faster dissolution rate does not always lead to significantly increased absorption [31], as the drug must still permeate the gastrointestinal tract. Therefore, intestinal permeability is another critical factor in oral absorption. Drugs may cross the epithelial barrier through various mechanisms, including passive diffusion and endocytosis. Rapidly dissolving drugs are more likely to be absorbed via molecular diffusion. However, as discussed earlier, NCs can also be absorbed via endocytosis, thus presenting absorption through both mechanisms. Conversely, the drug's absorption rate may also influence its dissolution. Drugs with high membrane permeability are absorbed quickly upon oral administration, thereby promoting the continuous dissolution of the drug from the NC particles, and vice versa. Hence, the interplay between dissolution and permeability plays a pivotal role in oral drug absorption. In this context, and in line with the work of Zhang et al. (2021), recent studies have explored more refined strategies to directly track the fate of NCs in vivo, aiming to clarify the extent to which intact particles contribute to systemic absorption [32]. For instance, quercetin hybrid NCs (QT-HNCs) embedded with ACQ fluorophores were developed to allow the discrimination of intact particles from their dissolved counterparts after oral administration. Unlike traditional tracking methods, this approach enabled researchers to follow the behaviour of integral NCs in real-time. Notably, a fraction of QT-HNCs remained in the gastrointestinal tract for up to 12–16 hours post-administration, resisting complete dissolution. These intact NCs were absorbed through the intestinal membrane, preferentially in the ileum rather than the jejunum, and subsequently distributed to various organs, with the liver identified as the primary site of accumulation. Importantly, mechanistic studies showed that both trans-epithelial transport and M cell-mediated pathways were involved in the uptake of integral QT-HNCs, although the presence of the mucus layer could hinder this process. Additionally, the authors also studied the influence of the particle size finding that the particle size significantly influenced both in vivo behaviour and cellular interactions ex vivo. Medium-sized QT-HNCs (~550 nm) showed optimal retention in the gastrointestinal tract and enhanced absorption compared to smaller (~280 nm) or larger (~1100 nm) counterparts. These findings not only support the use of ACQ fluorophores as a robust tool for visualising the in vivo behaviour of NCs, but also provide compelling evidence that intact NCs, under certain conditions, can traverse the intestinal barrier and contribute to drug absorption, in agreement with what was reported by Zhang et al. (2021). Building upon their previous findings, the authors further investigated the biological fate of QT-HNCs with two different particle sizes (≈280 nm and ≈550 nm) following oral and intravenous administration (Fig. 2) [33]. This study aimed to estimate the contribution of intact NC absorption to the overall enhancement of oral bioavailability. By combining in vivo fluorescence imaging and HPLC analysis, the authors were able to distinguish between integral NCs and dissolved drug, demonstrating that QT-HNCs remained in the body as intact particles for up to 48 hours and exhibited differential biodistribution patterns depending on particle size and administration route. Notably, despite inherent limitations in the quantification method, their results suggested that the absorption of integral NCs could significantly contribute to the improved bioavailability of poorly soluble drugs. Together, these complementary studies provide a deeper understanding of the mechanisms underlying the oral bioavailability enhancement of NC formulations. Altogether, these findings underscore the multifaceted nature of NCs performance in vivo and suggest that both dissolution and intact particle absorption must be considered when designing and optimising NC-based drug delivery systems.

Moreover, beyond solubility and permeability, additional factors must also be considered when evaluating the performance of NC-based formulations. Guo et al. (2015) studied the influence of the crystal form on the dissolution profile and bioavailability of NC formulations [34]. To this end, the authors obtained rod-shaped and spherical lovastatin (LOV) NCs and compared them in vitro and in vivo. The results showed that rod-shaped LOV NCs (LOV-RNs) and spherical lovastatin NCs (LOV-SNs) had similar hydrodynamic diameters (500.6 ± 21.0 nm vs 503.2 ± 20.4 nm, respectively) and the same crystalline state. However, the dissolution profile showed that LOV-RNs had a higher dissolution rate than LOV-SNs in no-sink conditions, which was attributed to a larger surface area and a lower diffusion layer thickness. The in vivo studies demonstrated a higher oral bioavailability for the rod-shaped NCs $AUC_{(0-24\ h)}$ and C_{max} of LOV for LOV-RNs were about 1.98-fold ($p < 0.05$) and 2.97-fold ($p < 0.05$) higher than those of the commercial control, Junning®. The corresponding values for LOV-SNs were 1.46-fold ($p > 0.05$) and 1.99-fold ($p > 0.05$) higher than Junning®. These results led the authors to conclude that the crystal form could have a significant effect on the oral bioavailability of drug NCs. Moreover, further studies corroborate and enrich these findings. Guo et al. (2019) studied the impact of particle shapes on the oral delivery of drug NCs in mucus permeation, transepithelial transport, and bioavailability, using LOV as a model drug [35]. The authors prepared spherical (SNCs), rod-shaped (RNCs), and flaky NCs (FNCs) with similar surface chemistry and crystalline state, and systematically compared their behaviour in mucus permeation, transepithelial transport, and in vivo absorption. Their results revealed that rod-shaped NCs exhibited superior diffusion through the mucus layer, likely due to their ability to align and rotate within the mesh network, reducing steric hindrance. Additionally, RNCs showed enhanced transcytosis across epithelial cells, which the authors attributed to shape-mediated differences in cellular uptake pathways. These advantages translated into improved oral bioavailability, with RNCs achieving significantly higher AUC values than their spherical and flaky counterparts. Overall,

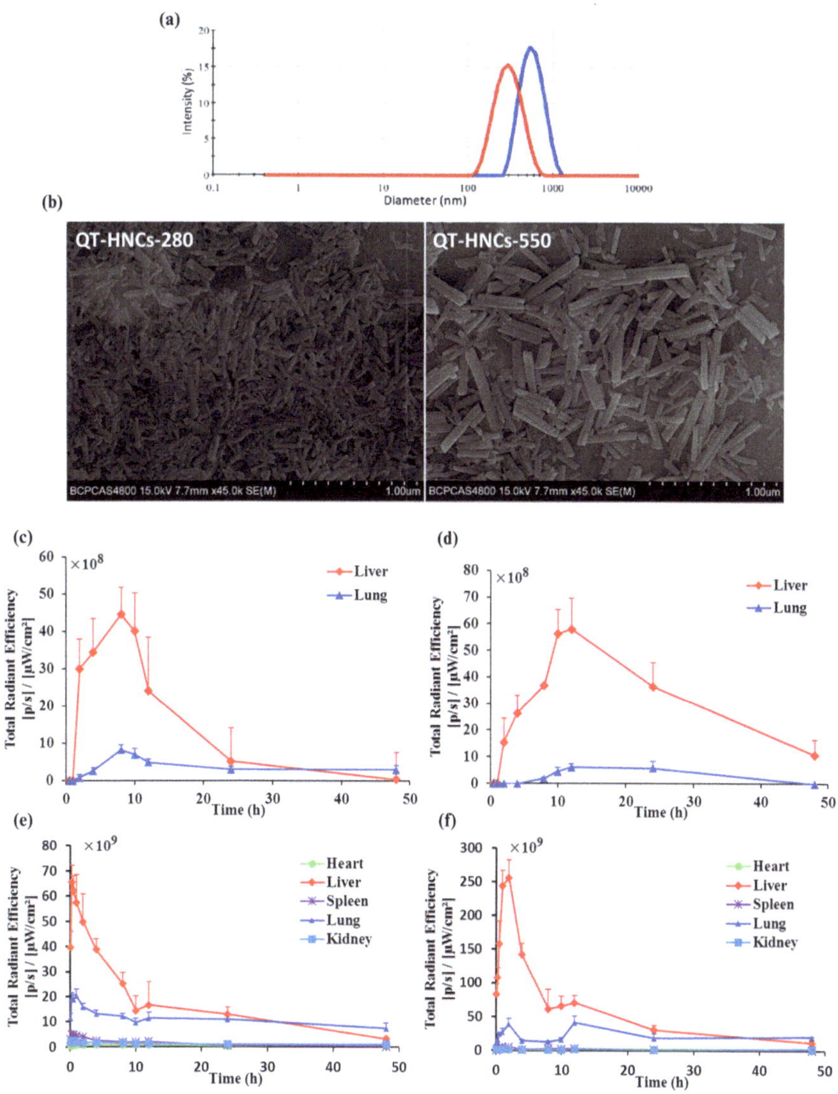

Fig. 2 (**a**) Particle size distribution and (**b**) SEM images of QT-HNCs of 280 and 550 nm. Fluorescence quantification of major organs based on total radiant efficiency following oral (**c**) and (**d**) and intravenous (**e**) and (**f**) administration of QT-HNCs of 280 nm (**c**) and (**e**) and QT-HNCs of 550 nm (**d**) and (**f**). (Reprinted from reference [33]. Copyright© (2021), with permission from Elsevier)

this study highlights that beyond solubility and particle size, NC morphology can play a critical role in determining the efficiency of oral drug delivery.

In line with the pivotal role of dissolution in oral NC absorption, Mu et al. (2016) conducted a detailed investigation into the influence of different stabilisers on the in vivo performance of spironolactone NCs [36]. The authors prepared several NC formulations of Spironolactone (SPN) using various stabilisers, including Pluronic F-127 (F127), Poloxamer 188 (F68), hydroxypropyl methyl cellulose E5 (HPMC-E5), and sodium deoxycholate (NaDC), and systematically compared their physicochemical properties, dissolution profiles, and pharmacokinetic behaviour. Although the formulations showed broadly comparable dissolution rates in simulated intestinal fluid (markedly higher than that of the coarse SPN), significant differences were observed under acidic conditions. Notably, the NaDC-stabilised NCs exhibited a delayed dissolution profile in HCl 0.1 M, which the authors attributed to particle aggregation at low pH, because of the transformation of NaDC (PKa 5.15) to insoluble deoxycholic acid at gastric pH. This effect likely limited the availability of dissolved drug for absorption in the upper gastrointestinal tract. In contrast, NCs stabilised with F127 demonstrated superior stability and a more consistent dissolution behaviour, ultimately resulting in enhanced oral bioavailability. Pharmacokinetic studies in rats confirmed that F127-stabilised NCs achieved the highest systemic exposure, with an AUC approximately four times greater than that of the coarse drug suspension. All formulations improved the oral bioavailability after nanonisation, but different stabilisers induced differences in pharmacokinetic profiles. These findings underscore the critical role of excipient selection in NC formulations, not only in terms of physical stabilisation but also in dictating the in vivo dissolution behaviour and subsequent absorption efficiency. By modulating the interaction of NCs with the gastrointestinal environment, stabilisers can markedly affect the fate of the particles during transit, reinforcing the notion that dissolution alone does not fully explain the bioavailability-enhancing effects of NCs. Instead, the complex interplay between formulation components, physicochemical stability, dissolution kinetics, and biological barriers must be carefully considered when designing oral NC delivery systems.

4 Advances in Oral Delivery of Drug Nanocrystals

In recent years, significant advances have been made in the development of oral NC-based formulations, driven by the pressing need to improve the bioavailability of poorly water-soluble drugs. While the preceding section addressed the mechanistic aspects of NCs absorption, this section centres on formulation strategies, technological innovations, and translational efforts aimed at enhancing the performance, stability, and scalability of orally administered NCs. Furthermore, commercially available oral products utilising NC technology are discussed to exemplify the successful industrial application of these systems.

A broad spectrum of formulation approaches has been investigated to overcome the physiological challenges posed by the GI tract, such as pH variability, enzymatic degradation, and mucosal barriers. These strategies include employing appropriate stabilisers to maintain particle stability in gastric fluids and designing solid dosage forms that enable rapid redispersion and preserve the nanometric particle size upon administration. Additionally, advancements in manufacturing technologies, ranging from bottom-up precipitation to top-down milling and hybrid techniques, have provided improved control over critical particle attributes and facilitated scale-up for industrial production. Beyond mere particle size reduction, various formulation strategies have been explored recently to enhance the oral bioavailability of NC-based drug delivery systems. One approach involves pH-sensitive encapsulation to protect NCs from premature dissolution in the stomach and promote targeted release in the intestine. This strategy is particularly beneficial not only for site-specific drug release but also for safeguarding acid-labile drugs from degradation within the harsh gastric environment. Premature exposure to low pH can lead to significant loss of drug potency prior to absorption. By embedding NCs within enteric matrices or coatings that dissolve selectively at intestinal pH, drug integrity is preserved, maximising therapeutic efficacy. For instance, authors developed ivermectin NCs via wet milling followed by spray drying, utilising Eudragit® L100–55 as an encapsulating agent. The resulting dry powder exhibited minimal release under gastric conditions (pH 1.2) but showed a marked increase in solubility at intestinal pH, illustrating the potential of combining NC technology with pH-responsive materials to optimise site-specific drug delivery [37]. More recently, Che et al. (2024) formulated nintedanib NCs coated with Eudragit L100, aiming to reduce drug solubility in the stomach, maintain supersaturation in the intestine, and improve oral absorption [38]. An in vivo study compared coated NCs, uncoated NCs, and an oral soft capsule formulation. The results demonstrated that the oral bioavailability of the coated and uncoated NCs increased by 2.58- and 1.43-fold, respectively, relative to the soft capsule. Encapsulation effectively curtailed premature release of BIBF in the stomach, allowing greater quantities of the drug to reach the intestine as NCs, sustaining supersaturation and thereby enhancing oral bioavailability.

Oral NC suspensions have been extensively studied. For example, an in vivo study of nitrendipine nanosuspension (~290 nm) in rats showed that the maximum plasma concentration (C_{max}) and AUC were approximately six times higher compared to marketed tablets [39]. In another human in vivo study, faster absorption from naproxen NC suspensions reduced the time to reach maximum concentration (T_{max}) by nearly 50% and increased AUC by approximately fivefold compared to conventional tablet (Anaprox®) and suspension (Naprosyn®) formulations under fed conditions [40]. Similarly, a study comparing oral administration of candesartan cilexetil NCs prepared with a combination of hydroxypropyl methylcellulose (HPMC) and Pluronic® F127 (50:50 w/w) against candesartan cilexetil free base revealed a 3.8-fold increase in oral bioavailability and a two-fold increase in C_{max} in a murine model [41]. Regarding incorporation into solid dosage forms, NCs have been successfully included in capsules. For example, Novartis compound A and itraconazole NCs were encapsulated and evaluated for bioavailability in Beagle dogs [42] and rats [43], showing significant improvements. Another study assessed the pharmacokinetics of glimepiride NCs in capsules in rats, demonstrating marked bioavailability enhancement compared to commercial microcrystal-loaded capsules [44]. An additional study compared oral albendazole NCs with a control formulation consisting of a physical mixture of albendazole and Poloxamer 188 in capsules, evaluating pharmacokinetics and efficacy. The NCs exhibited improved AUC, T_{max}, and C_{max} relative to the control. Efficacy studies showed no statistical difference between NCs and the commercial formulation at doses of 25 and 12.5 mg/kg. However, at 6.25 mg/kg (one-quarter of the reference dose), significant differences emerged: faecal nematode egg counts were reduced by 62.0 ± 21.1% with the control and by 100 ± 0% with the NCs [45]. Eplerenone NCs were produced by controlled crystallisation during lyophilisation, yielding particles with an average size of approximately 47 nm and a highly crystalline structure without detectable polymorphic transition. These NCs showed about a 17-fold increase in aqueous solubility and achieved 95% dissolution within 10 minutes, compared to only 29% for the standard powder form. In vivo rat studies revealed notable improvements in pharmacokinetic parameters (AUC, $Cmax$) and a reduced T_{max}, with no signs of systemic toxicity [46]. In another study, Khare et al. (2023) prepared NCs of Zileuton, an FDA-approved BCS Class II drug. To evaluate efficacy and mucosal immune profile, 10-week-old rats were dosed via capsules containing either Zileuton alone, NC-formulated Zileuton, or a physical mixture, administered by flexible oral gavage syringes. Over 15 days at 30 mg/kg, animals treated with NCs or the physical mixture received significantly lower amounts of active Zileuton compared to those given the micronised form. At treatment end, reduced levels of pro-inflammatory cytokines (IL-18, TNF-α, RANTES) were observed in the ileal mucosa, attributed to the NC formulation. Additionally, NCs exhibited an immunosuppressive profile both locally and systemically, comparable to the micronised form but with reduced immunotoxic potential [47].

Incorporation of NCs into rapidly disintegrating solid dosage forms represents another strategy for oral delivery. For example, Liu et al. (2017) developed orally fast-dissolving films (OFDFs) containing lutein NCs to improve bioavailability. The OFDFs consisted of 2.05% (w/v) HPMC, 1.03% (w/v) PEG 400, and 0.43% (w/v) Cremophor EL, disintegrating rapidly within 29 ± 8 s. Following administration to Sprague–Dawley rats, T_{max} decreased from 3 hours for the oral solution to less than 0.8 hours for OFDFs, while C_{max} increased from 150 ng/mL (solution) to 350 ng/mL (conventional OFDF) and 830 ng/mL (NC-loaded OFDF) [48]. Another study developed a mucoadhesive oral film based on NC suspensions, incorporating a layer of carvedilol NCs between a mucoadhesive layer and a backing layer. NCs were dispersed in a hydrogel prepared with HPMC and Carbopol 934P, using PEG 400 as a plasticiser. In vivo studies in rabbits showed a significant increase in relative bioavailability compared to commercial tablets [49]. 3D printing has also emerged as a promising approach. For instance, rapidly dissolving solid dosage forms were fabricated by 3D printing using PEG 1500 as the primary matrix former and propylene glycol as a plasticiser. The printed tablets contained 50% albendazole NCs and were compared in an in vivo pharmacokinetic study in dogs against 3D-printed tablets loaded with a solid dispersion of the same drug and pure drug as controls [50]. The solid dispersion exhibited a two-fold increase in relative bioavailability compared to control, whereas the NC-loaded tablets demonstrated an almost ninefold increase. Additionally, gastroretentive 3D-printed formulations containing NCs of niclosamide have been developed [51]. These achieved controlled release of NCs in simulated gastric medium, with in vivo canine studies demonstrating prolonged gastric retention times up to 180 min. These findings underscore the potential of 3D printing to produce gastroretentive, slow-release oral solid dosage forms loaded with NCs of poorly soluble drugs, ideal systems for treating gastric conditions such as *Helicobacter pylori* infections.

Another promising direction involves the development of multi-drug NC systems to enable synergistic or complementary pharmacological actions. In a recent study, a NC formulation of a baicalin–berberine (BA–BBR) complex was developed to simultaneously enhance the oral absorption of both components. The complex was first prepared and subsequently processed using high-pressure homogenisation to yield NCs with a mean size of 318.40 ± 3.32 nm and a polydispersity index (PDI) of 0.26 ± 0.03. In pharmacokinetic studies in rats, the AUC values for BA and BBR increased by 7.49- and 2.64-fold, respectively, when compared to a coarse suspension of the same complex. These findings suggest that NC formulation of drug–drug complexes can be a valuable strategy to promote co-absorption and enhance the bioavailability of multiple therapeutic agents simultaneously [52]. A comprehensive overview of the most recent preclinical and clinical studies exploring oral NCs can be found in Table 1. This table was compiled following a structured search strategy using the algorithm: TITLE-ABS-KEY("oral drug") AND TITLE-ABS-KEY ("nanocrystal*" OR "drug nanocrystal*") AND TITLE-ABS-KEY("in vivo") AND TITLE-ABS-KEY("animal model*" OR "rat model*" OR "mouse model*" OR "preclinical study" OR "bioavailability" OR "pharmacokinetics"), filtered for articles only. Abstracts were screened to identify studies focused on NCs development with in vivo evaluations.

Among the most compelling advances in translation are those NC formulations that have successfully navigated regulatory approval and are now commercially available for oral use. A recently published review article compiled a table summarising NC-based products currently approved for clinical use, highlighting that almost 30 NC-containing drug products have received FDA approval to date, 15 of which are intended for oral administration, making it the most prominent route [84]. This growing number of approved formulations underscores the technological maturity and regulatory acceptance of NCs as a viable strategy for enhancing the oral bioavailability of poorly soluble drugs. These marketed products span a wide range of therapeutic areas, including immunosuppression, anti-inflammatory, hypercholesterolemia, and pain management, among others, reflecting the versatility of NC technology in addressing solubility-limited absorption across diverse drug classes. In most cases, the key advantage of NC-based formulations lies in their ability to improve dissolution rate, achieve higher peak plasma concentrations, and reduce inter-individual variability compared to conventional oral dosage forms. Moreover, the commercial success of these formulations illustrates the feasibility of scaling up NC production using industrially relevant techniques such as wet media milling and high-pressure homogenisation, which have proven robust and reproducible at large scale.

Table 1 Comprehensive overview of the most recent preclinical and clinical studies exploring oral delivery of drugs NCs

API	Production method	Stabiliser	Size	Crystal morphology	In vivo studies	Key outcome	Year	Refs.
Breviscapine	HPH	Tween 80	472 ± 2 nm	Irregular rod-like tendency	PK studies in Sprague-Dawley rats. Nanocrystal powder showed and increased bioavailability by 4.73- fold compared to an orally administered raw powder of breviscapine	A nanocrystal powder formulation with increased saturation solubility and oral bioavailability was successfully developed. Furthermore, the modification with tween 80 showed enhanced cellular uptake and transport compared to the reference raw powder and the formulation without surface modification	2025	[53]
Albendazole	WMM	Tea saponins	186 nm PDI: 0.187 ± 0.006	Spherical	In vivo studies included everted gut-sac permeability experiments in Sprague Dawley rat's duodenum, jejunum, ileum and colon, showing markedly higher apparent permeability (Papp) for the nanocrystals (3.6–5.3-fold increase) PK studies in ICR mice showed that nanocrystals presented 4.65-fold increase in plasma AUC of the active metabolite and markedly higher plasma concentrations relative to the commercial tablet and untreated drug	Tea-saponin-stabilised albendazole nanocrystals improved saturation solubility, dissolution rate, intestinal permeability, and enhanced oral bioavailability	2025	[54]

Itraconazole	ASP and HPH	Cholic acid	225.7 ± 10.3 nm PDI: 0.189 ± 0.023	Rods	PK studies in Wistar rats showed the Cmax increased and AUC value for the nanocrystals was 1.33-fold higher compared to the drug pellets Sporanox®	Nanocrystals showed increased saturation solubility and oral bioavailability, as well as superior cell uptake in Caco-2 cells	2024	[55]
Quercetin	WMM	HPMC E15 Poloxamer 407 Poloxamer 188 TPGS Glycyrrhizin acid (GL)	202 ± 7 nm PDI: 0.21 ± 0.02	Irregular rod-like tendency	PK studies in Sprague Dawley (SD) rats showed that nanocrystals formulation exhibited a significant increase in Cmax and AUC compared to the powdered quercetin. However, there were variations depending on the stabiliser used	The extent at which bioavailability, dissolution rate and saturation solubility increase is affected by the choice of the stabiliser	2024	[56]
Etoposide	HPH Thin film hydration Sonoprecipitation	Hexadecyl trimethylammonium bromide (CTAB)	220.3 ± 14.2 nm	Spherical	PK studies in Sprague Dawley rats showed the lipid encapsulated nanocrystals had a bioavailability 1.96- and 10.92-fold higher compared to etoposide nanocrystals and coarse crystals, respectively	Etoposide nanocrystal-loaded lipid nanoparticles demonstrated increased intestinal permeability and drug absorption, as well as improved oral bioavailability and antitumour activity	2024	[57]
Abemaciclib-Flurbiprofen	ASP	Sodium deoxycholate	224.9 ± 2.53 nm PDI: 0.141 ± 0.014	Spherical	PK studies in Sprague Dawley rats showed nanocrystal formulation increased the drug half-life and AUC by 3.6 and 2.5-fold, respectively The study included drug efficacy evaluation using tumours induced BAL B/C	Lysine-polydopamine codrug nanocrystals showed a synergistic effect leading to increase oral absorption and enhanced antitumoural activity	2024	[58]

(continued)

Table 1 (continued)

API	Production method	Stabiliser	Size	Crystal morphology	In vivo studies	Key outcome	Year	Refs.
					mice and measuring apoptosis, necrosis, and proliferation of tumours by TUNEL, HE, and ki67 staining			
Nintedanib	Acid solubilisation and alkaline precipitation followed by WMM	SDS	290.80 ± 5.05 nm PDI: 0.103 ± 0.034	Spherical	PK studies in Sprague Dawley rats showed that the bioavailability of non-coated and coated nanocrystals was 1.43 and 2.58 times increased compared to a nintedanib soft capsule formulation	Eudragit L100 coated nintedanib nanocrystals demonstrated increased dissolution rate and retarded release by avoiding the acidic environment of the stomach and achieving drug release in the intestines	2024	[38]
Luteolin	HPH	Hydroxyethyl starch	191.1 ± 16.8 nm PDI: 0.239	Spherical	ICR mouse model; the formulation's effect was compared to free LUT and febuxostat (positive control). Key evaluated outcomes were serum uric acid levels, renal and hepatic function, serum/liver xanthine oxidase activity, urate transporter expression (GLUT9 and URAT1), and histopathology	The nanocrystals demonstrated a 69.93% reduction in serum uric acid levels and improved renal function, with superior efficacy and safety compared to coarse luteolin. The formulation showed enhanced oral bioavailability (355.7% relative to free luteolin) and increased intestinal absorption	2024	[59]

Zileuton	WMM	KollidonVA64 Dowfax2A1	571 ± 47 nm	Irregular	Pharmacological studies in Sprague Dawley rats assessed cytokine levels in ileal tissue extract and plasma samples and found that drug activity is dependent on particle size	Zileuton nanocrystals showed increased dissolution rate as well as enhanced anti-inflammatory activity coupled with an improvement in the safety profile and immunotoxicity compared to the powdered drug	2023	[60]
Niclosamine	WMM	P188	188.5 nm PDI: 0.186	Spherical	Gastroretentive assay in dog model	Nanocrystals loaded into 3D-printed Gastroretentive tablets showed improved controlled release, and enhanced dissolution of niclosamide in acidic conditions while maintaining cell viability	2023	[51]
Amiodarone	ASP And HPH	Pluronic F127	221 ± 1.2 nm PDI: 0.333	Spherical	PK studies in Sprague Dawley rats showed nanocrystals achieved a Cmax 1.6-fold greater than the powdered drug, leading to a 2.1-fold increase in bioavailability, as the time to reach peak plasma concentration was also shortened, indicated enhanced absorption	Amiodarone nanocrystals had increased saturation solubility and dissolution rate, as well as increased bioavailability compared to a commercially available product and powdered drug. Nanocrystals were also stable at room temperature for a period of 3 months	2022	[61]

(continued)

Table 1 (continued)

API	Production method	Stabiliser	Size	Crystal morphology	In vivo studies	Key outcome	Year	Refs.
Enzalutamide	ASP, WMM Spray drying and tableting	Hydroxypropyl methylcellulose acetate succinate (HPMCAS-HG)	Anti-solvent method: 199.1 nm PDI: 0.202 Wet milling method: 378.0 nm PDI: 0.141	Spherical	Pk studies in beagle dogs sing a randomised crossover design to compare ENZ/NS-WM tablets to marketed Xtandi® tablets (amorphous solid dispersion)	Nanocrystals showed improved dissolution rates, their bioavailability was inferior to Xtandi®, with 0.43-fold lower Cmax and AUC0-14d. The study concluded that supersaturation parameters (Cspring and AUSPC2-24 h) from "spring-parachute" testing are better predictors of in vivo bioavailability than dissolution rate alone	2022	[62]
Puerarin	ASP	F68	83.05 ± 1.96 nm, PDI: 0.047 ± 0.009	Spherical	In vivo FRET imaging in zebra fish model and oral PK studies in Sprague-Dawley rats	Orally administered Puerarin nanocrystals significantly enhance the drug's bioavailability and brain accumulation compared to conventional Puerarin suspension, showing improved absorption, higher plasma and brain concentrations, and greater anti-parkinsonian efficacy in animal models, including significant neuroprotection,	2022	[63]

						amelioration of dopamine depletion, and improved behavioural deficits in Parkinson's disease models		
Baicalin (BA)–Berberine (BBR)	HPH	P188	318.40 ± 3.32 nm PDI: 0.26 ± 0.03	Rectangular rods	PK studies in Sprague–Dawley rats exhibited an area under curve for BA and BBR in the BA-BBR complex nanocrystals of 622.65 ± 456.95 h·ng/ml and 167.28 ± 78.87 h·ng/ml, respectively, which were separately 7.49- and 2.64-fold than the complex coarse suspension	The results indicated that the developed and optimised BA-BBR complex nanocrystals could improve the dissolution rate and extent and oral bioavailability, as well as facilitate the co-absorption of the drug prescriptions BA and BBR	2022	[52]
Nintedanib	ASP	CMC-Na	325.30 ± 1.03 nm PDI: 0.14 ± 0.02	Rods	PK studies in Sprague-Dawley rats showed that the oral bioavailability of the nanocrystals exceeded by 2.51-fold that of the reference nintedanib oral solution	Nintedanib nanocrystals effectively increased saturation solubility, dissolution rate and improved oral bioavailability by increasing absorption through multiple intestinal pathways	2022	[64]
Eplerenone	Precipitation and freeze drying	Mannitol	46.8 ± 0.5 nm	Elongated parallelepiped	PK studies in Sprague Dawley rats showed that eplerenone nanocrystals attained rapid absorption and higher drug concentration in each time point compared to the powdered drug, resulting in a 1.5-fold higher bioavailability	Eplerenone nanocrystals exhibited a 17-fold increase in saturation solubility, higher dissolution rate and extent and improved oral bioavailability without acute toxicity detected	2021	[46]

(continued)

Table 1 (continued)

API	Production method	Stabiliser	Size	Crystal morphology	In vivo studies	Key outcome	Year	Refs.
Venetoclax	Precipitation and HPH	PVA	340 nm PDI: 0.160	Irregular	PK studies in Sprague-Dawley rats. The nanocrystals were compared to free drug suspension through the oral route where nanocrystals showed a 2.02-fold rise in oral bioavailability, with Cmax $\approx$ 792 ng mL^{-1} and AUC_0-∞ $\approx$ 12,714 ng mL^{-1}·h versus the free drug (Cmax $\approx$ 448 ng mL^{-1}, AUC_0-∞ $\approx$ 6274 ng mL^{-1}·h)	The nanocrystals demonstrated a 20-fold increase in saturation solubility and ~ 100% dissolution within 120 min, far surpassing the < 43% dissolution of the free drug. They also exhibited improved oral bioavailability	2021	[65]
Ricobendazole	WMM followed by spray-drying	P188	181.30 ± 5.93 nm PDI: 0.17 ± 0.02	Spherical	PK studies: crossover trial was conducted in healthy dogs. Results showed that AUC_0 for nanocrystals was 1.9-fold higher than that of the micronised formulation. In addition, Cmax increased by approximately 2.1-fold versus plain drug	The results of the study demonstrated that nanocrystal formulation substantially improves dissolution rate and the oral bioavailability of poorly soluble ricobendazole	2020	[66]
Glyburide	Precipitation and HPH	P188	400 nm	Irregular	PK studies in Sprague Dawley rats showed faster drug absorption into the gastrointestinal tract and higher Cmax and AUC for the nanocrystals compared to commercial formulation	Nanocrystal-loaded tablets had an improved dissolution rate and higher bioavailability than the powdered drug and the commercial reference product	2019	[67]

Lutein	ASP and ultrasonication	Soy phosphatidylcholine (SPC) PVP K30	110.7 nm PDI: 0.141	Spherical	PK studies in Sprague-Dawley rats showed that nanocrystals Cmax and AUC0–24 h were 3.24 and 2.28 times higher, respectively, than the coarse suspension, demonstrating improved dissolution and absorption	The lutein nanocrystals exhibited superior dissolution rate and oral bioavailability compared to the coarse drug	2018	[68]
Cinacalcet hydrochloride	precipitation–ultrasonication	Soluplus	244 ± 2 nm PDI: 0.168 ± 0.001	Spherical	4 × 4 crossover PK studies in beagle dogs showed that nanocrystals effectively maintain an increased bioavailability both in fed and fasted states, contrary to the commercial product Sensipar® whose bioavailability decreases during fasted state in 40%	The cinacalcet hydrochloride nanocrystals showed enhanced dissolution rate and bioavailability compared to commercially available product Sensipar®	2018	[69]
Berberine	Solvent precipitation	Brij-S20	148.0 ± 3.2 nm PDI: 0.310 ± 0.050	Prisms	PK studies in Sprague Dawley rats showed a double absorption curve due to enterohepatic circulation where Cmax and AUC values for the nanocrystals were 2 times greater than the powdered berberine	The nanocrystals showed improved dissolution rate and exhibited a sustained release. In addition, the formulated nanocrystals showed inhibitory activity on P-gp while increasing oral bioavailability	2018	[70]
Trans -resveratrol	Probe sonication	TPGS Lecithin PF-127	110.28 ± 12.55 nm PDI: 0.187 ± 0.03	Rods	PK studies in Sprague-Dawley (SD) rats showed in an increment in AUC (3.5-fold) and *C*max (2.2-fold) for the nanocrystal formulation when compared to the powdered drug	The nanocrystal formulation effectively increased trans-resveratrol activity and improved oral bioavailability	2017	[71]

(continued)

Table 1 (continued)

API	Production method	Stabiliser	Size	Crystal morphology	In vivo studies	Key outcome	Year	Refs.
Lutein	ASP	SPC SDS	377.9 ± 32.1 nm	Not presented	PK studies in Sprague Dawley rats showed reduce Tmax, greater Cmax and AUC for the nanocrystal loaded film, leading to a two-fold higher oral bioavailability	Lutein nanocrystal loaded dispersible film showed increased dissolution rate and onset of action, peak plasma concentration and greater bioavailability compared to a lutein oral solution	2017	[48]
Aceclofenac	WMM	PVA	484.7 ± 54.12 nm PDI: 0.108 ± 0.009	Irregular and angular shape	PK studies in Wistar albino rats comparing pure drug with nanocrystals. The nanocrystal formulation showed a significantly higher maximum plasma concentration (Cmax: 3.75 ± 0.28 μg/mL vs. 1.96 ± 0.17 μg/mL for pure drug) and area under the curve (AUC0–12: 9.0 ± 1.00 vs. 5.8 ± 0.87 μg·h/mL), indicating improved absorption and bioavailability	Wet-milled aceclofenac nanocrystals with PVA significantly enhanced solubility, dissolution rate, and oral bioavailability, making this a promising and cost-effective alternative to current therapies	2017	[72]
Diacerein	Sonoprecipitation	PVP K25	145.4 ± 4.8 nm PDI: 0.141 ± 0.02	Oval and spindle-like shape	PK analysis in Wister albino rats (plasma concentration-time profiles, Cmax, Tmax, AUC), ex vivo intestinal permeation using noneverted rat intestinal	The study showed that chitosan-coated DCN nanosuspensions significantly enhanced bioavailability (relative bioavailability of 172.1%)	2017	[73]

					sacs, and diarrhoea score tests (severity and frequency over 5 days)	and reduced diarrheal side effects compared to DCN suspension, supporting improved clinical efficacy and safer dosing recommendations		
Spironolactone	WMM	F127 F68 HPMC-E5 Sodium deoxycholate (NaDC)	370 nm with PDI < 0.2	Spherical	In vivo studies in Sprague-Dawley rats showed increased Cmax and AUC for all the nanocrystals formulations except the one stabilised with sodium deoxycholate. AUC total was 4.96-, 3.91-, 2.88- and 1.72-fold higher for F127, F68, HPMC-E5 and NaDC, respectively	Spironolactone nanocrystals showed enhanced dissolution rate and oral bioavailability compared to spironolactone solution Furthermore, the study found that ionic stabilisers (NaDC) can cause aggregation in the GI tract and are unsuitable alone for NC stabilisation	2016	[36]
Candesartan cilexetil	ASP	HPMC Pluronic® F 127	159 ± 8.1 nm PDI: 0.177 ± 0.043	Plates	PK studies in Sprague–Dawley rats showed enhanced gastrointestinal absorption of the nanosized drug with increased Tmax, Cmax and AUC for the nanocrystals compared to powdered drug	The nanocrystals in suspension showed an increased saturation solubility and dissolution rate, 4 times permeation rate and 3.8-fold oral bioavailability compared to the powdered form of the drug and demonstrated stability up to 3 months	2016	[41]
Hyperoside	HPH followed by lyophilisation	SDS PVP K-30	384 ± 21 nm PDI: 0.172 ± 0.027	Rods	PK studies in Sprague Dawley rats showed a 210.63% increase in oral bioavailability. PD studies	Hyperoside nanocrystals showed significantly enhanced dissolution rate, oral bioavailability, and	2016	[74]

(continued)

Table 1 (continued)

API	Production method	Stabiliser	Size	Crystal morphology	In vivo studies	Key outcome	Year	Refs.
					in ducks tested efficacy by inducing hepatitis B virus and measuring serum viral DNA and liver histopathology	superior anti-HBV activity in vivo, confirmed by higher inhibitory rates and improved liver histology compared to powdered drug		
Allisartan Isoproxil	WMM followed by spray-drying	PVP K30	304.0 ± 14.8 nm PDI: 0.21 ± 0.02	Irregular	PK studies in Sprague-Dawley rats showed increased Cmax and a 4.73-fold AUC increased compared to the plain drug	The study demonstrated that surfactant addition is essential for high-loading, redispersible nanocrystals of highly hydrophobic drugs, yielding improved dissolution and bio-availability	2016	[75]
Lovastatin	Sonoprecipitation WMM	PVP K30 F68 HPMC E5 Sodium deoxycholate	500.6 ± 21.0 nm	Rods Spherical	PK studies in beagle dogs showed that both rod-shaped and spherical-shaped nanocrystals had increased AUC over the reference product Junning®. Nonetheless, the rod-shaped lovastatin nanocrystals exhibited a superior in vivo performance over spherical-shaped ones	The study shows that nanocrystal shape is an important factor as the rod nanocrystals exhibited higher dissolution rate and oral bioavailability than spherical nanocrystals	2015	[34]
Ezetimibe	ASP And HPH	TPGS L-ascorbic acid-2-glucoside (AA2G)	359.1 ± 5.97 nm PDI: 0.008 ± 0.004	Spindle	PD studies in albino Wistar rats showed a superior reduction in TG, TC and HDL levels for the	The ezetimibe nanocrystals demonstrated increased saturation solubility, dissolution rate and	2015	[76]

					nanocrystal formulation stabilised with TPGS	antilipidemic activity compared to plain ezetimibe		
Paclitaxel	HPH	Pluronic F127 grafted chitosan	192.7 ± 9.2 nm PDI: 0.15	Rods	PK studies in Wistar rats showed the nanocrystals produced a higher plasma exposure: AUC had a 12.6-fold increase and Cmax a 13-fold increase compared to Taxol™	The paclitaxel nanocrystals provided a significant benefit as nanometre-scale particles increased surface area, leading to faster dissolution and a nine-fold rise in solubility, which together translated into a greater than 12-fold enhancement in oral bioavailability and superior antitumour efficacy	2015	[77]
Paclitaxel	HPH	PSS Tween 80	202.3 ± 13.28 nm PDI: 0.174	Rods	PK studies in Wistar rats showed that the orally administered nanocrystal formulation increased the AUC and Cmax 10 times compared to plain paclitaxel powdered drug	The paclitaxel nanocrystal formulation demonstrated increased dissolution rate directly translated to enhanced bioavailability in vivo without affecting the drug activity against cancer cells	2015	[78]
Diacerein	ASP Ultrasonication HPH	SDS	374.0 ± 1.95 nm PDI: 0.312 ± 0.01	Rods	A single-dose, two-period randomised cross-over design in human volunteers showed the nanocrystal formulation had increased Cmax and AUC compared to a commercially available capsule formulation. Leading to a relative bioavailability of 131.4%	Diacerein nanocrystals demonstrated increased saturation solubility, in vitro dissolution rate and relative bioavailability	2014	[79]

(continued)

Table 1 (continued)

API	Production method	Stabiliser	Size	Crystal morphology	In vivo studies	Key outcome	Year	Refs.
Atovaquone	Microprecipitation and HPH followed by freeze drying	Solutol® HS 15 Capryol™ 90	371.50 nm PDI: 0.19	Rods and cubes	PK studies in Sprague–Dawley rats showed that when compared to the commercially available product (Malarone®), the nanosuspension had a 4.6 and a 4.8-fold increased AUC and Cmax, respectively. PD studies were assessed by a 4-day suppressive test in plasmodium berghei-infected mice	Nanocrystals in suspension showed increased saturation solubility and dissolution rate compared to the reference product. They also showed increased bioavailability and produced superior efficacy, with a quarter-dose curing malaria and 100% survival, outperforming Malarone®	2014	[80]
Bexarotene	Nanoprecipitation and microfluidization	F68 Lecithin	279.0 ± 3.2 nm PDI: 0.104 ± 0.014	Spherical	PK study in Wistar rats showed. The plasma drug concentration profile of nanocrystals represented significant improvement in drug absorption compared to a bexarotene standard solution	The nanocrystal formulation significantly increased oral bioavailability of bexarotene and showed a sustained release profile	2014	[81]
Betulin	ASP	Tween 80	110 nm	Spherical	PK and PD studies in Sprague–Dawley rats showed increased Cmax and AUC for the betulin nanocrystals, resulting in a 1.21-fold increase in bioavailability and more potent anti-glycemic effect compared with raw betulin	Betulin nanocrystals showed increased saturation solubility and dissolution rate, leading to enhanced oral bioavailability when compared to the raw powdered drug	2014	[82]

Bicalutamide	ASP	Soluplus	168 nm	Spherical	Oral PK studies were performed in male Wistar rats; plasma drug levels were measured after administration of pure drug, plain precipitated drug, and BCT nanocrystals	The nanocrystals showed a ~ 3.5-fold increase in Cmax and a significantly higher AUC compared to pure drug, indicating much improved oral bioavailability. The formulation was physically stable for at least 6 months	2013	[83]

HPH high pressure homogenisation, *WMM* wet media milling, *ASP* anti-solvent precipitation

Among the FDA-approved NC-based oral formulations, several noteworthy examples demonstrate the breadth of therapeutic applications and the tangible benefits of NC technology. One of the earliest NC-containing products, Gris-Peg® (griseofulvin), approved in 1982, was developed to enhance the solubility and absorption of this antifungal agent, known for its poor water solubility. Similarly, Cesamet® (nabilone), introduced in 1985 as a synthetic cannabinoid used for chemotherapy-induced nausea and vomiting, employed NC technology to achieve improved bioavailability. Several analgesic and anti-inflammatory drugs have also benefitted from nanonisation. For example, Neprelan® (naproxen sodium), approved in 1996, and Zorvolex® (diclofenac), launched in 2016, demonstrated faster dissolution and more consistent plasma profiles compared to their conventional counterparts, enabling more predictable therapeutic effects. In the cardiovascular field, Verelan® PM (verapamil hydrochloride, 1998), Herbesser® (diltiazem hydrochloride, 2002), and Tricor® (2004) and Triglide® (2005), both fenofibrate formulations, illustrate how NCs can improve the pharmacokinetic profiles of drugs with low aqueous solubility, facilitating once-daily dosing and reducing food effects. Rapamune® (sirolimus), approved in 1999, is a macrolide immunosuppressant with extremely low solubility and a narrow therapeutic index. It is a clear example of how NC technology can be employed to enhance the oral bioavailability of complex molecules that would otherwise present major formulation challenges. In the central nervous system category, Ritalin LA® (methylphenidate hydrochloride, 2002), and Focalin XR® (dexmethylphenidate hydrochloride, 2005) as well as Zanaflex™ (tizanidine hydrochloride, 2002), demonstrate how NCs can be incorporated into extended-release formulations to optimise onset and duration of action. Other examples include Avinza® (morphine sulphate, 2002), a once-daily opioid formulation designed to maintain stable plasma levels while reducing dosing frequency, and Emend® (aprepitant, 2003), used to prevent chemotherapy-induced nausea and vomiting, which achieved enhanced oral absorption through NC formulation. Megace ES® (megestrol acetate, 2001), used for appetite stimulation in cancer and AIDS-related cachexia, also benefited from nanonisation, allowing for reduced dosing volumes and improved patient compliance.

5 Conclusion

NC technology has established itself as a powerful and versatile strategy for improving the oral delivery of poorly water-soluble drugs, a category that includes a large and growing number of therapeutic agents with significant formulation challenges. Many drugs suffer from low aqueous solubility, which limits their dissolution, absorption, and ultimately their clinical efficacy. By reducing particle size to the nanometre scale, NCs significantly enhance dissolution rate and saturation solubility, resulting in improved oral bioavailability and more consistent pharmacokinetic profiles. These benefits enable lower dosing, less frequent administration, and better patient compliance, critical advantages for oral therapies. This field remains an area of intense research and commercial interest, driven by the widespread prevalence of poorly soluble drugs and the need for effective delivery solutions. Although challenges remain in scaling up production and maintaining long-term stability, continuous advancements in formulation science and manufacturing technologies are expanding the potential applications of oral NCs. Overall, NCs represent a clinically validated and promising platform to overcome solubility and absorption barriers, ultimately facilitating the development of safer and more effective oral medications.

References

1. Market Research Future – Industry Analysis Report, Business Consulting and Research., https://www.marketresearchfuture.com/. Accessed 2 July 2025 (n.d.)
2. Home | American Pharmaceutical Review, https://www.americanpharmaceuticalreview.com/. Accessed 2 July 2025 (n.d.)
3. Oral Solid Dosage Pharmaceutical Market Size & Trends 2032., https://www.factmr.com/report/oral-solid-dosage-pharmaceutical-market. Accessed 2 July 2025 (n.d.)
4. V. Prasad, K. De Jesús, S. Mailankody, The high price of anticancer drugs: origins, implications, barriers, solutions. Nat. Rev. Clin. Oncol. **14**, 381–390 (2017)
5. P. Ma, J. Seguin, N.K. Ly, L.C. Henríquez, E. Plansart, K. Hammad, R. Gahoual, H. Dhôtel, C. Izabelle, B. Saubamea, C. Richard, V. Escriou, N. Mignet, Y. Corvis, Designing fisetin nanocrystals for enhanced in cellulo anti-angiogenic and anticancer efficacy. Int J Pharm X **4**, 100138 (2022). https://doi.org/10.1016/J.IJPX.2022.100138
6. M.S. Alqahtani, M. Kazi, M.A. Alsenaidy, M.Z. Ahmad, Advances in oral drug delivery. Front. Pharmacol. **12**, 618411 (2021). https://doi.org/10.3389/FPHAR.2021.618411/XML
7. A. Rossetti, D.A. Real, B.A. Barrientos, D.A. Allemandi, A.J. Paredes, J.P. Real, S.D. Palma, Significant progress in improving atorvastatin dissolution rate: Physicochemical characterization and stability assessment of self-dispersible atorvastatin/tween 80® nanocrystals formulated through wet milling and freeze-drying. Int. J. Pharm. **650**, 123720 (2024). https://doi.org/10.1016/J.IJPHARM.2023.123720
8. O. Article, S. Labib, M. Nasr, S. Badawy, Formulation and evaluation of atorvastatin calcium nanocrystals containing p-glycoprotein inhibitors for enhancing oral delivery. Int. J. Curr. Pharm. Res., 19–23 (2021). https://doi.org/10.22159/IJCPR.2021V13I3.42087
9. J. Lou, H. Duan, Q. Qin, Z. Teng, F. Gan, X. Zhou, X. Zhou, Advances in oral drug delivery systems: Challenges and opportunities. Pharmaceutics **15**, 484 (2023). https://doi.org/10.3390/PHARMACEUTICS15020484
10. M.R. Neutra, P.A. Kozlowski, Mucosal vaccines: The promise and the challenge. Nat. Rev. Immunol. **6**, 148–158 (2006)
11. J.B. Dressman, G.L. Amidon, C. Reppas, V.P. Shah, Dissolution testing as a prognostic tool for oral drug absorption: Immediate release dosage forms. Pharm. Res. **15**, 11–22 (1998). https://doi.org/10.1023/A:1011984216775/METRICS
12. L. Liu, C. Tian, B. Dong, M. Xia, Y. Cai, R. Hu, X. Chu, Models to evaluate the barrier properties of mucus during drug diffusion. Int. J. Pharm. **599**, 120415 (2021). https://doi.org/10.1016/J.IJPHARM.2021.120415
13. T. Lenhardt, G. Vergnault, P. Grenier, D. Scherer, P. Langguth, Evaluation of nanosuspensions for absorption enhancement of poorly soluble drugs: In vitro transport studies across intestinal epithelial monolayers. AAPS J. **10**, 435–438 (2008). https://doi.org/10.1208/S12248-008-9050-7/FIGURES/2

14. M. Elmeliegy, M. Vourvahis, C. Guo, D.D. Wang, Effect of P-glycoprotein (P-gp) inducers on exposure of P-gp substrates: Review of clinical drug–drug interaction studies. Clin. Pharmacokinet. **59**, 699–714 (2020). https://doi.org/10.1007/S40262-020-00867-1
15. M.N. Martinez, G.L. Amidon, A mechanistic approach to understanding the factors affecting drug absorption: A review of fundamentals. J. Clin. Pharmacol. **42**, 620–643 (2002). https://doi.org/10.1177/00970002042006005
16. M. Markovic, S. Ben-Shabat, A. Dahan, Prodrugs for improved drug delivery: Lessons learned from recently developed and marketed products. Pharmaceutics **12**, 1031 (2020). https://doi.org/10.3390/PHARMACEUTICS12111031
17. G. Tiwari, R. Tiwari, A.K. Rai, Cyclodextrins in delivery systems: Applications. J. Pharm. Bioallied Sci. **2**, 72 (2010). https://doi.org/10.4103/0975-7406.67003
18. D. Gupta, D. Bhatia, V. Dave, V. Sutariya, S.V. Gupta, Salts of therapeutic agents: Chemical, physicochemical, and biological considerations. Molecules **23**, 1719 (2018). https://doi.org/10.3390/MOLECULES23071719
19. A. Schittny, J. Huwyler, M. Puchkov, Mechanisms of increased bioavailability through amorphous solid dispersions: A review. Drug Deliv. **27**, 110 (2019). https://doi.org/10.1080/10717544.2019.1704940
20. M.R. Gigliobianco, C. Casadidio, R. Censi, P. Di Martino, Nanocrystals of poorly soluble drugs: Drug bioavailability and physicochemical stability. Pharmaceutics **10**, 134 (2018). https://doi.org/10.3390/PHARMACEUTICS10030134
21. M.B. McGuckin, J. Wang, R. Ghanma, N. Qin, S.D. Palma, R.F. Donnelly, A.J. Paredes, Nanocrystals as a master key to deliver hydrophobic drugs via multiple administration routes. J. Control. Release **345**, 334–353 (2022). https://doi.org/10.1016/J.JCONREL.2022.03.012
22. R.H. Müller, S. Gohla, C.M. Keck, State of the art of nanocrystals – special features, production, nanotoxicology aspects and intracellular delivery. Eur. J. Pharm. Biopharm. **78**, 1–9 (2011). https://doi.org/10.1016/J.EJPB.2011.01.007
23. B. Wahlang, Y.B. Pawar, A.K. Bansal, Identification of permeability-related hurdles in oral delivery of curcumin using the Caco-2 cell model. Eur. J. Pharm. Biopharm. **77**, 275–282 (2011). https://doi.org/10.1016/J.EJPB.2010.12.006
24. Y. He, D.N. Xia, Q.X. Li, J.S. Tao, Y. Gan, C. Wang, Enhancement of cellular uptake, transport and oral absorption of protease inhibitor saquinavir by nanocrystal formulation. Acta Pharmacol. Sin. **36**, 1151–1160 (2015). https://doi.org/10.1038/APS.2015.53;KWRD
25. G. Zhang, Y. Wang, Z. Zhang, Z. He, Y. Liu, Q. Fu, FRET imaging revealed that nanocrystals enhanced drug oral absorption by dissolution rather than endocytosis: A case study of coumarin 6. J. Control. Release **332**, 225–232 (2021). https://doi.org/10.1016/J.JCONREL.2021.02.025
26. M. Imono, H. Uchiyama, H. Ueda, K. Kadota, Y. Tozuka, In-situ dissolution and permeation studies of nanocrystal formulations with second-derivative UV spectroscopy. Int. J. Pharm. **558**, 242–249 (2019). https://doi.org/10.1016/J.IJPHARM.2018.12.086
27. R. Mauludin, R.H. Müller, C.M. Keck, Development of an oral rutin nanocrystal formulation. Int. J. Pharm. **370**, 202–209 (2009). https://doi.org/10.1016/J.IJPHARM.2008.11.029
28. L.D.O. Macedo, J.F. Masiero, N.A. Bou-Chacra, Drug nanocrystals in oral absorption: Factors that influence pharmacokinetics. Pharmaceutics **16**(9), 1141 (2024)
29. J. Li, Z. Wang, H. Zhang, J. Gao, A. Zheng, Progress in the development of stabilization strategies for nanocrystal preparations. Drug Deliv. **28**, 19–36 (2021). https://doi.org/10.1080/10717544.2020.1856224;PAGE:STRING:ARTICLE/CHAPTER
30. Y. Ding, T. Zhao, J. Fang, J. Song, H. Dong, J. Liu, S. Li, M. Zhao, Recent developments in the use of nanocrystals to improve bioavailability of APIs. Wiley Interdiscip. Rev. Nanomed. Nanobiotechnol. **16**, e1958 (2024). https://doi.org/10.1002/WNAN.1958
31. A. Dahan, J.M. Miller, The solubility-permeability interplay and its implications in formulation design and development for poorly soluble drugs. AAPS J. **14**, 244–251 (2012). https://doi.org/10.1208/S12248-012-9337-6/FIGURES/8

32. C. Shen, Y. Yang, B. Shen, Y. Xie, J. Qi, X. Dong, W. Zhao, W. Zhu, W. Wu, H. Yuan, Y. Lu, Self-discriminating fluorescent hybrid nanocrystals: Efficient and accurate tracking of translocation via oral delivery. Nanoscale **10**, 436–450 (2017). https://doi.org/10.1039/C7NR06052A
33. B. Shen, C. Shen, W. Zhu, H. Yuan, The contribution of absorption of integral nanocrystals to enhancement of oral bioavailability of quercetin. Acta Pharm. Sin. B **11**, 978–988 (2021). https://doi.org/10.1016/J.APSB.2021.02.015
34. M. Guo, Q. Fu, C. Wu, Z. Guo, M. Li, J. Sun, Z. He, L. Yang, Rod shaped nanocrystals exhibit superior in vitro dissolution and in vivo bioavailability over spherical like nanocrystals: A case study of lovastatin. Colloids Surf. B Biointerfaces **128**, 410–418 (2015). https://doi.org/10.1016/J.COLSURFB.2015.02.039
35. M. Guo, M. Wei, W. Li, M. Guo, C. Guo, M. Ma, Y. Wang, Z. Yang, M. Li, Q. Fu, L. Yang, Z. He, Impacts of particle shapes on the oral delivery of drug nanocrystals: Mucus permeation, transepithelial transport and bioavailability. J. Control. Release **307**, 64–75 (2019). https://doi.org/10.1016/J.JCONREL.2019.06.015
36. S. Mu, M. Li, M. Guo, W. Yang, Y. Wang, J. Li, Q. Fu, Z. He, Spironolactone nanocrystals for oral administration: Different pharmacokinetic performances induced by stabilizers. Colloids Surf. B Biointerfaces **147**, 73–80 (2016). https://doi.org/10.1016/J.COLSURFB.2016.07.051
37. L. Lopez-Vidal, P. Parodi, M.R. Actis, N. Camacho, D.A. Real, A.J. Paredes, F.J. Irazoqui, J.P. Real, S.D. Palma, Formulation and optimization of pH-sensitive nanocrystals for improved oral delivery. Drug Deliv. Transl. Res. **14**(5), 1301–1318 (2024)
38. J. Che, Y. Fu, Y. Li, Y. Zhang, T. Yin, J. Gou, X. Tang, Y. Wang, H. He, Eudragit L100-coated nintedanib nanocrystals improve oral bioavailability by reducing drug particle size and maintaining drug supersaturation. Int. J. Pharm. **658**, 124196 (2024). https://doi.org/10.1016/J.IJPHARM.2024.124196
39. D. Xia, P. Quan, H. Piao, H. Piao, S. Sun, Y. Yin, F. Cui, Preparation of stable nitrendipine nanosuspensions using the precipitation–ultrasonication method for enhancement of dissolution and oral bioavailability. Eur. J. Pharm. Sci. **40**, 325–334 (2010). https://doi.org/10.1016/J.EJPS.2010.04.006
40. E. Merisko-Liversidge, G.G. Liversidge, E.R. Cooper, Nanosizing: A formulation approach for poorly-water-soluble compounds. Eur. J. Pharm. Sci. **18**, 113–120 (2003). https://doi.org/10.1016/S0928-0987(02)00251-8
41. S. Jain, V.A. Reddy, S. Arora, K. Patel, Development of surface stabilized candesartan cilexetil nanocrystals with enhanced dissolution rate, permeation rate across CaCo-2, and oral bioavailability. Drug Deliv. Transl. Res. **6**, 498–510 (2016). https://doi.org/10.1007/S13346-016-0297-8/FIGURES/12
42. S. Bose, D. Schenck, I. Ghosh, A. Hollywood, E. Maulit, C. Ruegger, Application of spray granulation for conversion of a nanosuspension into a dry powder form. Eur. J. Pharm. Sci. **47**, 35–43 (2012). https://doi.org/10.1016/J.EJPS.2012.04.020
43. S. Kumar, J. Shen, D.J. Burgess, Nano-amorphous spray dried powder to improve oral bioavailability of itraconazole. J. Control. Release **192**, 95–102 (2014). https://doi.org/10.1016/J.JCONREL.2014.06.059
44. B. Du, G. Shen, D. Wang, L. Pang, Z. Chen, Z. Liu, Development and characterization of glimepiride nanocrystal formulation and evaluation of its pharmacokinetic in rats. Drug Deliv. **20**, 25–33 (2013). https://doi.org/10.3109/10717544.2012.742939;REQUESTEDJOURNAL:JOURNAL:IDRD20;WGROUP:STRING:PUBLICATION
45. A.J. Paredes, N. Litterio, A. Dib, D.A. Allemandi, C. Lanusse, S.S. Bruni, S.D. Palma, A nanocrystal-based formulation improves the pharmacokinetic performance and therapeutic response of albendazole in dogs. J. Pharm. Pharmacol. **70**, 51–58 (2018). https://doi.org/10.1111/JPHP.12834
46. M.A. Khan, M.M. Ansari, S.T. Arif, A. Raza, H.I. Choi, C.W. Lim, H.Y. Noh, J.S. Noh, S. Akram, H.A. Nawaz, M. Ammad, A.A. Alamro, A.A. Alghamdi, J.K. Kim, A. Zeb, Eplerenone nanocrystals engineered by controlled crystallization for enhanced oral bioavailability. Drug Deliv. **28**, 2510–2524 (2021). https://doi.org/10.1080/10717544.2021.2008051

47. S. Khare, R. Jog, A. Bright, D.J. Burgess, S.K. Chakder, K. Gokulan, Evaluation of mucosal immune profile associated with Zileuton nanocrystal-formulated BCS-II drug upon oral administration in Sprague Dawley rats. Nanotoxicology **17**, 583–603 (2023). https://doi.org/10.1080/17435390.2023.2289940;JOURNAL:JOURNAL:INAN20;WGROUP:STRING:PUBLICATION
48. C. Liu, D. Chang, X. Zhang, H. Sui, Y. Kong, R. Zhu, W. Wang, Oral fast-dissolving films containing lutein nanocrystals for improved bioavailability: Formulation development, in vitro and in vivo evaluation. AAPS PharmSciTech **18**, 2957–2964 (2017). https://doi.org/10.1208/S12249-017-0777-2/TABLES/6
49. P. Rana, R.S.R. Murthy, Formulation and evaluation of mucoadhesive buccal films impregnated with carvedilol nanosuspension: A potential approach for delivery of drugs having high first-pass metabolism. Drug Deliv. **20**, 224–235 (2013). https://doi.org/10.3109/10717544.2013.779331
50. L. Lopez-Vidal, M. Tinti, M.E. Melian, L. Canton, M. Lorenzutti, L. Schofs, M.L. Formica, A.J. Paredes, S.S. Bruni, N. Litterio, R. Faccio, S.D. Palma, J.P. Real, In vivo pharmacokinetic study and PBPK modeling: Comparison between 3D-printed nanocrystals and solid dispersions. Int. J. Pharm. **669**, 125063 (2025). https://doi.org/10.1016/J.IJPHARM.2024.125063
51. J.P. Real, D.A. Real, L. Lopez-Vidal, B.A. Barrientos, K. Bolaños, M.G. Tinti, N.J. Litterio, M.J. Kogan, S.D. Palma, 3D-printed Gastroretentive tablets loaded with Niclosamide nanocrystals by the melting solidification printing process (MESO-PP). Pharmaceutics **15**, 1387 (2023). https://doi.org/10.3390/pharmaceutics15051387
52. Z. Li, Y. Liu, J. Wang, X. Feng, E.O. Nwafor, Y. Zhang, R. Liu, W. Dang, Q. Zhang, C. Yu, J. Pi, Z. Liu, Baicalin-berberine complex nanocrystals orally promote the co-absorption of two components. Drug Deliv. Transl. Res. **12**, 3017–3028 (2022). https://doi.org/10.1007/S13346-022-01167-W/TABLES/3
53. L. Huang, W. Hu, P. Yue, Y. Liao, H. Chen, Y. Chen, Development and evaluation of Tween-80 modified breviscapine nanocrystals for enhanced oral absorption. J Drug Deliv Sci Technol **107**, 106819 (2025). https://doi.org/10.1016/J.JDDST.2025.106819
54. S. Lan, K. Chen, L. Feng, P. Sima, X. Ji, F. Wu, Y. Lin, Tea Saponins: a novel stabilizer for enhancing the oral bioavailability of Albendazole nanocrystals. AAPS PharmSciTech **26**, 1–13 (2025). https://doi.org/10.1208/S12249-024-03015-1/TABLES/3
55. S. Chang, Q. Yang, J. Liu, L. Yin, J. Han, L. Zong, X. Pu, The increased dissolution and oral absorption of Itraconazole by nanocrystals with an endogenous small-molecule surfactant as a stabilizer. Molecules **29**, 1769 (2024). https://doi.org/10.3390/MOLECULES29081769
56. Y. Zhu, F. Hu, C. Shen, B. Shen, H. Yuan, Quercetin nanocrystals for bioavailability enhancement: Impact of different functional stabilizers on in vitro/in vivo drug performances. Pharm. Dev. Technol. **29**(6), 551–558 (2024)
57. Y. Wang, P. Wang, H. Li, X. Han, H. Zhu, X. Jin, Nanocrystal-loaded lipid carriers for improved oral absorption and anticancer efficacy of etoposide: Formulation development, transport mechanism, in vitro and in vivo evaluation. Mol. Pharm. **21**, 1170–1181 (2024). https://doi.org/10.1021/ACS.MOLPHARMACEUT.3C00785/ASSET/IMAGES/LARGE/MP3C00785_0009.JPEG
58. T. Sun, F. Zhang, Y. Xu, X. Wang, J. Jia, L. Sang, J. Li, D. Wang, Z. Yu, Lysine-Polydopamine nanocrystals loaded with the Codrug Abemaciclib-Flurbiprofen for Oral treatment of cancer. ACS Omega **9**, 18137–18147 (2024). https://doi.org/10.1021/ACSOMEGA.3C10142/ASSET/IMAGES/LARGE/AO3C10142_0008.JPEG
59. H. Luo, X. Wang, M. Fang, H. Yu, L. Gui, Z. Wu, J. Sheng, F. Li, Combined hydroxyethyl starch Luteolin nanocrystals for effective anti-hyperuricemia effect in mice model. Int. J. Nanomed. **19**, 5139–5156 (2024). https://doi.org/10.2147/IJN.S464948/ASSET/0C292E2B-1FC3-4FC4-BC5C-35AA7ED8C7FC/ASSETS/GRAPHIC/DIJN_A_12303318_F0011_C.JPG
60. S. Khare, R. Jog, A. Bright, D.J. Burgess, S.K. Chakder, K. Gokulan, Evaluation of mucosal immune profile associated with Zileuton nanocrystal-formulated BCS-II drug upon oral

administration in Sprague Dawley rats. Nanotoxicology **17**, 583–603 (2023). https://doi.org/10.1080/17435390.2023.2289940/ASSET/FC94043B-A841-476E-BBD7-229EB88B2F6B/ASSETS/GRAPHIC/INAN_A_2289940_F0009_C.JPG

61. A.M. Awan, A. Farid, S.U. Shah, D. Khan, F. Ur Rehman, M.J. Dar, T. Iftikhar, S. Ghazanfar, C.M. Galanakis, A.S. Alamri, S.M.B. Asdaq, K.U. Shah, Nanocrytals-mediated oral drug delivery: Enhanced bioavailability of amiodarone. Pharmaceutics **14**, 1300 (2022). https://doi.org/10.3390/PHARMACEUTICS14061300
62. X. Guo, Y. Guo, M. Zhang, B. Yang, H. Liu, T. Yin, Y. Zhang, H. He, Y. Wang, D. Liu, J. Gou, X. Tang, A comparative study on in vitro and in vivo characteristics of enzalutamide nanocrystals versus amorphous solid dispersions and a better prediction for bioavailability based on "spring-parachute" model. Int. J. Pharm. **628**, 122333 (2022). https://doi.org/10.1016/J.IJPHARM.2022.122333
63. S. Xiong, W. Liu, D. Li, X. Chen, F. Liu, D. Yuan, H. Pan, Q. Wang, S. Fang, T. Chen, Oral delivery of Puerarin nanocrystals to improve brain accumulation and anti-parkinsonian efficacy. Mol. Pharm. **16**, 1444–1455 (2019). https://doi.org/10.1021/ACS.MOLPHARMACEUT.8B01012/ASSET/IMAGES/LARGE/MP-2018-01012D_0010.JPEG
64. Y. Zhu, Y. Fu, A. Zhang, X. Wang, Z. Zhao, Y. Zhang, T. Yin, J. Gou, Y. Wang, H. He, X. Tang, Rod-shaped nintedanib nanocrystals improved oral bioavailability through multiple intestinal absorption pathways. Eur. J. Pharm. Sci. **168**, 106047 (2022). https://doi.org/10.1016/J.EJPS.2021.106047
65. S.G. Kala, S. Chinni, Development and characterization of Venetoclax nanocrystals for oral bioavailability enhancement. AAPS PharmSciTech. **22**, 1–11 (2021). https://doi.org/10.1208/S12249-021-01968-1/FIGURES/7
66. A.J. Paredes, N.M. Camacho, L. Schofs, A. Dib, et al., Ricobendazole nanocrystals obtained by media milling and spray drying: Pharmacokinetic comparison with the micronized form of the drug. Int. J. Pharm. **585**, 119501 (2020). https://doi.org/10.1016/J.IJPHARM.2020.119501
67. H.S. Ali, A.F. Hanafy, a. alqurshi, engineering of solidified glyburide nanocrystals for tablet formulation via loading of carriers: downstream processing, characterization, and bioavailability. Int. J. Nanomed. **14**, 1893–1906 (2019). https://doi.org/10.2147/IJN.s194734
68. D. Chang, Y. Ma, G. Cao, J. Wang, X. Zhang, J. Feng, W. Wang, Improved oral bioavailability for lutein by nanocrystal technology: formulation development, in vitro and in vivo evaluation. Artif. Cells Nanomed. Biotechnol. **46**, 1018–1024 (2018). https://doi.org/10.1080/21691401.2017.1358732/ASSET/942F8629-A0DA-4003-9FD4-0EC23AA017B8/ASSETS/IMAGES/IANB_A_1358732_F0006_C.JPG
69. X. Xu, G. Chen, Y. Li, J. Wang, J. Yin, L. Ren, Enhanced dissolution and oral bioavailbility of cinacalcet hydrochlorde nanocrystals with no food effect. Nanotechnology **30**, 055102 (2018). https://doi.org/10.1088/1361-6528/AAEF46
70. W. Xiong, W. Sang, K.G. Linghu, Z.F. Zhong, W.S. Cheang, J. Li, Y.J. Hu, H. Yu, Y.T. Wang, Dual-functional Brij-S20-modified nanocrystal formulation enhances the intestinal transport and oral bioavailability of berberine. Int. J. Nanomed. **13**, 3781–3793 (2018). https://doi.org/10.2147/IJN.S163763/ASSET/65694EAD-C7A5-4E0C-8D1C-EE09D49EADC1/ASSETS/GRAPHIC/DIJN_A_12194064_F0007_C.JPG
71. S.K. Singh, V. Makadia, S. Sharma, M. Rashid, S. Shahi, P.R. Mishra, M. Wahajuddin, J.R. Gayen, Preparation and in-vitro/in-vivo characterization of trans-resveratrol nanocrystals for oral administration. Drug Deliv. Transl. Res. **7**, 395–407 (2017). https://doi.org/10.1007/S13346-017-0362-Y/FIGURES/6
72. R. Narayan, A. Pednekar, D. Bhuyan, C. Gowda, K.B. Koteshwara, U.Y. Nayak, A top-down technique to improve the solubility and bioavailability of aceclofenac: In vitro and in vivo studies. Int. J. Nanomed. **12**, 4921–4935 (2017). https://doi.org/10.2147/IJN.S141504/ASSET/24502B27-4FB6-4E21-B538-F825D654CFB3/ASSETS/GRAPHIC/DIJN_A_12193741_F0010_C.JPG
73. A.N. Allam, S.I. Hamdallah, O.Y. Abdallah, Chitosan-coated diacerein nanosuspensions as a platform for enhancing bioavailability and lowering side effects: Preparation, characterization,

and ex vivo/in vivo evaluation. Int. J. Nanomed. **12**, 4733–4745 (2017). https://doi.org/10.2147/IJN.S139706/ASSET/419B67A9-05D7-40D4-8BA3-74C8226913B5/ASSETS/GRAPHIC/DIJN_A_139706_SF0001_B.JPG
74. B. Shen, N. Wu, C. Shen, F. Zhang, Y. Wu, P. Xu, L. Zhang, W. Wu, Y. Lu, J. Han, Y. Wang, H. Yuan, Hyperoside nanocrystals for HBV treatment: process optimization, in vitro and in vivo evaluation. Drug Dev. Ind. Pharm. **42**, 1772–1781 (2016). https://doi.org/10.3109/03639045.2016.1173051;SUBPAGE:STRING:FULL
75. Y. Hou, J. Shao, Q. Fu, J. Li, J. Sun, Z. He, Spray-dried nanocrystals for a highly hydrophobic drug: Increased drug loading, enhanced redispersity, and improved oral bioavailability. Int. J. Pharm. **516**, 372–379 (2017). https://doi.org/10.1016/J.IJPHARM.2016.11.043
76. K.M.R. Srivalli, B. Mishra, Preparation and pharmacodynamic assessment of ezetimibe nanocrystals: Effect of P-gp inhibitory stabilizer on particle size and oral absorption. Colloids Surf. B Biointerfaces **135**, 756–764 (2015). https://doi.org/10.1016/J.COLSURFB.2015.08.042
77. S. Sharma, A. Verma, G. Pandey, N. Mittapelly, P.R. Mishra, Investigating the role of Pluronic-g-cationic polyelectrolyte as functional stabilizer for nanocrystals: Impact on paclitaxel oral bioavailability and tumor growth. Acta Biomater. **26**, 169–183 (2015). https://doi.org/10.1016/J.ACTBIO.2015.08.005
78. S. Sharma, A. Verma, B.V. Teja, P. Shukla, P.R. Mishra, Development of stabilized paclitaxel nanocrystals: In-vitro and in-vivo efficacy studies. Eur. J. Pharm. Sci. **69**, 51–60 (2015). https://doi.org/10.1016/J.EJPS.2014.11.012
79. I. Elsayed, A.A. Abdelbary, A.H. Elshafeey, Nanosizing of a poorly soluble drug: Technique optimization, factorial analysis, and pharmacokinetic study in healthy human volunteers. Int. J. Nanomed. **9**, 2943–2953 (2014). https://doi.org/10.2147/IJN.S63395/ASSET/06605E49-17FA-49FA-990D-6A32905F3B7F/ASSETS/GRAPHIC/DIJN_A_63395_F0007_C.JPG
80. V. Borhade, S. Pathak, S. Sharma, V. Patravale, Formulation and characterization of atovaquone nanosuspension for improved oral delivery in the treatment of malaria. Nanomedicine **9**, 649–666 (2014). https://doi.org/10.2217/NNM.13.61/SUPPL_FILE/INNM_A_12343475_SM0002.DOCX
81. L. Chen, Y. Wang, J. Zhang, L. Hao, H. Guo, H. Lou, D. Zhang, Bexarotene nanocrystal – oral and parenteral formulation development, characterization and pharmacokinetic evaluation. Eur. J. Pharm. Biopharm. **87**, 160–169 (2014). https://doi.org/10.1016/J.EJPB.2013.12.005
82. X. Zhao, W. Wang, Y. Zu, Y. Zhang, Y. Li, W. Sun, C. Shan, Y. Ge, Drug delivery preparation and characterization of betulin nanoparticles for oral hypoglycemic drug by antisolvent precipitation preparation and characterization of betulin nanoparticles for oral hypoglycemic drug by antisolvent precipitation. Drug Deliv. **21**, 467–479 (2014). https://doi.org/10.3109/10717544.2014.881438
83. V.B. Pokharkar, T. Malhi, L. Mandpe, Bicalutamide nanocrystals with improved oral bioavailability: in vitro and in vivo evaluation. Pharm. Dev. Technol. **18**, 660–666 (2013). https://doi.org/10.3109/10837450.2012.663391/ASSET/642B7794-34A6-45C6-A728-4C66FF6FCDAE/ASSETS/IMAGES/IPHD_A_663391_F0007_B.GIF
84. E.J. Catlin, L. Lopez-Vidal, R.F. Donnelly, A.J. Paredes, A dynamic duo comes of age: nanocrystals and microneedles for hydrophobic drug delivery. Expert Opin. Drug Deliv. (2025). https://doi.org/10.1080/17425247.2025.2531059

Nanocrystals for Transdermal Drug Delivery and Hair Follicle Targeting

Abraham M. Abraham, Adriparna Paul, Pedro S. Londoño, Aws Ahmad Rashed Al Jayyousi, Marta G. Fuster, Franco D. Battistini, and Alejandro J. Paredes

Abstract Nanocrystals, nanosized form of active pharmaceutical ingredients (APIs), have emerged as a promising strategy to overcome the formidable barrier of the *stratum corneum* and enhance both transdermal and follicular drug delivery. As matrix-free systems, nanocrystals offer exceptionally high drug loading, increased saturation solubility, rapid dissolution, and strong bioadhesion, enabling improved skin penetration, retention, and bioavailability of poorly water-soluble drugs. Their capacity to accumulate within hair follicles allows them to act as drug reservoirs, supporting sustained release and targeted therapy for conditions such as acne, psoriasis, eczema, leishmaniasis, skin cancer, and alopecia. This chapter discusses the application of nanocrystals in transdermal drug delivery and hair follicle targeting, while also highlighting key factors, including particle size, shape, surface charge, stabilisers, and delivery vehicles, which critically influence their performance.

Keywords Nanocrystals · Nanosuspensions · Solid drug nanoparticles · Transdermal · Skin · Dermal delivery

1 Transdermal Drug Delivery Systems

The large surface area of the skin (~1.8 m^2) provides a lucrative option of drug delivery as it is easily accessible and serves as a moderately painless route of administration of drugs [1]. It overcomes the drawbacks of oral and parenteral routes, including first-pass metabolism, gastrointestinal pH, and interactions with other drugs and digestive enzymes, thus allowing a higher bioavailability and

---

A. M. Abraham · A. Paul · P. S. Londoño · A. A. R. Al Jayyousi · M. G. Fuster · A. J. Paredes (✉)
School of Pharmacy, Queen's University Belfast, Belfast, UK
e-mail: a.paredes@qub.ac.uk

F. D. Battistini
Secretaria de Salud, Municipalidad de Cordoba, Cordoba, Argentina

A. J. Paredes (ed.), *Nanocrystals*, https://doi.org/10.1007/978-3-032-10238-6_4

sustained release. It allows a continuous intervention in case of quick removal, replacement, or modification of the medication, as it can be applicable in the form of a patch on the skin. Therefore, this method of drug delivery offers increased patience compliance. Even in several instances where oral drug intake is not feasible and intravenous route is considered invasive, transdermal drug delivery systems (TDDS) play a crucial alternative role. The market for TDDS has been increasing rapidly, and its net worth is projected to reach 88,422.40 million US$ by 2030. However, the current limited commercial success of TDDS owes to the defensive "brick and mortar" barrier called the *stratum corneum*, which needs to be efficiently permeated to reach the target site. To achieve this, the organisation of the skin layers and functions needs to be thoroughly understood [2].

1.1 Skin

The skin is made up of three different layers with varying structure and composition. The outermost, epidermis, consists of a non-viable hydrophobic part called the *stratum corneum* (SC), and the viable hydrophilic layer consisting of *stratum lucidum*, *stratum granulosum*, *stratum spinosum*, and *stratum basale*. The SC forms the main barrier between the environment and the internal body. It represents the last stage of epidermal differentiation and forms a protective layer of dead cells called corneocytes. With its high density (1.4 g/cm^3 in the dry state) and low hydration (15–20%), this layer prevents water loss and protects from pathogens and harmful chemicals [3]. The SC is composed of dead corneocytes filled with hydrophilic keratin proteins, arranged in a lamellar organisation with densely packed repeat units. A lateral packing of a rigid matrix of lipids such as cholesterol (25%), ceramides (50%), and saturated fatty acids is also closely associated to the SC, providing the "brick and mortar" barrier to transport through the skin [2]. Below the epidermis lies the dermis, which consists of a dense network of fibres like elastin and collagen, providing strength and elasticity. Blood vessels are usually present in this zone, providing nutrients to both epidermis and dermis. Moreover, macrophages, nerves, and lymphatic vessels are also present in this layer [4]. The hypodermis, the deepest layer of the skin, serves as a transitional zone between the skin and the tissues beneath it. It is mainly composed of adipose (fat) cells, which function to insulate the body against heat loss, cushion it from physical impact, and facilitate the transmission of nerves and blood vessels.

1.2 Routes of Drug Delivery Through the Skin

The 15–20 layers of corneocytes of the SC present the most significant challenge for drug penetration. Once the drug reaches the dermis, it can pass into systemic circulation [5]. Many active ingredients can penetrate the SC via different routes

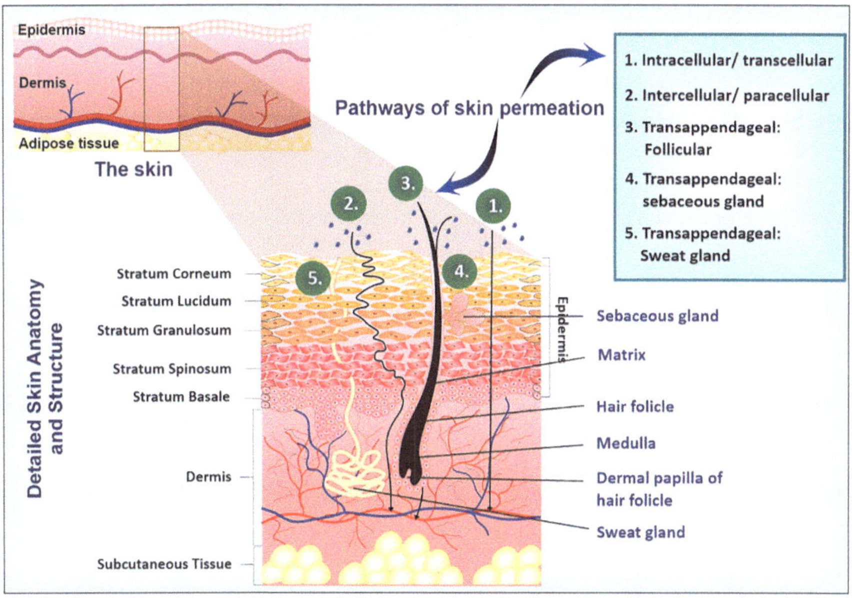

Fig. 1 Pathways of entry of different drug molecules through the *stratum corneum*. (Reprinted from reference [5])

such as intercellular, transcellular, and transappendageal routes. The intercellular or paracellular pathway is the more prevalent route, where drug molecules penetrate through the intercellular spaces between the corneocytes. It allows the passive diffusion of lipophilic or non-polar compounds, through the continuous lipid matrix. The transcellular or intracellular pathway allows the transport of small hydrophilic or polar solutes to pass through the corneocytes membrane. It is a less predominant route for the penetration of molecules through the SC, as the water content in this layer is around 13%. However, this pathway becomes more relevant in the dermis once the solutes have crossed the SC. The third route is the transappendageal or follicular route, in which drugs are delivered through the hair follicles, sweat glands, and sebaceous glands. These are invaginations which increase the surface area for the penetration and absorption of substances applied on the skin. Since these follicles are rooted deeper than the SC layer, it is relatively easier to transport drug molecules to the dermis via this route. Ionisable and large molecules, both, can be transported through this route. However, the small surface area covered by these structures on the skin makes their applications limited (Fig. 1) [5–7].

1.3 *Nanostructures for Transdermal Drug Delivery*

Permeation of molecules through the different layers of the skin, which are structurally and functionally different from one another, involves several steps. Penetration is the entry of the substance into a layer of the skin, followed by partitioning from the hydrophobic SC to the viable epidermis and further diffusion into the upper dermis from where the final absorption of molecules occurs into the systemic circulation [4]. The challenge to penetrate the skin in adequate therapeutic dosage without causing any inflammation or irritation is enormous. The use of penetration enhancers can often cause skin rashes and irritation.

Nanotechnology is a novel strategy to fluidise the SC by optimising nanocarriers to penetrate the skin. Simultaneously, the follicular and sebaceous routes could also be exploited for nanocarrier drug delivery systems. The incorporation of drugs into nanocarriers enhances penetration of drugs without causing significant disruption of the skin barrier and thereby irritation. Such nanocarriers such as liposomes, emulsions, and polymeric nanoparticles have been explored before. However, they have inadequate drug-loading, often show drug leakage, and rupture of skin during permeation. Furthermore, they require complex manufacturing excipients, which might cause toxicity issues. In this scenario, nanocrystals emerge as a potential solution to the skin penetration problem.

Nanocrystals are nano-sized drug crystals, with higher (~100%) therapeutic drug loading, low excipient content, and easier manufacturing methods. The persistent problem of low solubility of new drugs can also be addressed by preparing nanocrystals. The elevated surface area and dissolution pressure increase "apparent solubility" of nanocrystals and therefore enhance concentration gradient across the skin barrier facilitating passive diffusion. Nanocrystals also show higher adhesiveness to skin membranes and form a diffusional corona to enhance penetration. Delivering of nanocrystals in the lower *infundibulum* region of hair follicles further shows enhanced permeability due to less developed SC in this region. Furthermore, modifications of the morphology of the nanocrystals can contribute to enhancement in efficient and adequate transdermal delivery. To achieve nanocrystal formulations for efficient delivery, their properties need to be monitored and optimised carefully.

2 Nanocrystals Production Considerations for TDD

2.1 *Particle Size and Surface Charge*

The size of nanocrystals impacts the permeation through the skin as well as the safe targeting of cells. Nanocrystals of drugs show higher desired activity than micronised drug. Furthermore, a higher activity and skin permeation are shown by smaller nanocrystals. This is because of the amplified contact between the nanocrystals and the skin, causing greater transport of nanocrystals through the

SC. This close association also causes interaction with the lipids of the skin layer to promote formation of transient nano-channels for the promotion of drug transfer through the SC [8]. Improved skin retention and penetration have been achieved with smaller and medium-sized nanocrystals more than larger ones.

Surface charge, indicated by zeta potential, indicates stability of the formulation, and a value $>\pm 30$ mV is desirable for an electrostatically stabilised nanosuspension. While permeation of positively charged nanosystems with negatively charged cell membranes is through electrostatic interactions, negatively charged nanosystems show uptake by endocytosis or other non-specific processes of adsorption on the cell surface [9].

The penetration of nanocrystals through hair follicles is somewhat different from that of general transdermal delivery. It is observed that the main mechanism working in the entrance of nanocrystals into the *infundibulum* of hair follicles is the "ratchet mechanism" and "pawl" condition of the cuticular structure of the hair follicles. Varying sizes of nanocrystals result in only marginal differences in their penetration through hair follicles. Although slight variations in efficacy may occur when different excipients are used to formulate nanocrystals, no consistent correlation has been found between particle size and penetration efficacy with different excipient formulations.

2.2 Shape of Nanocrystals

Morphology of nanocrystals plays a crucial role in enhancement of transdermal delivery of the drug. It has been noted that anisotropic nanocrystals, such as nanorods and nanowires, have an improved penetration through the skin owing to the increased interfacial contact points. This leads to greater affinity for the lipid bilayers and potential formation of transient aqueous channels for the amplified permeation of drug molecules [8].

2.3 Surfactants

Surfactants interact with the lipids and proteins of the SC and penetrate this barrier. This might lead to disruption of cell membranes causing release of pro-inflammatory molecules. Ionic surfactants interact strongly with proteins via electrostatic forces and often cause skin irritation. In contrast, nonionic stabilisers form weaker van der Waals bonds with the skin and hence, these stabilisers are more suited for topically delivered nanosuspensions [7]. Other factors like skin hydration and pH of the different layers of the skin can play an important role in enhancing drug nanocrystal penetration [9].

3 Properties of Nanocrystals for Topical Applications

3.1 Dissolution Rate

A significant reduction in particle size increases the surface area-to-mass ratio of solid particles. This means that smaller particles have a larger surface available to interact with the solution molecules or the dissolution medium (Fig. 2).

The increased surface area leads to a greater quantity of material dissolved per unit of time. The increased dissolution rate dynamics are governed by the Noyes–Whitney equation (Eq. 1), where the velocity at which the drug dissolves is directly proportional to the surface area [10, 11].

$$\frac{dc}{dt} = A.D.\left(\frac{C_S - C_X}{h}\right) \tag{1}$$

dc/dt is the dissolution rate/velocity, A is the surface area, D is the diffusion coefficient, C_s the saturation solubility of the drug in the diffusion layer, Cx is the concentration in the bulk of the solution, and h is the diffusional distance [11, 12].

3.2 Saturation Solubility

It should be noted that Eq. 1 does not exclusively refer to the dissolution rate; it also introduces the concept of saturation solubility (Cs). Cs can be defined as the amount of drug dissolved until the saturation point is reached in a given medium and at a given temperature. For conventional drug substances and micronised powders, the Cs is typically regarded as an intrinsic thermodynamic property, independent of particle size. However, when the particle diameter decreases below a critical threshold of approximately 1–2 μm, Cs becomes dependent on particle size [10]. Thus, producing drug nanocrystals below this critical size directly results in an increase in the saturation solubility of the substance or an improved "apparent solubility". The

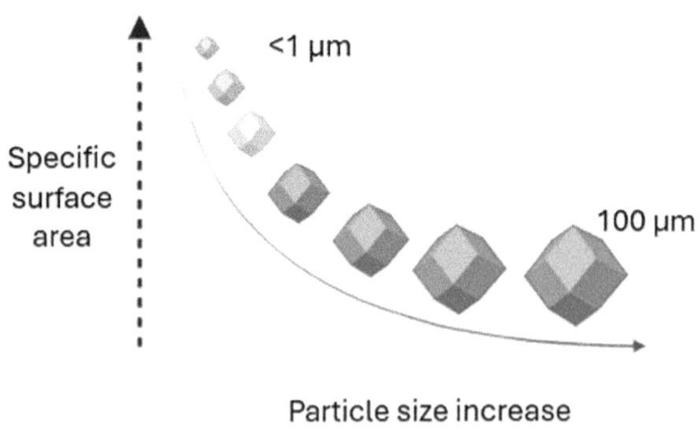

Fig. 2 Particle size reduction and surface area increase for drug nanocrystals. (Created with biorender.com)

relationship between particle size and solubility was first elucidated by Ostwald, who based his derivation on the thermodynamic framework established by Gibbs [13]. This work was later refined by Freundlich, who formulated Eq. 2, now known as the Ostwald–Freundlich equation [14].

$$\mathrm{Ln}\frac{S}{S_0} = \frac{2\gamma V_m}{rRT} \tag{2}$$

S is drug solubility of sufficiently small particles with radius (r). S_0 represents the equilibrium solubility, V_m is the molar volume, γ is the interfacial tension, R (8.31 J/mol.K) is the ideal gas constant, and T is the temperature of the system.

The Ostwald–Freundlich equation is derived from the Kelvin model, which describes how surface curvature influences vapour pressure in gas phase [15]. In the context of nanocrystals, their dissolution behaviour can be predicted using a similar mathematical form because both phenomena involve curvature-driven changes. Extremely small particles exhibit higher surface curvature, which increases vapour pressure and, in turn, enhances drug solubility [14, 15].

As mentioned earlier in the chapter, this elevated saturation solubility, combined with a faster dissolution rate, is particularly advantageous for topical drug delivery. Simultaneously, higher saturation solubility raises the concentration gradient at the site of application, which drives more effective passive diffusion across the skin or mucosal barriers. Furthermore, an increased dissolution rate ensures that the drug rapidly converts into its absorbable form upon contact with the skin's limited moisture, enabling a quicker onset of action. Together, these properties enhance the availability of the drug in its dissolved state, improving local bioavailability, promoting more efficient permeation, and contributing to greater therapeutic efficacy, particularly for poorly water-soluble drugs frequently used in topical formulations.

3.3 *Enhanced Bioadhesion*

Nanocrystals show enhanced adhesion to surfaces and bioadhesion to cells and tissues due to their increased surface-to-volume ratio, which provides more contact points (Fig. 3) [16]. The adhesion force is directly proportional to the surface area involved in the adhesion process [17]. Moreover, adhesion is influenced not only by particle size but also by crystal roughness and shape, which can further enhance adhesive performance [18].

Enhanced bioadhesion of nanocrystals offers significant advantages in topical drug delivery. Improved adhesion to skin and mucosal surfaces results in longer residence time at the site of application, which enhances the opportunity for drug absorption and therapeutic action. This close and prolonged contact with biological surfaces facilitates greater penetration through the skin or mucosa and supports the maintenance of a higher localised drug concentration. As a result, the need for

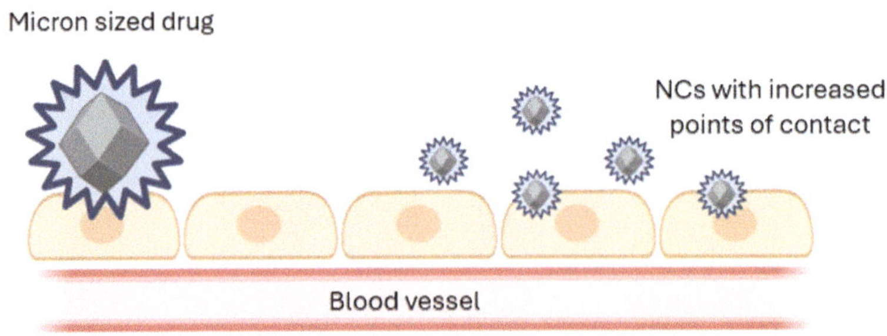

Fig. 3 Schematic of enhanced bioadhesiveness of nanocrystals. (Made with biorender.com)

frequent reapplication is reduced, contributing to better patient compliance and more consistent treatment outcomes.

3.4 *Increased Concentration Gradient*

Nanocrystals are a breakthrough in topical drug delivery due to their small size and large surface area, which increase drug loading, solubility, and skin penetration. Their high saturation solubility enhances the concentration gradient, a key factor driving passive diffusion and improving bioavailability for effective skin treatment [19, 20]. As per Fick's law, a high concentration gradient boosts the rate of diffusion across biological membranes—specifically, the *stratum corneum* and follicular epithelium. When applied directly onto the skin, nanocrystals create a concentrated layer of medication. Since the drug remains in its pure crystalline form rather than being dissolved in a medium, its thermodynamic activity reaches its peak, leading to more drug molecules ready to penetrate the skin [21]. Moreover, many formulations based on nanocrystals exhibit strong skin adhesion, prolonging their retention and enabling a sustained release of the medication. This is especially beneficial for managing chronic conditions like eczema or psoriasis, where maintaining consistent drug levels over time is crucial [22].

Some stabilisers and excipients, like hyaluronic acid or gum arabic, have been found to improve the formation of films and enhance dermal penetration by boosting skin hydration or interacting with follicular microenvironments [21, 23]. These agents can also steepen the concentration gradient further by keeping the nanocrystals concentrated at the application site.

3.5 *Improved Bioavailability*

One major drawback of traditional topical treatments is their limited bioavailability, particularly in case of hydrophobic drugs. Nanocrystals provide a solution by enhancing drug solubility, adhesion, and skin retention—all of which lead to higher

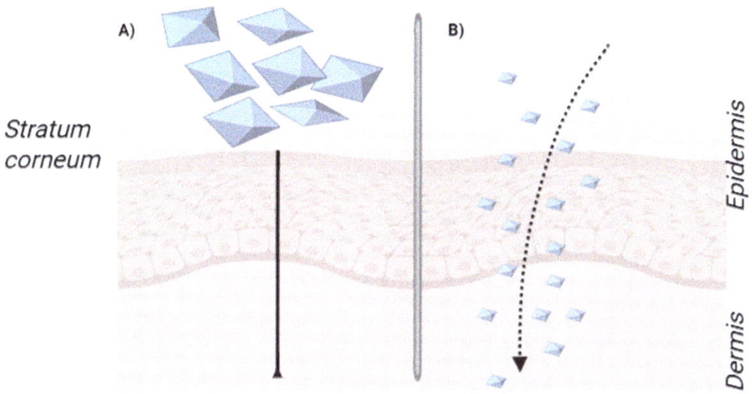

Fig. 4 (**A**) Macroparticles: poor penetration due to limited solubility and a low concentration gradient across the *stratum corneum*. (**B**) Nanocrystals: increased surface area and saturation solubility create a steeper concentration gradient, enhancing passive diffusion through the epidermis into the dermis, resulting in higher local bioavailability

concentrations of the drug within the skin layers [20, 24]. Thus, nanocrystals can effectively penetrate both the outer skin layer and the hair follicles, delivering larger amounts of the drug straight to the viable epidermis and dermis in comparison to large particles (Fig. 4). Research has shown that even small doses of nanocrystals can lead to significantly better skin drug absorption than conventional emulsions or suspensions [25, 26].

The hair follicle pathway is especially important for boosting local bioavailability. When nanocrystals are deposited into hair follicles, they serve as a reservoir for the drug, gradually releasing active ingredients over time, which promotes longer-lasting therapeutic effects. This reservoir function is particularly helpful for treating localised inflammatory issues like acne [21, 24]. Thus, nanocrystals are ideal for improving the dermal bioavailability while minimising systemic side effects, which is a primary objective in dermatological treatments [27].

4 Nanocrystals for Dermatological Applications

Skin conditions such as acne, psoriasis, eczema, leishmaniasis, and skin cancer impact millions of individuals around the world. They not only affect physical health but also take a toll on emotional and psychological well-being. Many of these ailments are long-lasting and tend to recur, necessitating ongoing treatment. Traditional therapies, e.g. incorporating drugs into creams, gels for topical application, often face challenges like poor solubility and limited skin absorption [8, 28]. The various advantages of nanocrystals, as mentioned in the previous section, make them a promising formulation principle to treat many skin conditions as illustrated in Fig. 5.

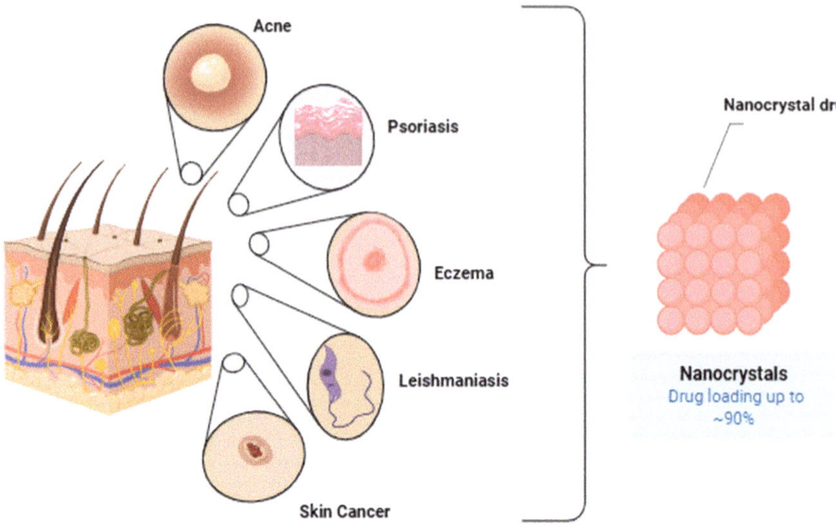

Fig. 5 Nanocrystals, with a high drug loading capacity (>90%), are utilised for enhanced penetration and localisation in treating acne, psoriasis, eczema, leishmaniasis, and skin cancer

4.1 Nanocrystals in Acne Management

Acne vulgaris is one of the most common skin conditions around the globe, especially affecting teenagers and young adults. Traditional treatments, such as topical retinoids, antibiotics, and benzoyl peroxide, can sometimes lead to irritation and may not penetrate the hair follicles very well. However, nanocrystals have the potential to significantly enhance these treatments by improving how the skin absorbs them and delivering the medication directly to sebaceous follicles. For example, tretinoin nanocrystals mixed into a hydrogel have shown to be more effective in combating acne compared to standard formulations, with less irritation and better drug absorption [29]. Likewise, azelaic acid nanocrystals, when crafted into an in situ hydrogel, demonstrated improved skin adherence and deeper penetration, which resulted in less inflammation and better lesion clearing in clinical scenarios [30]. Recently, minocycline nanocrystals have been developed to more effectively target the sebaceous glands, showing strong anti-inflammatory properties while minimising overall systemic absorption—a common issue with oral minocycline treatment [31].

4.2 Nanocrystals in Psoriasis Treatment

Psoriasis is a long-lasting autoimmune condition that results in red, scaly patches due to an overproduction of skin cells. While treatments like topical corticosteroids, vitamin D analogues, and systemic immunomodulators are often used, they can lead to side effects when used over a long period. However, nanocrystals present a good alternative by enabling localised delivery of medications at effective levels while reducing overall exposure to the body. Research shows that formulations based on apremilast nanocrystals have excellent skin penetration and provide a steady release of the drug, which means lower doses and better safety [32]. Additionally, diosmin nanocrystal gels have been found to influence inflammatory signalling pathways, particularly TLR7/8 and NF-κB, leading to significant improvements in psoriasis symptoms in animal studies [33]. Furthermore, nanocrystals can also facilitate the combination of different medications into a single gel matrix, enhancing their therapeutic effects in psoriasis [34].

4.3 Nanocrystals for Eczema and Atopic Dermatitis

Eczema, especially atopic dermatitis (AD), is characterised by ongoing inflammation, dryness, and itching. It weakens the skin barrier, making it more susceptible to secondary infections. Usually, treatment involves corticosteroids and emollients, but prolonged use of corticosteroids can lead to skin thinning and risks of systemic absorption. Nanocrystals present a promising platform for more effective delivery of corticosteroids directly to the epidermis, which helps to lower the dosages needed and thus lessen side effects. They boost the solubility of the drug and help avoid liver metabolism, thus improving the therapeutic index. Other studies highlight that phytochemicals, including flavonoids and polyphenols, made as nanocrystals, show remarkable antioxidant and anti-inflammatory effects in managing eczema [35]. These innovative formulations assist in repairing the skin barrier while alleviating itchiness and inflammation.

4.4 Nanocrystals in Leishmaniasis Treatment

Cutaneous leishmaniasis is a type of parasitic skin infection that can cause long-lasting, disfiguring ulcers. While traditional systemic treatments like amphotericin B are effective, they can also have considerable toxicity. Nanocrystals present a promising strategy to target treatment more precisely, lessen systemic exposure, and improve the effectiveness of the drugs. When formulated as nanocrystals, amphotericin B and paromomycin showed better skin retention and enhanced activity against leishmaniasis in infected areas [36]. Additionally, chalcone nanocrystals demonstrated improved bioavailability and effective parasite clearance in mouse models, suggesting that nanomilling can enhance both efficacy and safety [37]. Furthermore, hydroxymethylnitrofurazone nanocrystals displayed strong dual activity against *Trypanosoma cruzi* and Leishmania species, indicating their potential for wider antiparasitic applications [38].

4.5 Nanocrystals in Skin Cancer Applications

Skin cancer, including both basal cell carcinoma and melanoma, is one of the most frequently occurring cancers around the world. The treatment is often hampered by issues like poor drug solubility and systemic toxicity. However, nanocrystals can significantly enhance the delivery of anticancer drugs directly to tumour sites, while minimising any harm to healthy tissue. For instance, curcumin nanocrystals have demonstrated an ability to effectively penetrate skin layers, helping to slow down tumour growth by decreasing oxidative stress and encouraging apoptosis in cancer cells [39, 40]. Additionally, other compounds like genistein and 5-fluorouracil have also been developed as nanocrystals to improve their delivery when applied topically, while aiming to reduce systemic toxicity [41].

5 Nanocrystals for Hair Follicle Delivery

5.1 Anatomy and Physiology of Hair Follicles

Hair follicles are little structures in our skin that do much more than just grow hair. They help regulate temperature, produce sebum, and send sensory signals—plus, they offer a smart way to deliver targeted medication. These mini-organs are nestled in the dermis and reach into the layer beneath the skin. Their intricate design, cyclical behaviour, and biological functions make them especially important when it comes to targeted therapies [42, 43].

Each hair follicle has three main parts:

- The *infundibulum*, which is the outermost section, extends from the skin's surface to the opening of the sebaceous gland. This section is readily accessible and typically the initial site reached by topical medications.
- The *isthmus* stretches from the sebaceous gland to where the arrector pili muscle attaches. This section helps anchor the follicle and is surrounded by immune cells that keep watch.
- The bulb, found at the deepest level, includes the hair matrix and dermal papilla. This area is where the hair production happens and has a rich blood supply, making it a prime target for delivering medications effectively [42, 43].

The outer root sheath (ORS) and inner root sheath (IRS) wrap around the growing hair strand. The ORS, which is continuous with the outer skin layer, is a significant target for delivering drugs to the hair follicle. Meanwhile, the IRS helps direct the hair as it grows, but gradually breaks down in the upper part of the follicle. Keratinocytes lining these sheaths can hold onto nanocrystals, creating local storage spots for gradual drug release [42]. Additionally, sebaceous glands linked to the upper follicle produce a fatty substance called sebum. Sebum is helpful for delivering lipophilic drugs and can even ease the movement of nanocrystals by lowering surface tension and helping distribute the particles along the follicular duct. Hair follicles serve not only as structures for hair growth; they also host immune cells like Langerhans cells, mast cells, and macrophages—playing a crucial role in monitoring and responding to inflammation. This rich immune landscape makes hair follicles valuable not just for direct drug delivery but also for treatments that modify immune responses, particularly for conditions such as acne, psoriasis, and alopecia areata. The cycle of hair growth, comprising the phases of anagen (growth), catagen (regression), and telogen (rest), also influences how well drugs can be absorbed. During the anagen phase, the follicle is actively elongating and in full swing, creating the perfect opportunity for nanocrystal therapies. On the flip side, in the telogen phase, the follicle is smaller and not as effective at absorbing [42]. Finally, the infundibulum of the follicle can serve as a little storage area for drug systems like nanocrystals. This means that nanocrystals can build up in the follicle and then be gradually released, providing long-lasting delivery to nearby tissues. By leveraging these distinctive anatomical and physiological features, nanocrystals can be precisely engineered to optimise their deposition, penetration, and therapeutic effectiveness within hair follicles.

5.2 The Hair Follicle: A Strategic Route for Drug Delivery

Unlike traditional transdermal routes, the follicular pathway skips the *stratum corneum* barrier, which is the outermost and least permeable layer of the skin. This allows hair follicles to act as a "shunt" route, providing direct access to the deeper layers of the skin where drug concentration is higher and resistance is minimised [44, 45]. With their reservoir-like structure, hair follicles also function as drug depots, allowing for gradual drug release over time.

Nanocrystals present a promising formulation principle for targeting hair follicles, thanks to their small size, increased surface area, and pure drug composition (Fig. 6). They can reach deeply into the follicular ducts, particularly when they're under 300 nanometres in size, which helps with effective delivery in the follicular epithelium [21, 22]. For example, curcumin nanocrystals have shown the ability to settle inside the *follicular infundibulum* through fluorescence tracking, proving their deep penetration into the follicle structure [3]. Likewise, azelaic acid nanocrystals used in acne treatment have effectively reached the sebaceous glands, boosting the therapeutic benefits while minimising surface irritation. This reservoir effect, where nanocrystals remain within the follicle, results in prolonged drug release and improved therapeutic retention [25]. Additionally, some studies suggest that pairing nanocrystals with enhancers like hyaluronic acid or clay-based hydrogels can further enhance penetration, ensure stability, and extend their stay in the follicular canal [19].

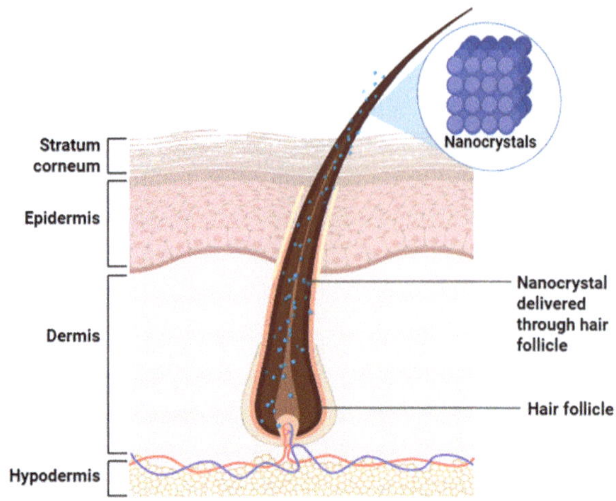

Fig. 6 Hair follicle targeting using nanocrystals

5.3 Advantages of Targeting Hair Follicles with Nanocrystals

Nanocrystals surpass other drug carriers like liposomes, solid lipid nanoparticles (SLNs), and polymeric micelles in several ways when it comes to follicular delivery:

- *Higher drug loading*: Nanocrystals consist almost entirely of the active pharmaceutical ingredient, unlike other carriers that require large amounts of lipids or surfactants.
- *Fewer excipients*: Reduces skin irritation and risk of allergic reactions.
- *Better penetration and retention*: Due to crystalline nature, small size, and surface charge adaptability.
- *Flexibility of formulation*: Nanocrystals can be incorporated into hydrogels, sprays, gels, and microneedle systems without stability issues [19, 21].

Their ability to be tailored to a desired size and surface profile gives nanocrystals a precise edge in achieving deeper, targeted delivery into hair follicles.

6 Nanocrystals for Hair Growth and Alopecia

6.1 The Trans-follicular Drug Delivery Pathway

Topically applied nanocrystals for the treatment and prevention of hair loss must target the hair follicles as an uptake route. A typical human hair follicle is a cavity formed by the outer and inner root sheaths, which support the structure of the follicle, the hair bulb at the base, containing the matrix, hair shaft, and dermal papilla, and the arrector pili muscle and the sebaceous glands as annexes to the main canal [46]. The hair follicle is a specialised structure with a determined life cycle. As mentioned in Sect. 5.1, it is constantly being renovated through the growth or anagen phase, the catagen phase when shedding occurs, and the telogen phase that involves hair shaft regeneration [47]. This comes into play when considering the histological morphologies, the hair follicles may present according to their developmental stage. It has been reported that human hair follicles can have a diameter between 172 μm and 82 μm for human terminal and vellus hair follicles, respectively. Furthermore, the follicles are evenly distributed every 50 μm on the skin surface [48]. This type of studies revealed that hair follicles morphology can be exploited to deliver drugs with nanoparticulated systems, as targeting the inside of the follicles is feasible and puts the drug delivery system in direct contact with diverse cell populations inside the follicles [49]. For drug delivery purposes, the hair follicles are also of great significance because they account for an increase in the skin surface area and a disruption of the epidermal barrier, also functioning as a reservoir [50]. The main sub-structure responsible for this is the *follicular infundibulum*, which is located at the most external part of the hair follicle (Fig. 7). The follicular infundibulum extends from the opening of the follicle at the epidermis and then goes down into the sebaceous gland, connecting the skin surface with the dermis without passing through the *stratum corneum* [51]. This is the main aspect that directly increases the surface area available for drug absorption under the skin, which in turn bypasses the biological protective barrier.

Studies have demonstrated that the absorption through the trans-follicular route is 2–4 times faster than through non-interrupted epidermis. The trans-follicular

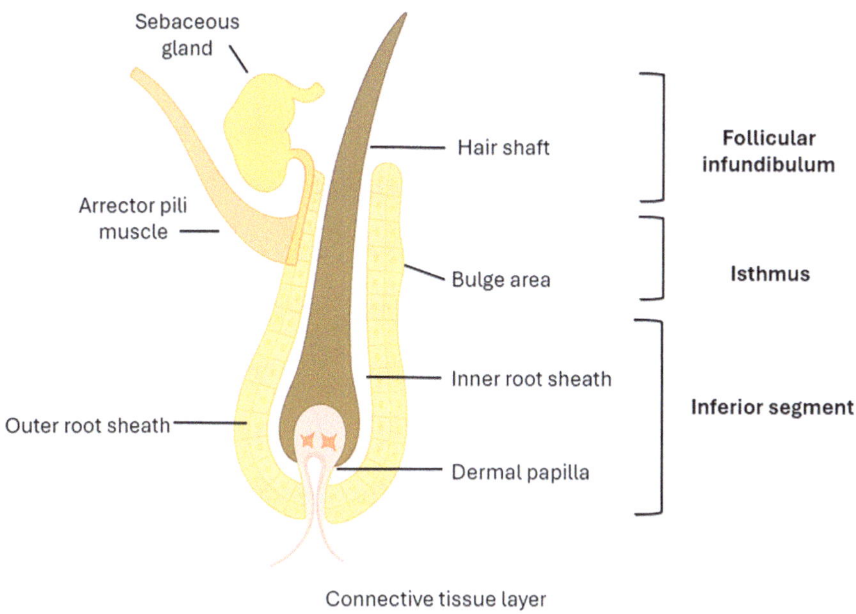

Fig. 7 Structure of the human hair follicle. (Created with the help of OpenAI. 2025. Digital illustration. ChatGPT. https://chat.openai.com)

pathway also favours lipophilic drug transdermal uptake, while hydrophilic drug absorption can be slowed down by the presence of sebum and the interaction with its lipids. The permeation in both cases occurs mainly by diffusion through the junctions of the internal and external root sheaths and into the surrounding connective tissue [51–53].

For this reason, targeting the trans-follicular delivery of therapeutic compounds is a very attractive approach as the hair follicles contribute greatly towards transdermal drug absorption. Furthermore, exploring this drug delivery route will help treat dermal diseases and disorders where the hair follicle is directly involved, such as alopecia [54].

6.2 Alopecia Disease and State of Current Treatments

Alopecia is a topical disorder characterised by sudden and often irreversible hair loss. Depending on the underlying physiological cause, it can be classified most commonly as alopecia areata, androgenic alopecia, or chronic telogen effluvium. Alopecia areata happens when patients have an autoimmune condition that triggers the attack of hair follicles by the immune system, whereas androgenic alopecia and chronic telogen effluvium can have hormonal and idiopathic causes [55]. The current efforts on the topical pharmacological treatment for androgenic alopecia are focused on the delivery of hair promoters, such as dutasteride, finasteride, and minoxidil, whereas clobetasol propionate and other topical corticosteroids are being used for the treatment of alopecia areata [56]. The conventional topical dosage forms such as lotions, foams, and ointments can present irritation as the most common side effect after regular application, which affects patient comfort and compliance. In addition, other side effects directly associated with the aforementioned APIs are also present, such as skin thinning, dry scalp, and hormonal imbalance, although these are most likely to occur when administered through the oral route [57, 58].

Topical approaches explored by nanotechnology include the formulation of nanosuspensions, polymeric nanoparticles, lipid nanoparticles, liposomes, and other vesicular systems such as ethosomes, niosomes, and transferosomes [58]. It has been shown that these systems can accumulate on hair follicles increasing local drug concentration provided they have a small enough particle size (< 300 nm), leading to minimal side effects while improving permeation. However, particulate systems still face technical challenges such as scalability issues, insufficient drug loading and encapsulation efficiency, and complex characterisation and production methods. These considerations can result in issues regarding batch-to-batch quality consistency, thus becoming a barrier for regulatory approval [58, 59].

6.3 *Nanocrystals for Alopecia Treatment Through Hair Promoter Delivery*

The API more extensively studied for prevention and treatment of alopecia is Minoxidil. This molecule acts as a vasodilator that increases blood flow to the dermal papilla in the hair follicle and promotes the anagen phase, revitalising the hair shaft. Recent studies have demonstrated that minoxidil nanocrystals produced through the wet media milling technique promote hair growth on dermal human fibroblast (HDF) and C57BL/6 mice, showing enhanced anagen/telogen growth when compared to commercially available minoxidil formulations [60, 61]. The study performed by Moradi was able to show that minoxidil nanocrystals with a particle size of 21.9 nm were able to reach the site of action at the dermal papilla inside the hair follicle and induce cell proliferation by direct contact [60]. In addition, in the study by Oaku et al., minoxidil nanocrystals with a particle size of 110 nm exhibited increased solubility and dispersibility compared to commercially available forms. The study compared the uptake route and found that nanocrystals prefer the transfollicular pathway, whereas commercially available formulations would rather permeate through the *stratum corneum*. The results from this study also indicated that relative to conventional dosage forms, minoxidil nanocrystals were retained in mice skin tissue, hair bulges, and hair bulbs up to 4 hours after application, presenting concentrations in tissue that were 2.3-, 2.6- and 3.9-fold higher, respectively. Additional key findings from the Oaku et al. study include earlier hair growth observed when compared to standard products, the use of Arabic gum as a penetration enhancer into the hair follicles, and the occurrence of aggregation after producing nanocrystals through the bottom-up approach opposite to wet media milling, since the study tested both techniques for producing nanocrystals [61].

Nanocrystal technology not only presents the aforementioned features, but it is also a highly versatile approach that can be combined with other techniques to offer an optimised effect. In the case of minoxidil, a study performed by Hamed et al. produced minoxidil nanocrystals with a size of 226.7 nm and loaded them into polymeric microneedles [62]. The results showed sustained drug release directly translated into hair growth on rats after 11 and 7 days when compared to the topical minoxidil foam showing initial hair growth after 18 days. Another study published by Pang reported the use of cellulose nanocrystals as carrier of deep eutectic solvents of minoxidil to target the dermal papilla, finding effective hair growth and confirming the targeting capabilities of nanocrystals for the follicular route [63].

Concerning other APIs used for the treatment of alopecia, nanocrystal technology has only been applied to produce clobetasol propionate and dutasteride nanocrystals. Clobetasol propionate nanocrystals were produced by Kumar et al. [64] to treat Alopecia Areata. This study aimed to produce nanocrystals that would induce a marked concentration gradient to promote passive dermal penetration. For this purpose, the clobetasol propionate nanocrystals with a particle size of 210.7 nm were produced by a combination of bottom-up and top-down techniques, namely, controlled precipitation and high-pressure homogenisation, respectively. These nanocrystals were embedded into a gel matrix to extend scalp contact time, and results showed both dermal and *stratum corneum* deposition and a fourfold increase in follicular delivery compared against a commercially available gel. This work makes a strong case for using the hair follicles as a reservoir of drug, from where sustained release can be obtained after forming an in-situ drug depot.

Regarding the delivery of the 5 α-reductase inhibitors, finasteride and dutasteride, the majority of the recent efforts to develop new dosage forms have been concentrated around dutasteride due to its higher potency [65]. Even though no study has been published about drug nanocrystals for the topical or transfollicular route uptake, studies have shown that nanoparticulated systems carrying dutasteride can easily penetrate and deliver drug into the hair follicles [66, 67].

There are two more studies addressing nanocrystals for the topical delivery of hair promoters. The first of them is an exploratory study reporting the development of a trigonelline nanocrystal-based topical cream by Shetty and collaborators, where the nanocrystals exhibited skin permeation and retention for sustained drug release and were effective in increasing hair growth and density on Wistar rats [68]. Furthermore, a proof-of-concept study performed by Pelikh et al. showed that curcumin nanocrystals' follicle targeting properties are independent of extrinsic factors other than the particle size [69].

Overall, these studies show that nanocrystal technology has delivered on its promise to increase bioavailability and therapeutic effect of poorly water-soluble drugs and is an interesting technology to keep exploring the delivery of active substances through the transfollicular route. In addition, nanocrystals are also being used to modify the drug release profile and increase target delivery dependent on particle size, which can result in lower side effects and better therapeutic outcomes. This way nanocrystal technology is promising towards safer and more effective treatments.

7 Surface Modification and Targeted Delivery: Strategies for Enhancing Nanocrystal Penetration and Targeting

The previous sections highlight how nanocrystals offer superior transdermal drug delivery with advantages over conventional nanoparticles, including up to 100% drug loading, enhanced saturation solubility, faster dissolution, and improved skin retention [16]. However, without deliberate design strategies, nanocrystals may still suffer from limited permeation due to the skin‘s formidable barrier, the *stratum corneum* [25]. Therefore, surface modification and formulation tailoring are crucial to modulate not only skin permeation but also the localisation within hair follicles, which act as promising depots for sustained release. A pivotal factor influencing dermal or transfollicular delivery is the vehicle system. Pelikh et al. [69] investigated the potential of curcumin nanocrystals as a strategy for TDD, with particular emphasis on hair follicle targeting. In their study, nanocrystals with an average size of ~300 nm were incorporated into various gel-based vehicles differing in viscosity, polarity, and lipophilicity. Using an ex vivo porcine skin model, they demonstrated that the penetration of nanocrystals into the hair follicles, specifically into the lower infundibulum, a region known for its higher permeability, was efficient and largely independent of the physicochemical properties of the vehicle. All formulations exhibited shear-thinning behaviour, which likely facilitated consistent follicular delivery during massage application. In contrast, passive dermal penetration of curcumin through the skin was significantly affected by the vehicle composition. Hydrogels containing humectants enhanced skin hydration and improved transdermal diffusion; however, these same excipients reduced the efficiency of follicular targeting. These findings suggest that nanocrystals in simple aqueous or shear-thinning gels are highly suitable for follicular drug delivery, while the addition of hydrating excipients may be more appropriate for enhancing transdermal absorption. Thus, the formulation strategy should be tailored according to the therapeutic objective either follicular targeting or passive dermal diffusion. This will be discussed in more detail in Sect. 8. Beyond vehicle influence, surface engineering of nanocrystals can be employed to manipulate biological interactions [70]. Coating nanocrystals with hydrophilic polymers or other surface modifiers may alter their interaction with skin barriers and improve targeting efficacy. In this study, nanocrystals of varying sizes were produced to investigate the effect of particle size on dermal penetration, with a focus on poorly soluble active ingredients relevant for TDD and hair follicle targeting. Additionally, the influence of different excipients and vehicles on penetration efficiency was assessed. Results confirmed that dermal penetration of poorly soluble actives increases as nanocrystals size decreases, with smaller nanocrystals promoting deeper skin layer permeation. Unexpectedly, many classical penetration enhancers, including hydrophilic excipients such as urea, glycerol, ethanol, and propylene glycol, as well as hydrogels, were found to reduce penetration efficacy. In contrast, lipid-based vehicles, namely, oleogels and creams, significantly enhanced penetration, likely by supporting the apolar pathway that facilitates transport of hydrophobic compounds through the skin. Hydrogels, despite their frequent historical use as nanocrystals carriers, were identified as suboptimal vehicles for dermal delivery of poorly soluble active ingredients. Optimal penetration was achieved when high concentrations of nanocrystals came into direct contact with the skin, either by applying concentrated suspensions or dispersing nanocrystals in lipid-rich, low-water-content creams that ensure intimate skin contact. These findings highlight the critical role of both nanocrystal size and vehicle composition in designing effective dermal drug delivery systems and suggest promising strategies for hair follicle targeting through tailored surface engineering and formulation approaches. Particle size is another decisive parameter for successful follicular delivery. Particles between 200 and 400 nm in diameter are best suited to exploit the mechanical ratchet effect induced by hair shaft movement, which draws particles into follicles [71]. Larger particles show reduced penetration due to steric hindrance, while particles below 100 nm may diffuse non-specifically, reducing targeted deposition. Moreover, adequate nanocrystal concentration must be achieved to reach the "saturation threshold" within follicles; otherwise, particles may simply remain on the skin surface. These findings highlight the critical role of both nanocrystal size and vehicle composition in designing effective dermal drug delivery systems and suggest promising strategies for hair follicle targeting through tailored surface engineering and formulation approaches. In addition to vehicle and size, stabilisers used during nanocrystals preparation also influence dermal delivery. Poloxamers, phospholipids, and other surfactants not only ensure redispersibility of nanocrystals but may also interact with skin lipids, improving permeability. For instance, poloxamer 188 has been employed in the formulation of diclofenac and rutin nanocrystals, improving both dermal and follicular penetration through subtle disruption of lipid packing [59]. These effects occur without severely compromising barrier function, maintaining skin integrity while enhancing drug transport.

More recently, innovative hybrid drug delivery systems have emerged, integrating nanocrystals into microneedle-based platforms, particularly in the form of core–shell architectures. These systems are designed to bypass the *stratum corneum*, the primary barrier to transdermal drug delivery, by creating microchannels that facilitate direct deposition of nanocrystals into the viable epidermis or upper dermis. The core–shell structure offers a dual function: the core, typically composed of drug-loaded nanocrystals, ensures high payload and controlled dissolution, while the shell, made from biodegradable polymers such as hyaluronic acid or polyvinylpyrrolidone (PVP), provides mechanical strength and biocompatibility. This strategy allows for precise localisation and sustained release of poorly soluble drugs, leveraging the enhanced bioavailability offered by nanocrystals and the minimally invasive penetration achieved through microneedles. While not the primary focus of this chapter, such approaches highlight the increasing convergence between physical enhancement methods and nanocrystals engineering to overcome traditional limitations of transdermal delivery [72]. Abdelghany et al. [73] successfully embedded curcumin nanocrystals into dissolving microneedles fabricated from biocompatible polymers. Upon application to ex vivo porcine skin, the microneedles fully dissolved, allowing direct delivery of the nanocrystals into skin layers. These findings illustrate the potential of combining nanocrystals with physical transdermal enhancement technologies to deliver hydrophobic drugs more effectively through the skin barrier. Although these systems are not specifically designed for hair follicle targeting, they represent an important frontier in nanocrystal-based drug delivery, where spatially controlled release and skin layer selectivity can be achieved through smart engineering of device–drug interactions.

Efforts to apply true active targeting strategies to nanocrystals have been limited, primarily due to their crystalline composition. However, insights can be drawn from analogous nanosystems. For example, Jeong et al. [74] developed hyaluronic acid (HA)-modified PLGA nanoparticles loaded with minoxidil for androgenic alopecia, exploiting CD44 receptor-mediated endocytosis for follicular uptake. While not nanocrystals per se, this study illustrates the therapeutic value of ligand-based targeting. Extrapolating from this model, one could envisage surface conjugation of nanocrystals with HA or other bioadhesive ligands to achieve similar follicular selectivity, particularly for pathologies like acne, alopecia, or localised inflammation. Despite these promising advances, the translation of nanocrystal-based systems into targeted transdermal therapies remains limited. One of the core challenges lies in the inherent physicochemical constraints of nanocrystals, which restrict the extent of chemical surface functionalisation possible without compromising crystallinity and drug stability. Surface engineering, in this case, often relies on the adsorption of stabilisers or co-formulated polymers, rather than true covalent binding. This presents limitations when compared to polymeric nanoparticles, which offer greater flexibility for ligand attachment and bioresponsive features. Another considerable hurdle is the complexity of skin as a biological barrier, which varies in structure, thickness, hydration, and follicular density across body sites and among individuals. These factors can significantly influence nanocrystal penetration and reproducibility of drug delivery. Additionally, hair follicle targeting poses its own set of challenges. Follicles represent a small fraction of the total skin surface area, and achieving sufficient nanocrystal accumulation in these structures without systemic absorption or superficial deposition requires precise control over size, surface charge, formulation viscosity, and application method [75].

8 Effects of the Vehicle on Nanocrystals in Transdermal Drug Delivery and Hair Follicle Targeting

Nanocrystals, with their distinctive physicochemical characteristics, enhanced surface area, and dissolution rate, guarantee better bioavailability and skin penetration, offering an attractive route for cutaneous and transcutaneous drug delivery, particularly to difficult microstructures such as hair follicles [76–78]. Nanocrystals have no carrier matrix as in encapsulated systems, which explains their high drug-loading capability and promising controlled-release ability [77, 79]. But the behaviour of nanocrystal-based formulation does not exclusively rely on size or surface property; vehicle used for nanocrystals formulation is determinant of the vehicle and the effect of nanocrystals formulations [69, 80].

The carrier, referred to as the vehicle that transports and delivers the nanocrystals to the target site, plays essential roles of drug stabilisation, skin permeability, and release profile as well as the overall activity against the intended infection. Depending on the composition and rheological properties of the vehicle, this may influence how the formulation behaves when in contact with the skin and the hair follicle. The vehicle design thus becomes a very important factor for the success of drug delivery [69, 81, 82]. This section dives into the complex role that vehicles play in delivering nanocrystals, highlighting how different formulation types affect both dermal and follicular penetration.

8.1 The Role of Vehicle in Nanocrystal Delivery

8.1.1 Aqueous Versus Lipid-Based Vehicles

The way nanocrystals are dispersed affects their behaviour on the skin. Aqueous vehicles, like suspensions or gels, are often the choice because they're biocompatible, easy to prepare, and generally safe. These systems can release drugs relatively quickly, making them great for short-term or surface-level treatments. However, their limited residence time on the skin can reduce the amount of drug that accumulates at the target site [77, 79, 83]. In contrast, lipid-based vehicles, such as emulsions, creams, or oleogels, tend to stick better to the skin and provide a longer contact time [69, 84]. Their lipophilic nature helps them interact more effectively with sebum and the lipids in the outer skin layer, allowing for deeper penetration and better targeting of hair follicles. Moreover, these vehicles can include penetration enhancers that tweak the skin barrier, making it easier for drugs to get through both the skin and into the follicles [24, 69, 76, 84]. The lipid content, thickness, and occlusive properties of these formulations are crucial for adjusting the hydration of the outer skin layer, which helps loosen the barrier and improve drug absorption. Therefore, for choosing aqueous or lipid-based formulations for nanocrystals is important to consider the drug target, the properties of the drug, and the release profile [69, 82, 84].

8.1.2 Gels and Hydrogels

Hydrogels are emerging as a promising vehicle for delivering nanocrystals due to their high-water content, flexibility, and excellent skin compatibility. Their semi-solid form allows for even distribution of medication and controlled release, making them ideal for treating sensitive or inflamed skin issues [77, 85]. When nanocrystals are incorporated into hydrogel matrices, they can serve as a reservoir that slowly releases the drug into the skin over time. The polymer structure of the hydrogel can be customised to adjust the time required for the drug to diffuse, its mechanical properties and moisture retention [85]. Furthermore, hydrogels provide a cooling sensation upon application, which can boost patient comfort and encourage adherence to treatment plans. These systems are also compatible with a variety of excipients and can be engineered to react to environmental changes, like shifts in temperature or pH, making them versatile options for both topical and transfollicular drug delivery [22, 76, 77, 85]. Furthermore, the gel-like texture of hydrogels makes hydrogels easy to apply, which can lead to better patient compliance.

8.1.3 Solid Lipid and Polymeric Carriers

Incorporating nanocrystals to solid carriers, like lipid nanoparticles or biodegradable polymers, has facilitated the way for hybrid delivery systems that boast impressive performance features [22, 27, 86, 87]. These innovative platforms bring many key benefits, among them: they enhance the physical stability of nanocrystals and allow for more precise, targeted drug release profiles. Solid lipid nanoparticles and polymeric matrices can either encapsulate or adsorb nanocrystals, shielding them from degradation while ensuring they are released when they encounter physiological conditions [22, 27, 86, 88]. Moreover, these systems can be tailored to have specific surface charges, shapes, or release kinetics that promote follicular targeting and cellular uptake. By integrating nanocrystals into solid or semi-solid scaffolds, we are opening up exciting possibilities for creating smart, responsive formulations that can release drugs in reaction to particular biological signals or environmental changes [22, 24, 69, 76, 81, 85]. These carriers can be designed to react to specific stimuli, such as pH or temperature, enabling targeted drug release. The adaptability of polymeric nanoparticles makes them ideal for a variety of applications in dermatology and for personalised treatments [22, 77, 85, 88].

8.1.4 Microneedles

Microneedle arrays present an exciting and minimally invasive way to enhance the penetration of nanocrystals into the skin and hair follicles. By creating tiny channels in the outer layer of the skin, microneedles effectively bypass the main barrier for drug penetration, all while causing minimal pain to the patient [76, 80, 89, 90]. When used with nanocrystal formulations, microneedles can significantly enhance the delivery of drugs to the deeper layers of the skin, such as the dermis and hair follicle openings. This technique allows for targeted and precise drug delivery, showing great potential for treating issues like alopecia or acne with improved effectiveness and fewer side effects throughout the body [22, 24, 69, 76]. Additionally, microneedle-assisted systems can be combined with hydrogel or patch technologies to create transdermal platforms that offer controlled and on-demand release of drugs [22, 85, 89, 90]. This topic on the combination of nanocrystals and microneedles is discussed in more details in Chap. 6.

9 Future Perspectives

As nanocrystal-based drug delivery systems advance, they present a complex landscape of promising opportunities and significant challenges that are poised to influence the future development of formulations for transdermal drug delivery and hair follicle targeting. The advancements in this field will bring together the effort of experts from pharmaceutical science, materials engineering, dermatology, and personalised medicine. The following perspectives outline the main pathways that will advance nanocrystal technology for skin and hair follicle targeting. One of the most promising areas for innovation is in enhancing the interactions between the delivery vehicles and the particles [69, 82]. While significant advances in stabilising nanocrystals were achieved, the next step will be to design vehicles that can effectively control drug release, improve targeting precision, and allow for deeper penetration into the skin. These vehicles are no longer just passive carriers; they play an active role in determining how nanocrystals behave biologically. By incorporating rheological modifiers, pH-responsive excipients, and lipid-skin mimicking systems, we can develop tailored delivery profiles that align with specific therapeutic objectives [77, 80, 81, 87].

Exploring ways to target specific areas within hair follicles is promoting new frontiers. By designing nanocrystals with specific sizes, surface modifications (like adding ligands or adjusting charges), and controlled release properties, we can achieve targeted delivery to many areas of the skin [69, 81]. This level of precision could greatly enhance treatment outcomes for issues such as androgenetic alopecia, acne, and infections related to hair follicles, all while reducing unwanted side effects [24, 76, 86, 88]. From an industrial standpoint, scaling up production and manufacturing poses significant challenges. Although various methods, such as high-pressure homogenisation, wet milling, and precipitation, are available for producing nanocrystals at the laboratory scale, translating these techniques into robust, cost-effective, and regulatory-compliant industrial processes remains a significant challenge [10, 78]. Future initiatives will need to prioritise continuous manufacturing systems, real-time quality checks, and standardising particle size and stability to ensure consistency from one batch to the next [81].

Safety and long-term biocompatibility are crucial for the widespread acceptance of nanocrystal-based therapies [78]. While existing research generally supports the safety of applying nanocrystals topically, there is still a lack of thorough investigations into their long-term use, potential for bioaccumulation, immune responses, and effects on the skin barrier [77]. To tackle these issues, we need extensive long-term studies in living organisms, human trials, and the creation of standardised safety evaluation protocols specifically designed for nanoscale systems. Another exciting avenue is the combination of nanocrystals with smart delivery systems. Innovative technologies like wearable patches, biosensor-guided administration, and materials that respond to stimuli (such as changes in pH, temperature, or light) open up the possibility for real-time, personalised treatment plans. These systems can adjust drug release based on physiological or environmental signals, enhancing therapeutic accuracy and easing the treatment process [80].

In line with the growing trend towards personalised medicine, nanocrystal formulations show incredible promise for tailoring treatments based on individual genetics, lifestyle choices, or specific skin conditions. By adjusting aspects like drug loading, release rates, and the composition of the delivery vehicle, these formulations can be customised to meet the unique needs of each patient [78, 83, 91]. This is especially important in dermatology, where variations in skin type and barrier function can greatly influence how well a drug is absorbed and how effective it is [81].

Looking ahead, prospective studies will explore combination therapies, where nanocrystals are paired with other treatment methods to create synergistic effects [86]. Light-activated nanocrystals, in particular, hold exciting potential for addressing photoresponsive conditions like acne, psoriasis, or skin cancer, as they can harness the advantages of localised drug release along with external activation [80].

Lately, there has been a noticeable shift towards focusing on sustainability and the environmental effects of pharmaceutical formulations. Researchers are increasingly prioritising the development of eco-friendly, biodegradable, and non-toxic methods for delivering nanocrystals [77, 84]. They are looking into natural polymers, plant-based lipids, and biodegradable surfactants as greener alternatives to synthetic excipients [87]. By applying green chemistry principles in both the creation and disposal of nanocrystal formulations, environmental issues can be tackled while also making these products more acceptable to patients and consumers alike.

10 Conclusion

Nanocrystals are a highly sophisticated option for targeted drug delivery, especially in transdermal drug delivery and hair follicle targeting. Thanks to their unique physicochemical properties like a high surface area, better solubility, and customisable particle features, they can penetrate the skin more effectively or gather preferentially hair follicles. This makes nanocrystals a fantastic choice for delivering drugs that have poor solubility, enhancing local drug concentration while keeping systemic exposure to a minimum. However, the effectiveness of nanocrystal-based formulations is not based just on the particles themselves but also on how the delivery vehicle is designed and chosen. The vehicle is not just a simple carrier; it plays a crucial role in determining stability, release rates, penetration depth, and overall therapeutic effectiveness. By engineering vehicles to extend skin contact time, interact with skin lipids, adjust hydration, and improve retention in the follicular unit, we can significantly enhance delivery outcomes [69, 77, 81]. As the field continues to grow, we are seeing an exciting blend of material science, pharmaceutical technology, and dermatological research that is leading to advanced delivery systems. Innovations like smart materials, carriers that respond to stimuli, and techniques involving microneedles are really pushing the boundaries of what nanocrystal-based treatments can do. These new platforms not only promise better effectiveness but also aim to make things easier and more convenient for patients. Furthermore, the incorporation of nanocrystals into personalised and precision medicine will represent a radical change. By customising the design of particles and the composition of delivery vehicles to match individual skin types, disease profiles, and treatment goals, we can boost the precision of treatments while reducing side effects. This approach is set to transform dermatological care, providing more targeted, effective, and patient-focused solutions [22, 80, 81, 86, 91, 92]. Sustainability is becoming a key focus in the development of future nanocrystal formulations. By using biodegradable polymers, natural excipients, and eco-friendly production methods, the pharmaceutical and cosmetic industries are showing a growing commitment to environmental responsibility. These initiatives not only help lessen the ecological impact of dermatological products but also improve their safety and acceptance among the public [77, 84, 87]. In summary, nanocrystal technology is leading the way in innovative topical drug delivery. With careful design of both the particles and the delivery systems, these technologies have the potential to transform the treatment of skin and hair issues. With ongoing research and the strengthening of interdisciplinary collaboration, nanocrystals are increasingly positioned to drive the next generation of dermatological therapies by enhanced safety, efficacy, and long-term sustainability.

References

1. A.M. Abraham, O.E. Afolabi, Q.K. Anjani, R.F. Donnelly, PowderInjector microneedles: smart cavity microneedles as a novel intradermal delivery system for high doses of niacinamide. Int. J. Pharm. **683**, 125992 (2025). https://doi.org/10.1016/j.ijpharm.2025.125992
2. R. Banerjee, Overcoming the stratum corneum barrier: a nano approach. Drug Deliv. Transl. Res. **3**, 205–208 (2013). https://doi.org/10.1007/s13346-013-0149-8
3. T. Schmitt, R.H.H. Neubert, State of the art in stratum corneum research. Part II: Hypothetical stratum corneum lipid matrix models. Skin Pharmacol. Physiol. **33**, 213–230 (2020). https://doi.org/10.1159/000509019
4. A. Alkilani, M.T. McCrudden, R. Donnelly, Transdermal drug delivery: Innovative pharmaceutical developments based on disruption of the barrier properties of the stratum corneum. Pharmaceutics **7**, 438–470 (2015). https://doi.org/10.3390/pharmaceutics7040438
5. V. Phatale, K.K. Vaiphei, S. Jha, D. Patil, M. Agrawal, A. Alexander, Overcoming skin barriers through advanced transdermal drug delivery approaches. J. Control. Release **351**, 361–380 (2022). https://doi.org/10.1016/j.jconrel.2022.09.025
6. D.H. Lee, S. Lim, S.S. Kwak, J. Kim, Advancements in skin-mediated drug delivery: Mechanisms, techniques, and applications. Adv. Healthc. Mater. **13**, 2302375 (2024). https://doi.org/10.1002/adhm.202302375
7. P.K. Parmar, J. Wadhawan, A.K. Bansal, Pharmaceutical nanocrystals: A promising approach for improved topical drug delivery. Drug Discov. Today **26**, 2329–2349 (2021). https://doi.org/10.1016/j.drudis.2021.07.010
8. V. Giradkar, A. Mhaske, R. Shukla, Nanocrystals: A multifaceted regimen for dermatological ailments. Part. Part. Syst. Charact. **41**, 2300147 (2024). https://doi.org/10.1002/ppsc.202300147
9. L. Liu, W. Zhao, Q. Ma, Y. Gao, W. Wang, X. Zhang, Y. Dong, T. Zhang, Y. Liang, S. Han, J. Cao, X. Wang, W. Sun, H. Ma, Y. Sun, Functional nano-systems for transdermal drug delivery and skin therapy. Nanoscale Adv. **5**, 1527–1558 (2023). https://doi.org/10.1039/D2NA00530A
10. J.U.A. Junghanns, R.H. Müller, Nanocrystal technology, drug delivery and clinical applications. Int. J. Nanomedicine **3**(3), 295–310 (2008)
11. A.A. Noyes, W.R. Whitney, The rate of solution of solid substances in their own solutions. J. Am. Chem. Soc. **19**, 930–934 (1897). https://doi.org/10.1021/ja02086a003
12. R. Mauludin, R.H. Müller, C.M. Keck, Development of an oral rutin nanocrystal formulation. Int. J. Pharm. **370**, 202–209 (2009). https://doi.org/10.1016/j.ijpharm.2008.11.029

13. G. Kaptay, On the size and shape dependence of the solubility of nano-particles in solutions. Int. J. Pharm. **430**, 253–257 (2012). https://doi.org/10.1016/j.ijpharm.2012.03.038
14. A. Mihranyan, M. Strømme, Solubility of fractal nanoparticles. Surf. Sci. **601**, 315–319 (2007). https://doi.org/10.1016/j.susc.2006.09.037
15. D.R. Ely, R.E. García, M. Thommes, Ostwald–Freundlich diffusion-limited dissolution kinetics of nanoparticles. Powder Technol. **257**, 120–123 (2014). https://doi.org/10.1016/j.powtec.2014.01.095
16. M.B. McGuckin, J. Wang, R. Ghanma, N. Qin, S.D. Palma, R.F. Donnelly, A.J. Paredes, Nanocrystals as a master key to deliver hydrophobic drugs via multiple administration routes. J. Control. Release **345**, 334–353 (2022). https://doi.org/10.1016/j.jconrel.2022.03.012
17. M. Cooley, A. Sarode, M. Hoore, D.A. Fedosov, S. Mitragotri, A. Sen Gupta, Influence of particle size and shape on their margination and wall-adhesion: implications in drug delivery vehicle design across nano-to-micro scale. Nanoscale **10**, 15350–15364 (2018). https://doi.org/10.1039/C8NR04042G
18. S. Roh, A.H. Williams, R.S. Bang, S.D. Stoyanov, O.D. Velev, Soft dendritic microparticles with unusual adhesion and structuring properties. Nat. Mater. **18**, 1315–1320 (2019). https://doi.org/10.1038/s41563-019-0508-z
19. M. Ruggeri, R. Sánchez-Espejo, L. Casula, et al., Clay-based hydrogels as drug delivery vehicles of curcumin nanocrystals for topical application. Pharmaceutics **14**, 2836 (2022). https://doi.org/10.3390/pharmaceutics14122836
20. X. Miao, J. Zhao, H. Xiang, X. Shi, Hyaluronidase promote transdermal diffusion of small sized curcumin nanocrystal by dissolving microneedles delivery. Pharmaceutics **15**, 788 (2023). https://doi.org/10.3390/pharmaceutics15030788
21. O. Pelikh, C.M. Keck, Hair follicle targeting and dermal drug delivery with curcumin drug nanocrystals – essential influence of excipients. Nano **10**, 2323 (2020). https://doi.org/10.3390/nano10112323
22. Y. Lei et al., Advances in polymeric nano-delivery systems targeting hair follicles for the treatment of acne. Drug Deliv. **31**, 2372269 (2024). https://doi.org/10.1080/10717544.2024.2372269
23. Y. Oaku, S. Shiroyama, H. Otake, Y. Yajima, A. Abe, N. Yamamoto, N. Nagai, Gum Arabic enhances hair follicle-targeting drug delivery of Minoxidil nanocrystal dispersions. Biol. Pharm. Bull. **47**, 2083–2091 (2024). https://doi.org/10.1248/bpb.b24-00697
24. H.L. Ji, J. Li, G. Yang, W. Zhang, Y. Shen, B. Xu, J. Liu, J. Wen, W. Song, Hair follicle-targeted delivery of Azelaic acid micro/nanocrystals promote the treatment of acne vulgaris. Int. J. Nanomed. **19**, 5173–5191 (2024). https://doi.org/10.2147/IJN.S459788
25. H. Xiang, S. Xu, J. Li, S. Pan, X. Miao, Particle size effect of curcumin nanocrystals on transdermal and Transfollicular penetration by hyaluronic acid-dissolving microneedle delivery. Pharmaceuticals **15**, 206 (2022). https://doi.org/10.3390/ph15020206
26. S.W. Chaiprateep, R.W. Eckert, C. Raab, S. Sengupta, C.M. Keck, Influence of dose, particle size and concentration on dermal penetration efficacy of curcumin. Pharmaceutics **15**, 2645 (2023). https://doi.org/10.3390/pharmaceutics15112645
27. M.A.S.A. Parveen, P.V. Thanikachalam, R.K. Khar, P. Kesharwani, Nanocrystals as an emerging nanocarrier for the management of dermatological diseases. Colloids Surf. B Biointerfaces **225**, 113231 (2023). https://doi.org/10.1016/j.colsurfb.2023.113231
28. P.A.F. Duarte, F.L. dos Santos, I.C.K. Guerreiro, A.M. Fuentefria, R.V. Contri, Drug nanocrystals: A review of their application in the topical treatment of skin infections. Int. J. Nano Dimension (2025). https://doi.org/10.57647/j.ijnd.2025.1604.25
29. P. Shinu, Development, evaluation, and in vitro anti-acne activity of Tretinoin nanocrystals gel. Indian J. Pharm. Educ. Res. **57**, 1012–1020 (2023). https://doi.org/10.5530/ijper.57.4.114
30. S.M. Tomić, I. Pepić, D. Šimić, J. Filipović-Grčić, Efficacy and safety of Azelaic acid nanocrystal-loaded in situ hydrogel in the treatment of acne vulgaris. Pharmaceutics **13**, 567 (2021). https://doi.org/10.3390/pharmaceutics13040567

31. S.M. Abudoleh, J.M. Abu Ershaid, D. Lafi, N.A. Dahshan, A. Talhouni, A. Abuirmeileh, Minocycline nanocrystals: A new approach for treating acne with reduced systemic side effects. Pharmaceutics **17**, 727 (2025). https://doi.org/10.3390/pharmaceutics17060727
32. M.A.L.H. Goonoo, I. Chummun, N. Karuri, K. Badu, F. Gimié, J. Bergrath, M. Schulze, M. Müller, A. Bhaw-Luximon, Nanomedicine-based strategies to improve treatment of cutaneous leishmaniasis. R. Soc. Open Sci. **9**, 220058 (2022). https://doi.org/10.1098/rsos.220058
33. Y. Shahine, S.A.A. El-Aal, A.M. Reda, E. Sheta, N.M. Atia, O.Y. Abdallah, S.S.A. Ibrahim, Diosmin nanocrystal gel alleviates imiquimod-induced psoriasis in rats via modulating TLR7,8/NF-κB/micro RNA-31, AKT/mTOR/P70S6K milieu, and Tregs/Th17 balance. Inflammopharmacol **31**, 1341–1359 (2023). https://doi.org/10.1007/s10787-023-01198-w
34. N.S. Singh, T. Behl, B.C. Sarkar, H.R. Saha, K. Garg, S.K. Singh, S. Arora, M.S. Amran, A.A.H. Abdellatif, A.L. Bilgrami, G.M. Ashraf, M.S. Rahman, Promising strategies of colloidal drug delivery-based approaches in psoriasis management. Pharmaceutics **13** (2021). https://doi.org/10.3390/pharmaceutics13111978
35. B.E.K. Radhakrishnan, E.B. Noftall, C.A. Giacomantonio, H.P.V. Rupasinghe, Recent advances in phytochemical-based topical applications for the management of Eczema, a review. Int. J. Mol. Sci. **25**, 5375 (2024). https://doi.org/10.3390/ijms25105375
36. R.M. Barbosa, M.M.S. Obata, et al., Development of Ag-ZnO/AgO nanocomposites effectives for Leishmania braziliensis treatment. Pharmaceutics **14**, 2642 (2022). https://doi.org/10.3390/pharmaceutics14122642
37. W.P.-L. Borsodi, J.C.V. Menezes, D. Escrivani-Oliveira, N. Arruda-Costa, A.J.M. Silva, L.M. Cabral, P.G. Steel, A.J. Sousa-Batista, B. Rossi-Bergmann, Nanocrystallization effectively improves the oral efficacy of an Antileishmanial Chalcone. Pharmaceutics **17**, 399 (2025). https://doi.org/10.3390/pharmaceutics17040399
38. C.B. Scarim, A. de Souza, D.S.S. Marins, E.G. dos Santos, et al., Synthesis, characterization, and activity of Hydroxymethylnitrofurazone nanocrystals against Trypanosoma cruzi and Leishmania spp. Drugs Drug Candidates **1**, 43–55 (2022). https://doi.org/10.3390/ddc1010005
39. A. Bagde, K. Patel, A. Mondal, S. Kutlehria, N. Chowdhury, A. Gebeyehu, N. Patel, N. Kumar, M. Singh, Combination of UVB absorbing titanium dioxide and quercetin Nanogel for skin cancer chemoprevention. AAPS PharmSciTech **20**, 240 (2019). https://doi.org/10.1208/s12249-019-1424-x
40. K.P. Patel, V. Patel, M.S. Rajput, R. Patel, A. Rajput, Nanocrystals: an emerging paradigm for cancer therapeutics. Future J. Pharm. Sci. **10**, 4 (2024). https://doi.org/10.1186/s43094-024-00579-4
41. K.K.S. Yeligar, M. Chandrakar, M.S. Jangde, Nanotechnology-based delivery of Genistein to overcome physicochemical hindrance and enhance therapeutic response in skin cancer. BioNanoSci **13**, 1339–1358 (2023). https://doi.org/10.1007/s12668-023-01118-w
42. S. Tripathi, S. Dwivedi, S. Shrivastav, A.K. Tripathi, Anatomy and physiology diversity of hair: A multidisciplinary approach to understanding structure and function, in *Hair Care Products: Efficacy, Safety and Global Regulation*, ed. by A. Dwivedi, A.B. Pant, S. Poojan, M. Kotak, A. Tripathi, (Springer Nature Singapore, 2024), pp. 5–21. https://doi.org/10.1007/978-981-97-6740-3_2
43. B. Erdoğan, Anatomy and physiology of hair, in *Hair and Scalp Disorders*, vol. 13, (IntechOpen, 2017), pp. 1–7
44. C.L. Fang, I.A. Aljuffali, Y.C. Li, J.Y. Fang, Delivery and targeting of nanoparticles into hair follicles. Ther. Deliv. **5**(9), 991–1006 (2014)
45. A. Rancan, Vogt, getting under the skin: What is the potential of the Transfollicular route in drug delivery? Ther. Deliv. **5**, 875–877 (2014). https://doi.org/10.4155/tde.14.56
46. B. Buffoli, F. Rinaldi, M. Labanca, E. Sorbellini, A. Trink, E. Guanziroli, R. Rezzani, L.F. Rodella, The human hair: from anatomy to physiology. Int J Dermatology **53**, 331–341 (2014). https://doi.org/10.1111/ijd.12362

47. E. Aamar, E. Avigad Laron, W. Asaad, S. Harshuk-Shabso, D. Enshell-Seijffers, Hair-follicle mesenchymal stem cell activity during homeostasis and wound healing. J. Invest. Dermatol. **141**, 2797–2807.e6 (2021). https://doi.org/10.1016/j.jid.2021.05.023
48. A. Vogt, S. Hadam, M. Heiderhoff, H. Audring, J. Lademann, W. Sterry, U. Blume-Peytavi, Morphometry of human terminal and vellus hair follicles. Exp. Dermatol. **16**, 946–950 (2007). https://doi.org/10.1111/j.1600-0625.2007.00602.x
49. J. Lademann, N. Otberg, U. Jacobi, R.M. Hoffman, U. Blume-Peytavi, Follicular penetration and targeting. J. Investig. Dermatol. Symp. Proc. **10**, 301–303 (2005). https://doi.org/10.1111/j.1087-0024.2005.10121.x
50. U. Blume-Peytavi, A. Vogt, Human hair follicle: reservoir function and selective targeting: Human hair follicle. Br. J. Dermatol. **165**, 13–17 (2011). https://doi.org/10.1111/j.1365-2133.2011.10572.x
51. T. Ogiso, T. Shiraki, K. Okajima, T. Tanino, M. Iwaki, T. Wada, Transfollicular drug delivery: Penetration of drugs through human scalp skin and comparison of penetration between scalp and abdominal skins in vitro. J. Drug Target. **10**, 369–378 (2002). https://doi.org/10.1080/10611860210000001814
52. E. Abd, H.A.E. Benson, M.S. Roberts, J.E. Grice, Follicular penetration of caffeine from topically applied nanoemulsion formulations containing penetration enhancers: In vitro human skin studies. Skin Pharmacol. Physiol. **31**, 252–260 (2018). https://doi.org/10.1159/000489857
53. E. Abd, H. Benson, M. Roberts, J. Grice, Minoxidil skin delivery from nanoemulsion formulations containing eucalyptol or oleic acid: Enhanced diffusivity and follicular targeting. Pharmaceutics **10**, 19 (2018). https://doi.org/10.3390/pharmaceutics10010019
54. N. Afika, A.F. Saniy, A.A. Fawwaz Dharma, C.K. Ko, R. Kamran, A.D. Permana, Trilayer dissolving microneedle for transdermal delivery of minoxidil: a proof-of-concept study. J. Biomater. Sci. Polym. Ed. **35**, 1750–1770 (2024). https://doi.org/10.1080/09205063.2024.2350187
55. M. Ahmad, *Alopecia* (InTech, 2018). https://doi.org/10.5772/intechopen.73642
56. S. Gregoire, B. McIntosh, K. Sanchez, U. Biba, A. Mostaghimi, Local corticosteroids for alopecia Areata: A narrative review. Dermatol Ther (Heidelb) **15**, 1607–1631 (2025). https://doi.org/10.1007/s13555-025-01421-2
57. R.E. Elnady, M.S. Abdon, H.R. Shaheen, R.M. Eladawy, Y.O. Azar, S.M. Al Raish, The future of alopecia treatment: Plant extracts, nanocarriers, and 3D bioprinting in focus. Pharmaceutics **17**, 584 (2025). https://doi.org/10.3390/pharmaceutics17050584
58. M.N. Pereira, C.Y. Ushirobira, M.S. Cunha-Filho, G.M. Gelfuso, T. Gratieri, Nanotechnology advances for hair loss. Ther. Deliv. **9**, 593–603 (2018). https://doi.org/10.4155/tde-2018-0025
59. R.H. Muller, C.M. Keck, Challenges and solutions for the delivery of biotech drugs – a review of drug nanocrystal technology and lipid nanoparticles. J. Biotechnol. **113**, 151–170 (2004). https://doi.org/10.1016/j.jbiotec.2004.06.007
60. A. Moradi, S.M. Seifati, M. Darroudi, S. Golmohammadzadeh, M.D. Ashkezari, Nanocrystal structure of Minoxidil: A safe stimulator for HairGrowth factor for C57BL/6 mice. CMC **32**, 4343–4355 (2025). https://doi.org/10.2174/0109298673283413240116065844
61. S.S. Oaku, H. Otake, Y. Yajima, A. Abe, N. Yamamoto, N. Nagai, Gum Arabic enhances hair follicle-targeting drug delivery of Minoxidil nanocrystal dispersions. Biol. Pharm. Bull. **47**, 2083–2091 (2024). https://doi.org/10.1248/bpb.b24-00697
62. R. Hamed, A.D. Abu Kwiak, Y. Al-Adhami, A.M. Hammad, R. Obaidat, O.H. Abusara, R.A. Huwaij, Microemulsions as lipid nanosystems loaded into Thermoresponsive in situ microgels for local ocular delivery of prednisolone. Pharmaceutics **14** (2022). https://doi.org/10.3390/pharmaceutics14091975
63. X. Pang, S. Han, K. Zheng, L. Jiang, J. Wang, S. Qian, Cellulose nanocrystal-stabilized Pickering emulsion gels as vehicles for follicular delivery of minoxidil. Int. J. Biol. Macromol. **277**, 134297 (2024). https://doi.org/10.1016/j.ijbiomac.2024.134297

64. P. Kumar, V. Kumar, S. Sahoo, S.K. Singh, Enhancing alopecia areata management: Nanocrystal-driven strategy for targeting hair follicles. Int. J. Pharm. **675**, 125557 (2025). https://doi.org/10.1016/j.ijpharm.2025.125557
65. A.K. Gupta, M. Talukder, G. Williams, Comparison of oral minoxidil, finasteride, and dutasteride for treating androgenetic alopecia. J. Dermatol. Treat. **33**, 2946–2962 (2022). https://doi.org/10.1080/09546634.2022.2109567
66. S. Ramkar, P.K. Suresh, Finasteride-loaded nano-lipidic carriers for follicular drug delivery: Preformulation screening and box-Behnken experimental design for optimization of variables. Heliyon **8**, e10175 (2022). https://doi.org/10.1016/j.heliyon.2022.e10175
67. O.A. Ahmed, W.Y. Rizg, Finasteride nano-transferosomal gel formula for management of androgenetic alopecia: ex vivo investigational approach. Drug Des. Devel. Ther. **12**, 2259–2265 (2018). https://doi.org/10.2147/DDDT.S171888
68. K.S. Shetty, M. Koland, M. Ashwini, Trigonelline nanocrystal-based topical creams for the treatment of hair loss. J. Adv. Pharm. Technol. Res. **15**, 297–303 (2024). https://doi.org/10.4103/JAPTR.JAPTR_26_24
69. O. Pelikh, R.W. Eckert, S.R. Pinnapireddy, C.M. Keck, Hair follicle targeting with curcumin nanocrystals: Influence of the formulation properties on the penetration efficacy. J. Control. Release **329**, 598–613 (2021). https://doi.org/10.1016/j.jconrel.2020.09.053
70. O. Pelikh, P.-L. Stahr, J. Huang, M. Gerst, P. Scholz, H. Dietrich, N. Geisel, C.M. Keck, Nanocrystals for improved dermal drug delivery. Eur. J. Pharm. Biopharm. **128**, 170–178 (2018). https://doi.org/10.1016/j.ejpb.2018.04.020
71. J. Lademann, H. Richter, U.F. Schaefer, U. Blume-Peytavi, A. Teichmann, N. Otberg, W. Sterry, Hair follicles—A long-term reservoir for drug delivery. Skin Pharmacol. Physiol. **19**, 232–236 (2006). https://doi.org/10.1159/000093119
72. L.K. Vora, P.R. Vavia, E. Larrañeta, S.E.J. Bell, R.F. Donnelly, Novel nanosuspension-based dissolving microneedle arrays for transdermal delivery of a hydrophobic drug. J. Interdiscip. Nanomed. **3**, 89–101 (2018). https://doi.org/10.1002/jin2.41
73. S. Abdelghany, I.A. Tekko, L. Vora, E. Larrañeta, A.D. Permana, R.F. Donnelly, Nanosuspension-based dissolving microneedle arrays for intradermal delivery of curcumin. Pharmaceutics **11**, 308 (2019). https://doi.org/10.3390/pharmaceutics11070308
74. W.Y. Jeong, S. Kim, S.Y. Lee, H. Lee, D.W. Han, S.Y. Yang, K.S. Kim, Transdermal delivery of Minoxidil using HA-PLGA nanoparticles for the treatment in alopecia. Biomater Res **23**, 16 (2019). https://doi.org/10.1186/s40824-019-0164-z
75. A. Patzelt, J. Lademann, Recent advances in follicular drug delivery of nanoparticles. Expert Opin. Drug Deliv. **17**, 49–60 (2020). https://doi.org/10.1080/17425247.2020.1700226
76. Y. Gu, Q. Bian, Y. Zhou, Q. Huang, J. Gao, Hair follicle-targeting drug delivery strategies for the management of hair follicle-associated disorders. Asian J. Pharm. Sci. **17**, 333–352 (2022). https://doi.org/10.1016/j.ajps.2022.04.003
77. A.S. Alnaim, Nanocrystals in dermal drug delivery: A breakthrough for enhanced skin penetration and targeted skin disorder treatments. Pharmaceutics **16**, 1561 (2024). https://doi.org/10.3390/pharmaceutics16121561
78. V. Anand, D. Kaushik, Nanocrystals for improved drug delivery: a review. IJPSR **16**, 900–907 (2025). https://doi.org/10.13040/IJPSR.0975-8232.16(4).900-07
79. K.A. Ghadge et al., Nanocrystals in drug delivery –a review. IJIRT **8**, 462–469 (2022)
80. L. Busch, D. Asadzadeh, A.L. Klein, P. Suriyaamporn, M.K. Vollrath, C.M. Keck, M.C. Meinke, The penetration efficiency of a dissolved model drug into hair follicles depends on the concentration of added nanoparticles. Drug Deliv. Transl. Res. **15**, 1444–1452 (2025). https://doi.org/10.1007/s13346-024-01718-3
81. Y. Yanamadala, C.M.R. Muthumula, S. Khare, K. Gokulan, Strategies to enhance nanocrystal formulations for overcoming physiological barriers across diverse routes of administration. IJN **20**, 367–402 (2025). https://doi.org/10.2147/IJN.S494224

82. A. Patzelt, H. Richter, L. Dähne, P. Walden, K.-H. Wiesmüller, U. Wank, W. Sterry, J. Lademann, Influence of the vehicle on the penetration of particles into hair follicles. Pharmaceutics **3**, 307–314 (2011). https://doi.org/10.3390/pharmaceutics3020307
83. S.B. Lohan, S. Saeidpour, M. Colombo, S. Staufenbiel, M. Unbehauen, A. Wolde-Kidan, R.R. Netz, R. Bodmeier, R. Haag, C. Teutloff, R. Bittl, M.C. Meinke, Nanocrystals for improved drug delivery of dexamethasone in skin investigated by EPR spectroscopy. Pharmaceutics **12**, 400 (2020). https://doi.org/10.3390/pharmaceutics12050400
84. C. Aguirre, A. Sintov, Microemulsions and nanoparticles as carriers for dermal and transdermal drug delivery. J. Pharmacol. Pharm. Res. **2** (2019). https://doi.org/10.31038/JPPR.2019241
85. E.J. Delgado-Pujol, G. Martínez, D. Casado-Jurado, J. Vázquez, J. León-Barberena, D. Rodríguez-Lucena, Y. Torres, A. Alcudia, B. Begines, Hydrogels and Nanogels: Pioneering the future of advanced drug delivery systems. Pharmaceutics **17**, 215 (2025). https://doi.org/10.3390/pharmaceutics17020215
86. P. Kayal, R. Raghul, U.K. Sahoo, N. Jawahar, Nanocarrier-based approaches for enhanced management of androgenetic alopecia: Advancements and future prospects. Int. J. App. Pharm., 13–27 (2025). https://doi.org/10.22159/ijap.2025v17i3.53645
87. W. Samprasit, P. Suriyaamporn, P. Sriamornsak, P. Opanasopit, B. Chamsai, Resveratrol-loaded lipid-based nanocarriers for topical delivery: Comparative physical properties and antioxidant activity. OpenNano **19**, 100211 (2024). https://doi.org/10.1016/j.onano.2024.100211
88. A.C. Santos, M. Pereira-Silva, C. Guerra, D. Costa, D. Peixoto, I. Pereira, I. Pita, A.J. Ribeiro, F. Veiga, Topical Minoxidil-loaded nanotechnology strategies for alopecia. Cosmetics **7**, 21 (2020). https://doi.org/10.3390/cosmetics7020021
89. N. Diao, H. Qu, W. Wang, M. Cao, X. Liu, W. Yang, Y. Cao, T. Sun, H. Pei, C. Guo, D. Chen, Preparation and evaluation of a soluble microneedle loaded with resveratrol nanocrystals. J. Drug Deliv. Sci. Technol. **94**, 105463 (2024). https://doi.org/10.1016/j.jddst.2024.105463
90. T. Waghule, G. Singhvi, S.K. Dubey, M.M. Pandey, G. Gupta, M. Singh, K. Dua, Microneedles: A smart approach and increasing potential for transdermal drug delivery system. Biomed. Pharmacother. **109**, 1249–1258 (2019). https://doi.org/10.1016/j.biopha.2018.10.078
91. C. Fornaguera, M. García-Celma, Personalized nanomedicine: A revolution at the nanoscale. JPM **7**, 12 (2017). https://doi.org/10.3390/jpm7040012
92. Y. Ji, H. Li, J. Li, G. Yang, W. Zhang, Y. Shen, B. Xu, J. Liu, J. Wen, W. Song, Hair follicle-targeted delivery of Azelaic acid micro/nanocrystals promote the treatment of acne vulgaris. Int. J. Nanomed. **19**, 5173–5191 (2024). https://doi.org/10.2147/IJN.S459788

Parenteral Administration of Drug Nanocrystals: The Role of Long-Acting Injectables

Caroline C. Sousa, Diogo R. M. Moreira, Lucia Lopez-Vidal, and Alejandro J. Paredes

Abstract Parenteral administration is an effective route for fast-acting treatments and conditions that demand high bioavailability or require avoidance of hepatic first-pass metabolism. However, developing injectable drugs with poorly water-soluble drugs finds several challenges, including precipitation risk, stability issues, and possible toxic effects. Injectable nanocrystals (NCs) provide a pathway to enhance bioavailability by addressing these limitations and allow the drug to be released in a controlled manner without organic solvents. Parenteral NCs showed superior pharmacokinetic profiles, prolonged therapeutic effects, and optimised patient adherence to chronic condition treatments. Intramuscular administration of NCs allows controlled and sustained drug release by forming depots that maintain therapeutic levels for weeks or months, with granulomatous inflammation playing a key role in modulating release. Despite their potential, NCs face obstacles in sterilisation, standardisation for industrial scale, and approval by regulatory agencies, requiring advances in nanotechnology, materials science, and quality control. Current technologies, such as artificial intelligence and targeted delivery systems, can directly impact the optimisation of the development and approval of these drugs. This chapter explores the scientific, clinical, and industrial implications of NCs, emphasising their revolutionary potential in treating acute and chronic diseases while highlighting the need for an interdisciplinary alliance to overcome existing challenges.

C. C. Sousa
Instituto Gonçalo Moniz, Fundação Oswaldo Cruz (Fiocruz), Salvador, Brazil
e-mail: caroline.sousa@fiocruz.br

D. R. M. Moreira
School of Pharmacy, Queen's University Belfast, Belfast, UK

Instituto Gonçalo Moniz, Fundação Oswaldo Cruz (Fiocruz), Salvador, Brazil
e-mail: diogo.magalhaes@fiocruz.br

L. Lopez-Vidal · A. J. Paredes (✉)
School of Pharmacy, Queen's University Belfast, Belfast, UK
e-mail: l.lopezvidal@qub.ac.uk; a.paredes@qub.ac.uk

A. J. Paredes (ed.), *Nanocrystals*, https://doi.org/10.1007/978-3-032-10238-6_5

Keywords Poorly water-soluble drugs · Parenteral route · Injectable nanocrystals · Long-acting treatment · Chronic disease treatments

1 Parenteral Administration Route: Challenges in Injectable Formulations of Poorly Soluble Drugs

Parenteral administration (PA) is an essential therapeutic approach commonly used in clinical practice, especially in fast-acting treatment. This route is preferred over other routes of administration not only in emergency cases, but also because of its high (absolute) bioavailability, drug delivered directly to the bloodstream, and local effect in a specific area that needs to be treated. Drugs with a narrow or low bioavailability or of enzymatic sensitivity, which need to avoid the gastrointestinal system and hepatic first-pass metabolism, are also suitable in the PA. Furthermore, PA is widely applied in long-term treatments, such as schizophrenia, diabetes, HIV/AIDS, and cancers [1]. Different treatments can require different interventions. PA can be achieved through various routes, including subcutaneous, intravenous, intramuscular, intravitreal, intra-articular, intrathecal, intra-epidural, intracisternal, intra-arterial, intracardiac, intrapleural, intraperitoneal, and intraosseous routes. However, the most commonly used routes are intramuscular, subcutaneous, intravenous, and intradermal (Fig. 1). These injection-based administration routes follow distinct principles and are associated with specific dosage forms to achieve the desired therapeutic effect [2]. The choice of route must consider multiple factors, including the patient's medical condition, required dose, drug safety profile, formulation characteristics, and overall efficacy. Among the commonly used routes,

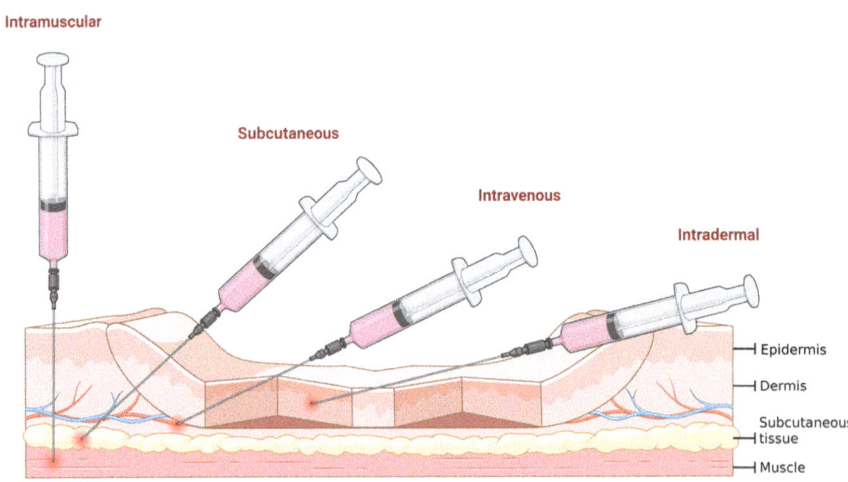

Fig. 1 Schematic representation of the most commonly used parenteral administration routes: intramuscular, subcutaneous, intravenous, and intradermal. (Image created using BioRender 2025)

intramuscular injection is particularly desirable for achieving prolonged therapeutic drug exposure due to depot formation at the injection site, allowing drug release over several weeks or even months. When intramuscular injection is not feasible, subcutaneous injection serves as an effective alternative, also enabling extended exposure. In contrast, intravenous injection is preferred when a rapid onset of action and systemic drug distribution are required, while still allowing for sustained exposure depending on the formulation. Finally, intradermal administration can provide relatively prolonged drug exposure, although the absorptive surface in the dermis is more limited compared to muscle tissue. Although the parenteral route offers general advantages in therapy, it also presents challenges, such as its limitation with poorly water-soluble drugs. The development of these injectable formulations requires the exploration of approaches to modify important issues, such as poor drug stability, local precipitation after injection, risk for erratic drug release and absorption, and unwelcome inflammation or immune response. To solve these problems, several strategies can be employed to improve poor drug water solubility; these may include pH adjustment, salt formation, micellar solubilisation, and the use of co-solvents and surfactants [2, 3].

1.1 pH Adjustment

Drugs such as ibuprofen, a widely used anti-inflammatory, were widely studied to enhance their solubility by adjusting the pH. Its weak acid properties allow it to improve the dissolution rate with a pH above 4.41. The pH modification is an interesting alternative for ionisable drugs, for example. These types of drugs can be protonated in the presence of basic groups or deprotonated if an acid group is present. In addition to buffers for pH adjustment, it is common to use co-solvents to enhance solubility, and this results in quick development and production of formulations. Despite the pH being simple and easy to improve solubilisation in poorly water-soluble drugs, some disadvantages can be associated, such as low tolerability and toxicity because of non-physiological pH. Drug precipitation is also a risk after injection due to the lower drug solubility and could cause emboli [4–6].

1.2 Salt Formation

Another common alternative to improve acidic and basic drug aqueous solubility is salification. Salt formation is the most preferred approach to developing liquid formulations for parenteral administration due to its effectiveness in increasing solubility and dissolution rates, making drugs more suitable for injections. Salt formation solubility is improved by modifications in the drug's physicochemical properties, such as wettability and diffusion layer thickness, which influence how quickly the dissolution occurs in aqueous environments [7]. Salt forms of ionisable drugs can enhance the dissolution rate by increasing the drug's apparent solubility, as described by the Noyes–Whitney equation [8]. The solubility of a salt form can be several orders of magnitude higher than that of the corresponding free base or free acid, greatly improving drug absorption. Furosemide, a diuretic drug, nowadays is used in intravenous and oral treatment of hypertension and oedema. This is possible because of its favourable physicochemical characteristics that allow changing a water-insoluble drug to a dissolved salt form [9, 10].

Salification as a dissolution rate improvement strategy is a good alternative to insoluble drugs. However, this method is limited to weak acidic and base drugs; in addition, solubility and stability must be in balance to avoid salt deposition after administration due pH differences. Even if the salt drug has all the best characteristics, the composition cannot be optimised to avoid hydrolysis or another type of degradation, a common issue in this form [9].

1.3 Co-solvents and Surfactants

Co-solvents are miscible organic solvents that enhance the solubility of active pharmaceutical ingredients (APIS). Utilising co-solvents is a straightforward and effective strategy to increase the solubility of poorly soluble drugs in parenteral formulations. Propylene glycol, ethanol, polysorbate, glycerine, and sorbitol are commonly used solvents in pharmaceutical preparations due to their low toxicity. Jouyban-Gharamaleki et al. demonstrated that a mixture of propylene glycol and ethanol (70:30) increased the solubility of ketoconazole by 20 times (50 mg/mL compared to 2.5 mg/mL in water), and raising the temperature (from 25 °C to 45 °C) further enhanced solubility and stability, confirmed by laser monitoring methods [11]. In an aqueous environment, the addition of co-solvents reduces intermolecular interactions between solvents, thereby lowering surface tension, dielectric constant, and solubility parameter.

Some compounds require additional techniques to enhance solubility, even when incorporating co-solvents into the formulation. In 2020, Thapa et al. examined the efficacy of the curcumin micellisation process using 1% ethanol or 1% DMSO as co-solvents. Curcumin, a well-known Class IV molecule with anticancer effects, demonstrated better performance with 1% ethanol, achieving curcumin encapsulation efficiency of up to 90% while maintaining greater formulation stability over time [12]. Becerril-Vega J et al. presented promising results in the development of an albendazole prodrug (ABZSO) and co-solvents (dimethylacetamide, PEG 400 and ethanol) to improve the solubility of parenteral albendazole [13]. Both approaches improved the solubility and pharmacokinetic profile of albendazole (ABZ), with the prodrug showing the highest AUC and C_{max}, resulting in sustained release. In contrast, a co-solvent-based formulation demonstrated immediate drug dissolution with a rapid plasma peak ($T_{max} < 1$ hour) and bioavailability up to 3 times higher than the oral route. Such strategies are applied in the clinic using prodrugs in prolonged treatments and co-solvents for emergencies [12, 13].

Liposomes are colloidal carrier systems composed of polar lipids and water, forming self-organised particulate structures. Unlike micelles—which consist of aggregates of surfactants with a hydrophobic core and a hydrophilic outer layer—liposomes favour the dispersion of the drug in the solvent. These small vesicles, measuring between 50 and 200 nm, are coated by a membrane composed of one or more bilayers of phospholipids or sphingolipids, often incorporating cholesterol. In parenteral formulations, in addition to the drug characteristics, liposomes can be modified on the surface to increase target specificity, such as by incorporating fatty acids with pH-sensitive lipid chains, which direct the action to acidic microenvironments characteristic of methicillin-resistant *Staphylococcus aureus* (MRSA) infections. The micellisation method is widely used in parenteral preparations, as it incorporates surfactants that form micelles capable of increasing the solubility of poorly soluble drugs in aqueous media, improving their bioavailability and therapeutic efficacy. In addition, this method contributes to the system's stability by encapsulating the drug and protecting it from chemical degradation, such as hydrolysis and oxidation. Another relevant benefit is the reduction of toxicity since it eliminates organic solvents, allowing a controlled and sustained release of the active ingredient [14–16].

Although co-solvents and surfactants are widely used in methodologies aimed at improving the solubility and stability of parenteral formulations, their usage has significant limitations. Co-solvents based on ethanol, PEG, and propylene glycol can cause vascular irritation, haemolysis, and pain at the injection site at high concentrations, along with the risk of drug precipitation after contact with biological fluids. Surfactants such as polysorbates can lead to cellular toxicity, protein denaturation, and immunogenic reactions. In preparations containing liposomes, for example, surfactants can destabilise lipid bilayers, and in micelles, low critical micelle concentrations can result in premature disintegration in biological media. Regulatory issues also tend to limit the approval of nanoformulated drugs due to concerns about safety and the challenge of balancing careful optimisation of excipient concentrations, efficacy, and biocompatibility [17, 18].

2 Injectables Nanocrystals: Extended-release Suspensions

Over the past few years, there has been a growing interest in employing parenteral delivery systems based on drug nanocrystals (NCs) formulations for biomedical purposes. More specifically, NCs have been explored for therapeutic purposes. This increasing interest in NCs as potential treatments is mainly due to their ability to enhance bioavailability, reduce toxicity, and efficiently target specific sites within the body. Compared to other parenteral approaches, NCs offer several advantages, such as the reduced reliance on organic solvents during manufacturing, customisable release profiles, and high versatility in tailoring the final product's characteristics. As a result, injectable NCs are emerging as promising candidates for extended-release suspensions.

Due to the simple and easy-to-handle manufacturing process, drug NCs are often preferred for improving the solubility, dissolution rate, and, consequently, bioavailability of poorly water-soluble drugs. Notably, here the definition of treatment can encompass both prophylactic and therapeutic interventions. Injectable NCs-based formulations are particularly promising for parenteral treatments, due to precise crystal size control. For instance, smaller particles (<200 nm) are favoured in intravenous administration for efficient drug delivery. Additionally, the nanosizing process has applications in controlled drug release and precise drug delivery [6, 19].

To overcome the challenges of low aqueous solubility and to reduce the toxicity associated with treatments for certain types of cancers, the development of drug NCs has been utilised alongside the antineoplastic drug paclitaxel (PTX). Below, we summarise a few examples of successfully employing drug NCs.

Although PTX is an excellent chemotherapeutic agent, its commercially available formulations, such as Taxol® and Abraxane®, present significant limitations from both therapeutic and pharmaceutical perspectives. Taxol® contains excipients like Cremophor® and dehydrated ethanol, which can induce severe side effects, including hypersensitivity reactions and peripheral neurotoxicity. In contrast, Abraxane®, a suspension of albumin particles in saline, involves a high-cost manufacturing process. These limitations have led to the development of PTX NCs as a promise alternative for cancer treatment. Park et al. have developed albumin-coated NCs of PTX to improve the therapeutic approach for solid tumours [20]. The PTX NCs were formulated by crystallisation in various surfactant-containing media; however, the formulation stabilised with poloxamer 407 resulted in the smallest particle size and a rod-shaped morphology that allows evasion from macrophage phagocytosis. Liu et al. formulated stable PTX NCs with TGPS as a stabiliser. This formulation demonstrated substantial specificity in inhibiting P-glycoprotein activity in cancer cells, which is crucial for overcoming multidrug resistance in cancer [21]. Furthermore, these NCs exhibited a higher capacity for drug loading compared to Taxol® and Abraxane®, which is an important aspect in decreasing infusion time, as well as changes in administration routines and patient adherence.

Drug NCs also have the versatility to improve drug delivery at both pharmacokinetic and pharmacodynamic levels [22]. For instance, Fuster et al. have introduced an approach for improving cancer treatment through the drug delivery NCs of curcumin, a well-known compound endowed with anticancer properties. The idea behind this strategy was to enhance curcumin's specificity to cancer cells by covalently attaching it to folic acid on the surface of the NCs. The nanoformulation displayed an enhanced cellular uptake in cancer cell lines and showed a promising selective cytotoxicity between cancerous cells and healthy cells. Moreover, by reducing macrophage uptake, these drug NCs were hypothesised to display a prolonged circulation in vivo and a decreased probability in targeting the immune system [23].

Park et al. demonstrated an improvement of Montelukast (MKT), a small-molecule drug used for chronic asthma and seasonal allergies, by changing its amorphous form into MKT NCs. This work had a goal for Alzheimer's parenteral treatment. Different types of stabilisers were used in the development process of the

NCs, among which formulations containing polysorbate 80 (P80) exhibited crystals with small particle sizes (between 200 and 500 nm), optimal homogeneity, and an improved dissolution rate. As a result, the pharmacokinetic profile was prolonged for up to 4 weeks after a single subcutaneous injection, as well as achieving higher maximum plasma concentrations compared to microcrystalline suspensions of MKT. Consistent to this result, the histopathological analysis also revealed a granulomatous inflammation induced by MKT NC injection, which was temporary and fully reversible after 28 days regardless of particle size [24].

Once intravenously administered, drug NCs are expected to dissolve more uniformly within the plasma, thus enhancing the overall pharmacokinetic properties and maximising tissue distribution of the drug. As a result of improved PK profile, an often-observable toxic reaction in the local injection is avoided in comparison to other parenteral formulations. A showcase of this is depicted for itraconazole, an antifungal drug. Pharmacokinetics profiles of itraconazole NCs have resulted in a remarkable superior sustained plasma concentration, while the micro-suspension solution was swiftly eliminated. Consistent to this, drug content from itraconazole NCs was detected in rat spleen 7 days following a single intravenous administration. These findings were confirmed by a Phase II study, which examined that the intravenous nanosuspension of itraconazole has exhibited a better tolerability and efficacy compared with the solution. It is rationalised that this is due to the lower peak plasma concentrations, an extended half-life, and an improvement in safety when high doses are employed. All these parameters allowed a greater drug accumulation in target organs, such as the kidneys, the fungal invasion site, and significantly reduced fungal colony counts [25, 26]. Here, it is important to highlight the underlying pharmacokinetic aspects that explain why drug NCs can exhibit an improved profile compared to other systems. A fascinating characteristic of nanoparticles is how particle size can influence the drug path through the body and its further distribution. In general, it is accepted that nano- and microcrystals may exhibit distinct tissue targeting profiles upon interaction with biological systems, as demonstrated by following intraperitoneal (I.P.) administration in rodents. While NCs can exhibit a rapid in vitro dissolution due to their large surface area, in in vivo studies in rodents, the microcrystals may achieve high systemic exposure at the injection site because of the particle sedimentation. However, the importance of tissue particle sedimentation upon microcrystal suspensions for the pharmacokinetic profile has not yet been fully understood, as it is modulated by local physiological factors [27, 28]. In contrast, NCs can be phagocytosed and/or enter the lymphatic system, as their size and surface charge result in an overall dilution factor. This is metabolised in the lymph nodes, leading to lower systemic drug levels. Furthermore, as soon as NCs are inside the phagocytic cells (namely, monocytes, macrophages, and neutrophils), the drug can also circulate in the bloodstream and accumulate in the mononuclear phagocyte system-rich organs, such as the liver, spleen, and lymph nodes. These findings support the overall idea of developing nanoformulations containing drug NCs for a gradual and controlled release system [3].

By discussing recent advances in nanotechnology approaches for treatment, there is no doubt that it is essential to mention in detail the breakthrough that injectable antiretroviral (ARV) nanoformulations have achieved. This was not only remarkable for the pharmaceutical and biomedical point of view, but it has also shifted the paradigm for patient adherence and health benefits globally. In fact, the development of ARV nanoformulations has coined the scientific term "long-acting" (LA) for the non-scientific community. This is best known as LA management of human immunodeficiency virus (HIV)/AIDS. The LA ARV nanoformulations for treatment and prophylaxis of HIV/AIDS are nowadays a reliable and efficient therapeutic option for the currently available oral medicines, which can benefit a broad range of HIV/AIDS patients and community. In general, the ARV NCs can exhibit an improvement in solubility and bioavailability and are able to form a drug depot at the injection site. This "depot" consists of a repository for drug in the muscle, which is necessary to release the drug that can be absorbed by the vasculature. Yin et al. observed that, after IM administration of prodrug paliperidone palmitate microsuspension, the depot-infiltrating is majority CD28+ macrophage cells seen 2 weeks after injection [29, 30].

Cabotegravir (CAB) and rilpivirine (RPV) are therefore a great alternative for HIV treatment, namely, for PREP. CAB NCs were manufactured via a wet bead milling process using mannitol, polysorbate 20, and polyethylene glycol 3350 as stabilisers. This formulation resulted in a particle size of approximately 200 nm. In a study with 72 healthy volunteers, this nanosuspension was administrated once by IM or SC route, and the results exhibit that the plasma concentration-time profiles were similar for both routes, with comparable maximal concentrations (C_{max}) and total plasma exposures. Furthermore, it was calculated a geometric mean concentrations 4 weeks post-dose remained over four times the protein-adjusted 90% inhibitory concentration (0.66 μg/mL) for IM (dose of 800 mg) and SC (dose of 400 mg) injections. These findings indicated that this is well comparable to the mean trough concentration observed in a 10-day monotherapy study of oral CAB 5 mg in HIV-infected antiretroviral-naive adults [6, 31].

Still in HIV/AIDS therapy, recent investigations have been carried out in vivo and human volunteers, employing nanosuspensions incorporating RPV NCs. Verloes et al. demonstrated that a single IM injection of RPV LA provides a drug depot, with a slow release over 28 days, resulting in almost stable RPV plasma concentrations over time and a prolonged and apparent half-life. To produce the RPV NCs, Poloxamer 338 was used as a wetting agent, and wet bead milling was selected as a method. Additionally, the injection volume was reduced, achieving a concentration of 300 mg/mL RPV. Furthermore, multiple injections (1200/600/600 mg) resulted in an average plasma concentration of 79 ng/mL over the 28-day interval, preserving therapeutic concentrations [32, 33].

Cobb et al. developed stable ProTide formulations of tenofovir (TFV), namely, M1TFV and M2TFV, for IM LA administration. One of these formulations (M1TFV) demonstrated enhanced drug and active metabolite delivery into macrophages and CEM-ss CD4+ T cells, which may play a role in the LA drug release. As a result, PK and efficacy evaluations showed significant improvements in the drug's apparent half-life and biodistribution. All these data support that both formulations could enable sustained release and extend the dosing intervals in HIV treatment.

Although IM administration can be painful due to the needle insertion at the injection site, NC formulations using aqueous solution avoid local irritation because they eliminate the need for organic solvents [34].

3 Intramuscular Nanocrystals: Characteristics and Applications

The intramuscular site has unique characteristics, which may influence the overall drug bioavailability. The muscular tissue comprises muscle fibres and can be organised into bundles or fasciculi, surrounded by a continuous network of connective tissue structures termed perimysium and epimysium. Muscle tissues are richly vascularised, supporting quick absorption of drugs and vaccines. When compared with other administration routes, the intramuscular route ensures a prompt onset of action, avoids first-pass metabolism, and provides a substantial space to retain a large volume of drugs, such as in the LA depot. Furthermore, due to the high vascularisation in muscle, it is possible that the fast distribution to lymph nodes through lymphatic vessels [35, 36].

Clearly, NCs dissolution requires a combination of parameters, such as the microenvironment, immune system response, drug dissolution rate and physicochemical properties, to release the drug in the bloodstream to produce biological effects. There is an interesting relationship between NCs size in the interaction with biological systems (Fig. 2). A prevailing hypothesis is that NCs smaller than 100 nm perform like a solution due to prompt dissolution, while larger particles exhibit higher cellular uptake, resulting in posterior drug release. In the muscle tissue, the undissolved drug particles form a depot through the interaction between the surrounding tissues and the foreign delivery system, leading to granuloma formation.

The injection site is initially characterised by an acute inflammation, followed by chronic inflammatory reactions, granuloma tissue formation, and fibrosis. The granulation tissue includes neovascularisation within the depot, proliferating fibroblasts, and various infiltrating inflammatory cells such as neutrophil granulocytes, lymphocytes, and phagocytic cells. Among these, macrophages (CD68+) play a particularly relevant role, not only by internalising NCs, but also by modulating drug distribution [37]. These macrophages may secrete extracellular vesicles containing NCs, contributing to a gradual release of the drug into the bloodstream. In parallel, a host tissue response involving further cell recruitment and extravasation can promote sustained migration of NCs-loaded cells from the injection site, potentially lasting for several months. Additionally, lymphatic transport and retention within lymph nodes are expected to influence systemic distribution and drug absorption. Altogether, these immune-mediated and physiological mechanisms modulate the pharmacokinetics of LA injectable formulations based on drug NCs.

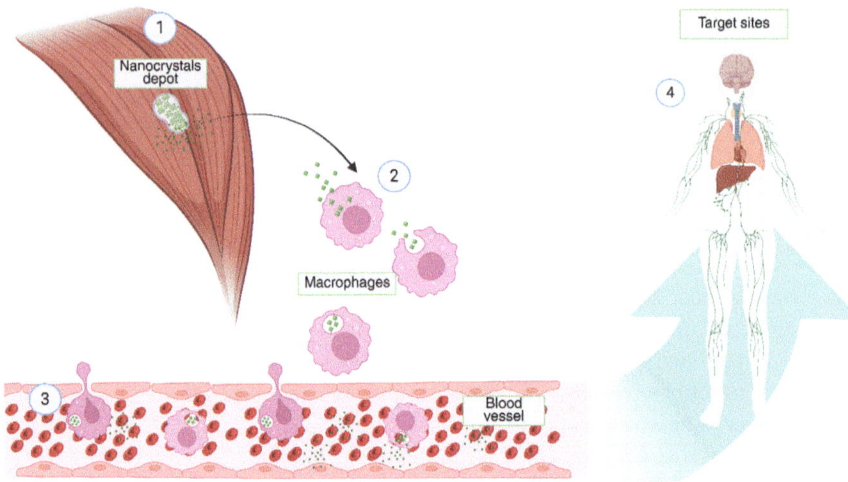

Fig. 2 Proposed pathways of the interplay between drug nanocrystals (NCs) and tissue physiology in the depot process. Upon intramuscular injection (1), drug NCs are deposited and surrounded by infiltrating cells, including macrophages, fibroblasts, and inflammatory cells. Macrophages (2) play a key role by internalising NCs and potentially secreting NC-loaded vesicles into the bloodstream. Additional mechanisms, such as cell migration and lymph node retention (3), are also involved in the systemic distribution of NCs. (Image was created using BioRender 2025)

More specifically, Darville et al. observed a time-dependent recruitment of various cell types following paliperidone palmitate (PP) administration. Infiltrating neutrophils were observed to migrate to the injection site as early as 4 hours post-injection for PP. As the inflammatory process progressed, the nanocrystals became enveloped by a moderate to marked chronic granulomatous inflammatory reaction, involving CD68+ macrophages and granulocytes. This reaction led to the formation of a depot after the efflux of the formulation vehicle, accompanied by a low presence of T lymphocytes. Subsequently, angiogenesis and vascularisation of the granuloma were observed at 48 hours and 7 days, respectively [38]. Certainly, the association of phagocytic cells and their migration to lymphoid tissues significantly influences pharmacokinetic parameters, particularly the half-life and plasma concentration. NCs are generally predominantly distributed by the mononuclear phagocyte system (MPS) organs, particularly macrophages. The deposition or dissolution of NCs is directly associated with the size, where those larger than 500 nm accumulate in the liver, captured by Kupffer cells. This NC capture is also observed in lymph nodes, due to lymph migration. Once phagocytised, NCs dissolve over up to 15–20 days. Subsequently, the free drug is released from the cells and re-enters circulation, altering the PK [39].

PK alterations may arise from flip-flop kinetics, where the drug absorption rate influences the elimination half-life. For instance, in oral administration, the drug follows a simple and linear pathway, where the elimination rate (k_{el}) is significantly lower than the absorption rate (k_a), allowing k_{el} to be determined. In contrast, with

IM administration, k_{el} exceeds k_a, eliminating the half-life controlled by the absorption rate. Fu et al. conducted a study comparing IM administration of NCs-based cannabidiol (CBD) with oral administration of CBD-NCs and CBD oil solution. The findings demonstrated that the IM formulation achieved a higher C_{max} compared to both oral CBD-NCs oil solution, with values of 239.41 ± 16.92, 151.40 ± 35.78, and 135.94 ± 38.15 ng/mL, respectively. Additionally, the IM route significantly enhanced the bioavailability of CBD-NC, which was up to 7.8-fold greater than the other two treatments [29, 39–41].

In summary, IM NC formulations provide several advantages for improving poorly water-soluble drugs. The relationship between the drug's physicochemical properties, the particle size, and the host's response results in enhanced pharmacokinetic and pharmacodynamic performance.

4 Revolutionising Chronic Disease Treatment: How Drug NCs of LA Injectables Improve Adherence and Outcomes

Parenteral formulations on NCs-based technology have shown an advantageous approach for the development of LA injectable formulations for the treatment of chronic conditions such as cancer, schizophrenia, mania, and bipolar disorder. These formulations can achieve sustained therapeutic drug levels for long periods and, at the same time, reduce the frequency of dosage, which is reflected in patient adherence.

Globally, Risperidone was the first antipsychotic available as an LA injectable treatment. To enhance the schizophrenia treatment, paliperidone, the active metabolite of risperidone, was introduced, first, as an extended-release oral formulation (Invega®, Janssen), and later as an LA injectable form of PP (PLAI) formulated as NCs of PP in aqueous solution (Invega Sustenna® and Xeplion®). The enhancement of the formulation allowed the monthly administration via intramuscular in the deltoid or gluteal muscle. After clinical studies, it was observed that PLAI is effective in controlling acute symptoms of schizophrenia and delays the time to relapse. The regimen can be initiated with a dose of 150 mg on day 1, followed by 100 mg on day 8, utilising both deltoid muscles as a site of injections. Invega Trinza and Invega Hafyera™ are two more alternatives available with an extended dosing interval of three and 6 months, respectively. These medical approaches enable patients to maintain therapeutic plasma levels of the medication, feel comfortable, and adhere to the treatment regimen. Another advantage of these formulations is the absence of refrigeration requirements and the need for reconstitution, thus facilitating clinical management. Such characteristics represent significant innovations in psychiatric treatment, offering both efficacy and convenience [42, 43].

The efficacy of both CAB and RPV has been previously demonstrated in clinical trials, where they achieved plasma concentrations sufficient to suppress the virus copies. Cabenuva®, a long-acting injectable therapy for HIV treatment, consists of a combination of two antiviral agents CAB and RPV. This formulation leverages nanocrystal technology to achieve controlled release of these antiretroviral drugs over 1 or 2 months, depending on the dosage. In two trial studies (Phase 3) with 1182 patients, it was observed the decrease of HIV-1 RNA copies in Cabenuva® group, maintaining <50 copies/mL for 48 weeks, with results comparable with current oral therapy. Cabenuva® approval garnered considerable enthusiasm, as conventional oral treatment for HIV continues to face social stigma. This innovative treatment can potentially reduce the stigmas associated with daily medication use, thereby improving patient adherence. Furthermore, it represents an effective and safe long-term therapeutic option [44, 45].

The development of LA injectables based on NCs technology represents a significant advancement in the treatment of chronic conditions such as schizophrenia and HIV. Invega Sustenna®, Invega Trinza®, and Invega Hafyera® show significant benefits, including sustained therapeutic drug levels, which are reflected in the reduction of frequency of doses and enhanced patient adherence. Furthermore, the Invega formulations eliminate the necessity for refrigeration and reconstitution, simplifying clinical management and improving accessibility for patients and healthcare providers. Likewise, Cabenuva®, a revolutionary LA injectable therapy for HIV, employs NCs technology to provide a controlled release of antiretroviral drugs for up to 2 months, resulting in a safe and effective alternative to daily oral regimens. These innovations significantly improve patients' treatment adherence and quality of life by addressing challenges such as social stigma and complex dosing schedules. The progress of these therapies in clinical trials highlights their potential success in revolutionising the management of chronic diseases, incorporating scientific innovation with patient-centred care. In scientific research, the development of LA injectable formulations plays an important role in modern medicine.

5 Challenges in Developing Parenteral Nanosuspensions: Sterilisation, Standardisation, and Regulatory Considerations

The development of nanosuspensions for parenteral administration faces significant challenges before reaching patients, including sterilisation, industrial standardisation, and regulatory aspects. Sterilisation is a critical stage since parenteral formulations must be sterile and non-pyrogenic. The most used methods, such as autoclaving and dry heat sterilisation, risk degrading the formulations due to particle aggregation and thermal degradation. Alternatives like filters with porosity smaller than 0.22 μm or γ-irradiation may be viable options during the sterilisation process, but they require rigorous quality control to ensure the stability and efficacy of the final product [37, 46].

The industrial standardisation of NCs is a limiting factor that relies on precise control of specific parameters such as particle size, homogeneous distribution, and long-term product stability. Wet media milling and controlled precipitation techniques are well-known and widely adopted; however, optimising these processes is essential for product quality and reproducibility. Equally important is identifying suitable excipients like stabilisers and surfactants, as they directly influence the preservation of the nanosuspension properties during storage.

Future scenarios for parenteral drugs based on NCs include studying sustained and targeted release platforms, such as the addition of specific ligands to increase selectivity in target tissues and cells, both for therapy and diagnostics [47, 48]. The rapid advancement of artificial intelligence associated with computational modelling can optimise planning and production time, reducing costs and development time [49]. Regulatory agencies are stringent regarding data on safety, efficacy, and quality when approving new drugs. Detailed information about NCs, such as particle size, surface charge, and release profile, is crucial for meeting regulatory requirements. Additionally, long-term toxicity and environmental impact are also considered in the approval process [50].

6 Conclusion

Parenteral administration of nanocrystals represents significant progress in treating acute and chronic diseases. It offers solutions for poorly water-soluble drugs and has potential as an LA injectable. Throughout this chapter, a number of aspects of NCs were discussed, such as their advantages over other types of pharmaceutical preparations concerning bioavailability, drug toxicity, and controlled release capacity. Finally, how these aspects can be applied in the clinic routine, for example, in treating and/or preventing schizophrenia and HIV/AIDS, was discussed here. Nowadays, the basic aspects of NCs for parental administration are well-understood; however, less is known about intramuscular administration using the drug depot strategy. A full understanding of the temporal drug distribution in tissue and the underlying plasma pharmacokinetics remains less known. These are unmet challenges for the future. Specifically, identifying these characteristics could provide a model to explain the scope and limitations of administering intramuscular NCs as therapeutics. Via this reasoning, the target selectivity of NCs is another issue. Whether this is necessary to adjust any parenteral or intramuscular therapy remains unclear. The regulatory stages and the thoroughness of toxicological studies are essential barriers to reaching a final product. However, these complementary areas of knowledge, such as nanotechnology, materials science, immunology, and pharmacology, can ensure the excellence of the final product safely and effectively. Moreover, optimising the sterilisation process and industrial production are potential perspectives for the development of NCs.

References

1. A.B. Jindal, A.R. Bhide, S. Salave, D. Rana, D. Benival, Long-acting parenteral drug delivery systems for the treatment of chronic diseases. Adv. Drug Deliv. Rev. **198**, 114862 (2023). https://doi.org/10.1016/J.ADDR.2023.114862
2. C. Camacho Vieira, L. Peltonen, A.P. Karttunen, A.J. Ribeiro, Is it advantageous to use quality by design (QbD) to develop nanoparticle-based dosage forms for parenteral drug administration? Int. J. Pharm. **657**, 124163 (2024). https://doi.org/10.1016/J.IJPHARM.2024.124163
3. K.C. Aluri, K. Sigfridsson, A. Xue, D. Ramsden, Pharmacokinetics of nano- and microcrystal formulations of low solubility compounds after intramuscular injection to mice. J. Pharm. Pharmacol. **77**, 635–644 (2025). https://doi.org/10.1093/JPP/RGAE118
4. E. Ahire, S. Thakkar, M. Darshanwad, M. Misra, Parenteral nanosuspensions: A brief review from solubility enhancement to more novel and specific applications. Acta Pharm. Sin. B **8**, 733–755 (2018). https://doi.org/10.1016/J.APSB.2018.07.011
5. J. Nerurkar, J.W. Beach, M.O. Park, H.W. Jun, Solubility of (±)-ibuprofen and S (+)-ibuprofen in the presence of Cosolvents and Cyclodextrins. Pharm. Dev. Technol. **10**, 413–421 (2005). https://doi.org/10.1081/PDT-54446
6. M.B. McGuckin, J. Wang, R. Ghanma, N. Qin, S.D. Palma, R.F. Donnelly, A.J. Paredes, Nanocrystals as a master key to deliver hydrophobic drugs via multiple administration routes. J. Control. Release **345**, 334–353 (2022). https://doi.org/10.1016/J.JCONREL.2022.03.012
7. P.C. Acharya, C. Fernandes, D. Suares, S. Shetty, R.K. Tekade, Solubility and solubilization approaches in pharmaceutical product development, in *Dosage Form Design Considerations: Volume I*, (2018), pp. 513–547. https://doi.org/10.1016/B978-0-12-814423-7.00015-0
8. A.A. Noyes, W.R. Whitney, The rate of solution of solid substances in their own solutions. J. Am. Chem. Soc. **19**, 930–934 (1897). https://doi.org/10.1021/JA02086A003/ASSET/JA02086A003.FP.PNG_V03
9. D.P. Elder, R. Holm, H.L. De Diego, Use of pharmaceutical salts and cocrystals to address the issue of poor solubility. Int. J. Pharm. **453**, 88–100 (2013). https://doi.org/10.1016/J.IJPHARM.2012.11.028
10. L.H. Nielsen, S. Gordon, R. Holm, A. Selen, T. Rades, A. Müllertz, Preparation of an amorphous sodium furosemide salt improves solubility and dissolution rate and leads to a faster Tmax after oral dosing to rats. Eur. J. Pharm. Biopharm. **85**, 942–951 (2013). https://doi.org/10.1016/J.EJPB.2013.09.002
11. V. Jouyban-Gharamaleki, A. Jouyban, H. Zhao, F. Martinez, E. Rahimpour, Solubility study of ketoconazole in propylene glycol and ethanol mixtures at different temperatures: A laser monitoring method. J. Mol. Liq. **337**, 116060 (2021). https://doi.org/10.1016/J.MOLLIQ.2021.116060
12. R.K. Thapa, F. Cazzador, K.G. Grønlien, H.H. Tønnesen, Effect of curcumin and cosolvents on the micellization of Pluronic F127 in aqueous solution. Colloids Surf. B. **195**, 111250 (2020). https://doi.org/10.1016/J.COLSURFB.2020.111250
13. J. Becerril-Vega, A. Hernández-Campos, I. González-Hernández, M. Flores-Ramos, R. Castillo, G. Leyva-Gómez, L. Mayet-Cruz, H. Jung-Cook, Development and pharmacokinetic evaluation of two parenteral formulations of Albendazole using prodrug and Cosolvent approaches. AAPS PharmSciTech **24**, 1–10 (2023). https://doi.org/10.1208/S12249-023-02613-9/TABLES/5
14. L.M. Ickenstein, P. Garidel, Lipid-based nanoparticle formulations for small molecules and RNA drugs. Expert Opin. Drug Deliv. **16**, 1205–1226 (2019). https://doi.org/10.1080/17425247.2019.1669558
15. S.S. Makhathini, R.S. Kalhapure, M. Jadhav, A.Y. Waddad, R. Gannimani, C.A. Omolo, S. Rambharose, C. Mocktar, T. Govender, Novel two-chain fatty acid-based lipids for development of vancomycin pH-responsive liposomes against Staphylococcus aureus and

methicillin-resistant Staphylococcus aureus (MRSA). J. Drug Target. **27**, 1094–1107 (2019). https://doi.org/10.1080/1061186X.2019.1599380

16. V.O. Nyandoro, E.A. Ismail, A. Tageldin, M.A. Gafar, X.Q. Peters, R. Mautsoe, C.A. Omolo, T. Govender, Potential of nanocarrier-mediated delivery of vancomycin for MRSA infections. Expert Opin. Drug Deliv. **22**, 347 (2025)
17. M. Rodriguez-Aller, D. Guillarme, J.L. Veuthey, R. Gurny, Strategies for formulating and delivering poorly water-soluble drugs. J Drug Deliv Sci Technol **30**, 342–351 (2015). https://doi.org/10.1016/J.JDDST.2015.05.009
18. K.U. Khan, M.U. Minhas, S.F. Badshah, M. Suhail, A. Ahmad, S. Ijaz, Overview of nanoparticulate strategies for solubility enhancement of poorly soluble drugs. Life Sci. **291**, 120301 (2022). https://doi.org/10.1016/J.LFS.2022.120301
19. L. Peltonen, Design space and QbD approach for production of drug nanocrystals by wet media milling techniques. Pharmaceutics **10**, 104 (2018). https://doi.org/10.3390/PHARMACEUTICS10030104
20. J. Park, B. Sun, Y. Yeo, Albumin-coated nanocrystals for carrier-free delivery of paclitaxel. J. Control. Release **263**, 90–101 (2017). https://doi.org/10.1016/J.JCONREL.2016.12.040
21. Y. Liu, L. Huang, F. Liu, Paclitaxel nanocrystals for overcoming multidrug resistance in cancer. Mol. Pharm. **7**, 863–869 (2010)
22. L. Lopez-Vidal, M. Tinti, M.E. Melian, L. Canton, M. Lorenzutti, L. Schofs, M.L. Formica, A.J. Paredes, S.S. Bruni, N. Litterio, R. Faccio, S.D. Palma, J.P. Real, In vivo pharmacokinetic study and PBPK modeling: Comparison between 3D-printed nanocrystals and solid dispersions. Int. J. Pharm. **669**, 125063 (2025). https://doi.org/10.1016/J.IJPHARM.2024.125063
23. M.G. Fuster, J. Wang, O. Fandiño, G. Víllora, A.J. Paredes, Folic acid-decorated nanocrystals as highly loaded Trojan horses to target cancer cells. Mol. Pharm. **21**, 2781–2794 (2024). https://doi.org/10.1021/ACS.MOLPHARMACEUT.3C01186/ASSET/IMAGES/LARGE/MP3C01186_0010.JPEG
24. J.S. Park, M.S. Kim, M.Y. Joung, H.J. Park, M.J. Ho, J.H. Choi, J.H. Seo, W.H. Song, Y.W. Choi, S. Lee, Y.S. Choi, M.J. Kang, Design of montelukast nanocrystalline suspension for parenteral prolonged delivery. Int. J. Nanomedicine **17**, 3673–3690 (2022). https://doi.org/10.2147/IJN.S375888
25. B. Rabinow, J. Kipp, P. Papadopoulos, J. Wong, J. Glosson, J. Gass, C.S. Sun, T. Wielgos, R. White, C. Cook, K. Barker, K. Wood, Itraconazole IV nanosuspension enhances efficacy through altered pharmacokinetics in the rat. Int. J. Pharm. **339**, 251–260 (2007). https://doi.org/10.1016/J.IJPHARM.2007.02.030
26. A.M.E. Jansen, R. Ter Heine, J.P. Donnelly, N. Blijlevens, R.J.M. Brüggemann, Repurposing antifungals: population pharmacokinetics of itraconazole and hydroxy-itraconazole following administration of a nanocrystal formulation. J. Antimicrob. Chemother. **78**, 1219–1224 (2023). https://doi.org/10.1093/JAC/DKAD072
27. K.L. Swetha, M. Paul, K.S. Maravajjala, S. Kumbham, S. Biswas, A. Roy, Overcoming drug resistance with a docetaxel and disulfiram loaded pH-sensitive nanoparticle. J. Control. Release **356**, 93–114 (2023). https://doi.org/10.1016/J.JCONREL.2023.02.023
28. M.S. Kim, M.J. Ho, M.Y. Joung, Y.S. Choi, M.J. Kang, Effect of dispersion medium on pharmacokinetic profile of Rotigotine crystalline suspension following subcutaneous injection. Pharmaceutics **14**, 2630 (2022). https://doi.org/10.3390/PHARMACEUTICS14122630
29. J. Yin, A.C. Collier, A.M. Barr, W.G. Honer, R.M. Procyshyn, Paliperidone palmitate long-acting injectable given intramuscularly in the deltoid versus the gluteal muscle: Are they therapeutically equivalent? J. Clin. Psychopharmacol. **35**, 447–449 (2015). https://doi.org/10.1097/JCP.0000000000000361
30. M. Ullah Nayan, B. Sillman, M. Hasan, S. Deodhar, S. Das, A. Sultana, et al., Advances in long-acting slow effective release antiretroviral therapies for treatment and prevention of HIV infection. Adv. Drug Deliv. Rev. **200**, 115009 (2023). https://doi.org/10.1016/J.ADDR.2023.115009

31. C. Trezza, S.L. Ford, W. Spreen, R. Pan, S. Piscitelli, Formulation and pharmacology of long-acting cabotegravir. Curr. Opin. HIV AIDS **10**, 239–245 (2015). https://doi.org/10.1097/COH.0000000000000168
32. R. Verloes, S. Deleu, N. Niemeijer, H. Crauwels, P. Meyvisch, P. Williams, Safety, tolerability and pharmacokinetics of rilpivirine following administration of a long-acting formulation in healthy volunteers. HIV Med. **16**, 477–484 (2015). https://doi.org/10.1111/HIV.12247
33. P.E. Williams, H.M. Crauwels, E.D. Basstanie, Formulation and pharmacology of long-acting rilpivirine. Curr. Opin. HIV AIDS **10**, 233–238 (2015). https://doi.org/10.1097/COH.0000000000000164
34. D.A. Cobb, N. Smith, S. Deodhar, A.N. Bade, N. Gautam, B.L.D. Shetty, J.E. McMillan, Y. Alnouti, S.M. Cohen, H.E. Gendelman, B. Edagwa, Transformation of tenofovir into stable ProTide nanocrystals with long-acting pharmacokinetic profiles. Nat. Commun. **12**, 1–16 (2021). https://doi.org/10.1038/s41467-021-25690-5
35. P.P. Purslow, The structure and role of intramuscular connective tissue in muscle function. Front. Physiol. **11**, 519145 (2020). https://doi.org/10.3389/FPHYS.2020.00495/XML/NLM
36. J. Lee, D. Kim, J. Byun, Y. Wu, J. Park, Y.K. Oh, In vivo fate and intracellular trafficking of vaccine delivery systems. Adv. Drug Deliv. Rev. **186**, 114325 (2022). https://doi.org/10.1016/J.ADDR.2022.114325
37. Z. Ma, H. Zhang, Y. Wang, X. Tang, Development and evaluation of intramuscularly administered nano/microcrystal suspension. Expert Opin. Drug Deliv. **16**, 347–361 (2019). https://doi.org/10.1080/17425247.2019.1588248
38. N. Darville, M. Van Heerden, A. Vynckier, M. De Meulder, P. Sterkens, P. Annaert, G. Van Den Mooter, Intramuscular administration of paliperidone palmitate extended-release injectable microsuspension induces a subclinical inflammatory reaction modulating the pharmacokinetics in rats. J. Pharm. Sci. **103**, 2072–2087 (2014). https://doi.org/10.1002/JPS.24014
39. Y. Yanamadala, C.M.R. Muthumula, S. Khare, K. Gokulan, Strategies to enhance nanocrystal formulations for overcoming physiological barriers across diverse routes of administration. Int. J. Nanomedicine **20**, 367–402 (2025). https://doi.org/10.2147/IJN.S494224
40. V.T.T. Nguyen, N. Darville, A. Vermeulen, Pharmacokinetics of long-acting aqueous nano–/microsuspensions after intramuscular administration in different animal species and humans – a review. AAPS J. **25**, 1–25 (2023). https://doi.org/10.1208/S12248-022-00771-5/METRICS
41. X. Fu, S. Xu, Z. Li, K. Chen, H. Fan, Y. Wang, Z. Xie, L. Kou, S. Zhang, Enhanced intramuscular bioavailability of Cannabidiol using nanocrystals: Formulation, in vitro appraisal, and pharmacokinetics. AAPS PharmSciTech **23**, 1–12 (2022). https://doi.org/10.1208/S12249-022-02239-3/TABLES/2
42. P. Chue, J. Chue, A review of paliperidone palmitate. Expert. Rev. Neurother. **12**, 1383–1397 (2012). https://doi.org/10.1586/ERN.12.137
43. C.-L. Chang, K.A. Forces, G. Hospital, Y.-C. Huang, P. Yang, Improving of cognition and quality of life in schizophrenia with one-month and three-month paliperidone palmitate treatment (2021). https://doi.org/10.21203/RS.3.RS-483945/V1
44. S. Yuan, D.S. Wang, H. Liu, S.N. Zhang, W.G. Yang, M. Lv, Y.X. Zhou, S.Y. Zhang, J. Song, H.M. Liu, New drug approvals for 2021: Synthesis and clinical applications. Eur. J. Med. Chem. **245**, 114898 (2023). https://doi.org/10.1016/J.EJMECH.2022.114898
45. L. Peters, M. Dyer, E. Schroeder, M.S. D'Souza, Invega Hafyera (Paliperidone palmitate): Extended-release injectable suspension for patients with schizophrenia. J. Pharm. Technol. **39**, 88–94 (2023). https://doi.org/10.1177/87551225231153541
46. D.D. Nakmode, B. Singh, S. Abdella, Y. Song, S. Garg, Long-acting parenteral formulations of hydrophilic drugs, proteins, and peptide therapeutics: mechanisms, challenges, and therapeutic benefits with a focus on technologies. Drug Deliv. Transl. Res. **15**, 1156–1180 (2024). https://doi.org/10.1007/S13346-024-01747-Y
47. J.V. Jokerst, D. Van de Sompel, S.E. Bohndiek, S.S. Gambhir, Cellulose nanoparticles are a biodegradable photoacoustic contrast agent for use in living mice. Photoacoustics **2**, 119–127 (2014). https://doi.org/10.1016/J.PACS.2014.07.001

48. K. Chaudhary, M.M. Patel, P.J. Mehta, Long-acting Injectables: Current perspectives and future promise. Crit. Rev. Ther. Drug Carrier Syst. **36**, 137–181 (2019). https://doi.org/10.1615/CRITREVTHERDRUGCARRIERSYST.2018025649
49. Y. He, Z. Ye, X. Liu, Z. Wei, F. Qiu, H.F. Li, Y. Zheng, D. Ouyang, Can machine learning predict drug nanocrystals? J. Control. Release **322**, 274–285 (2020). https://doi.org/10.1016/J.JCONREL.2020.03.043
50. M.L. Chen, M. John, S.L. Lee, K.M. Tyner, Development considerations for nanocrystal drug products. AAPS J. **19**, 642–651 (2017). https://doi.org/10.1208/S12248-017-0064-X/TABLES/3

Microneedle-mediated Administration of Drug Nanocrystals

Lucia Lopez-Vidal, Elise J. Catlin, Ryan F. Donnelly, and Alejandro J. Paredes

Abstract This chapter explores the combination of nanocrystals (NCs) and microneedles (MNs) as a versatile and promising strategy for drug delivery. Initially developed to enhance percutaneous administration, MNs have evolved into diverse formats—solid, coated, dissolving, hollow, and hydrogel-forming, each with distinct advantages depending on therapeutic goals. The combination of MNs with NCs offers a powerful platform to improve the delivery of poorly water-soluble drugs by enhancing dissolution rates while bypassing conventional administration barriers. This chapter outlines key MN fabrication techniques and discusses the rationale behind NC-loaded MNs. The common characterization methods used to evaluate drug loading, morphology, mechanical performance, and delivery efficiency are summarized. A comprehensive overview of recent studies is presented, highlighting successful applications across dermatological, infectious, ophthalmic, HIV-related, and neurodegenerative diseases. Collectively, the literature supports NC-loaded MNs as a patient-friendly approach that enables controlled dissolution, improved bioavailability, and sustained therapeutic levels. Overall, the integration of NC technology into MN systems represents a significant advancement in transdermal drug delivery, holding substantial potential to address limitations associated with low drug solubility and enhance therapeutic outcomes.

Keywords Microarray patches · Enhanced bioavailability · Nanomedicine · Nanosuspensions · Transdermal · Media milling

1 Introduction

Microneedles (MNs) are a minimally invasive delivery system that has emerged as a promising tool to overcome the skin's outer barrier for therapeutic and diagnostic purposes. Conceptually introduced by Gerstel and Place in 1976, MNs were

L. Lopez-Vidal · E. J. Catlin · R. F. Donnelly · A. J. Paredes (✉)
School of Pharmacy, Queen's University Belfast, Belfast, UK
e-mail: l.lopezvidal@qub.ac.uk; ecatlin01@qub.ac.uk; r.donnelly@qub.ac.uk; a.paredes@qub.ac.uk

A. J. Paredes (ed.), *Nanocrystals*, https://doi.org/10.1007/978-3-032-10238-6_6

originally envisioned to administer drugs percutaneously, laying the groundwork for modern microneedle-based technologies [1]. Structurally, MNs consist of micron-scale projections affixed to a supporting baseplate, designed to transiently disrupt the stratum corneum and create microchannels that allow for the passage of active pharmaceutical ingredients (APIs) [2]. This mechanism facilitates both transdermal and intradermal delivery while avoiding the limitations associated with conventional administration routes such as oral degradation or the invasiveness of hypodermic injections. Compared to traditional methods, MNs offer a unique balance between therapeutic efficacy and patient compliance. They bypass first-pass metabolism, reduce the need for trained personnel, and minimize the discomfort, anxiety, and poor adherence often associated with injectable formulations [3]. The micron-scale dimensions (typically 50–2000 μm in height) are sufficient to penetrate the skin's outer layers without stimulating dermal nerves due to the short length of the individual needles [4]. These versatile platforms are increasingly recognized as patient-friendly alternatives to conventional needles, offering both localized and systemic therapeutic effects without compromising comfort or safety. A plethora of benefits have been identified when using this promising technology, including reduction of infection transmission, a decrease in the prevalence of needle-stick injuries, and limited microbial access when paralleled to hypodermic needles [5]. The schematic and microscopic images in Fig. 1 provide a comprehensive visualization of MN array patches, highlighting their overall design functionality, giving great detail with respect to anatomical position during skin insertion and different geometric structures. Given their versatility, MNs have evolved into several distinct formats, each tailored to overcome specific challenges associated with drug delivery across the skin. The design and material composition of MNs play a crucial role in determining their performance, drug release kinetics, and patient acceptability. In the literature, five different types of MNs have been described, namely, solid, coated, dissolving, hollow and hydrogel-forming MNs [3, 6]. A clear understanding of the different types is essential to selecting the most appropriate system for a given therapeutic or diagnostic purpose.

2 Types of Microneedles

Over the past few decades, MNs technology has undergone significant advances, evolving from its earliest forms in the 1970s (such as solid and hollow metallic MNs) to the advanced systems available today. The introduction of coated MNs in the mid-1970s marked a pivotal step in drug delivery, with the active compound applied directly onto the MNs surface for controlled release. The development of dissolving MNs in 2010 further expanded the possibilities, providing a novel method for vaccine administration through a self-dissolving system [6]. This was followed by the emergence of hydrogel-forming MNs, enabling sustained drug release by swelling the needle material upon insertion into the skin. The idea was first described in 2008 [7] and first utilized in 2012 [4]. As it was mentioned before, various types of

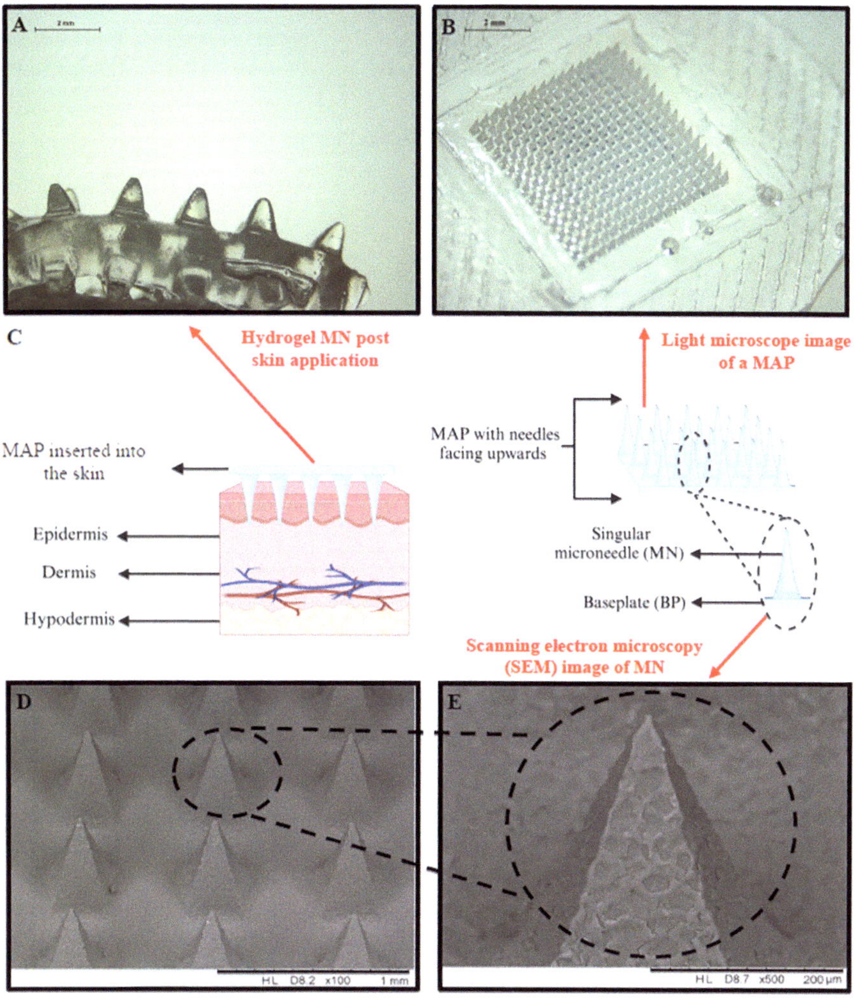

Fig. 1 A collection of schematic and microscopic images of MNs and MNs array patches. Light microscope image of a swollen hydrogel MAP post insertion into porcine skin (**a**), Light microscopic image of a dissolving MAP (**b**), Schematic illustration highlighting the structure of a MAP and position when inserted into skin (**c**), SEM image of hydrogel MAP (**d**), and SEM showing the sharp morphology of an individual MN (**e**). (With images in section C created Biorender.com 2024)

MNs have been developed, each with specific characteristics and applications for drug delivery (Fig. 2). Solid MNs are typically used to create micropores in the skin prior to the application of a topical formulation, in a two-step process known as "poke and patch." This type does not deliver the drug directly but facilitates its passive diffusion through the temporarily disrupted skin barrier [9]. Although solid MNs have been employed in studies for the slow diffusion of macromolecules such as bovine serum albumin, they are currently more widely used in dermatological

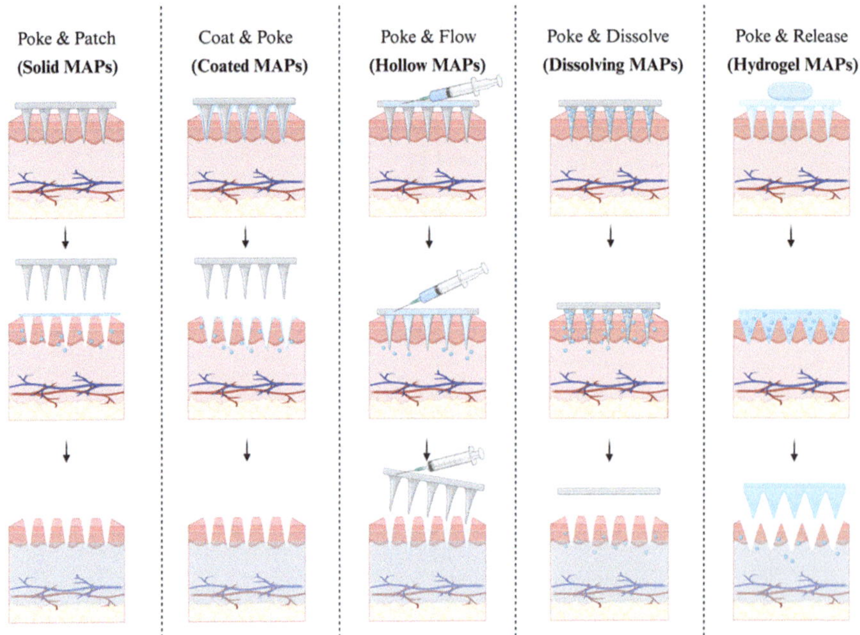

Fig. 2 Schematic illustration enabling visualization of the mode of action of drug delivery using solid, coated, hollow, dissolving, and hydrogel MNs, respectively. (This figure was reprinted from ref. [8] published under a Creative Commons Attribution 4.0 International (CC BY 4.0) license)

applications, such as facial rejuvenation, with commercial devices including Dermaroller® (Dermaroller GmbH, Wolfenbüttel, Germany) and Dermapen™ (DermapenWorld, Australia). Coated MNs, typically made from materials such as metal or silicon, are covered with a thin layer of drug on their surface. Upon insertion into the skin, this layer dissolves rapidly in the interstitial fluid, enabling the precise administration of low-dose, highly potent, water-soluble compounds [2]. This method, often referred to as "coat and poke," eliminates the need for multiple application steps and allows for immediate drug release upon skin penetration. This has been particularly advantageous for the delivery of highly potent, hydrophilic drugs that require low doses. However, its application is significantly constrained by the limited drug-loading capacity, typically less than 1 mg per patch, as only small amounts of formulation can adhere to the needle surface [10]. Additionally, achieving a uniform and reproducible coating across all microneedles presents a technical challenge, which can lead to variability in the dose delivered. Despite these limitations, coated MNs have been investigated extensively for the delivery of a variety of molecules, including small hydrophilic drugs, proteins, peptides, and genetic materials such as DNA [11, 12]. Examples reported in the literature include calcein, vitamin B, sulforhodamine, and different DNA constructs [13, 14]. One notable application was the development of Zolmitriptan-coated MNs for acute migraine treatment. Although long-term safety studies were completed in 2019, the project was ultimately discontinued following the failure of Zosano Pharma Corporation to gain FDA approval in 2022. Thus, while coated MNs present certain formulation and manufacturing challenges, they remain a promising platform for the transdermal delivery of potent therapeutic agents requiring precise low-dose administration. On the other hand, dissolving MNs are entirely composed of biodegradable materials such as sugars, carbohydrates or synthetic polymers, with the drug incorporated. Once inserted into the skin, the MNs dissolve in response to interstitial fluid, releasing the active compound in a controlled manner depending on the polymer composition [15]. This system allows single-step application, avoids potential needle stick injury, which prevents the spread of transmittable diseases, and removes the requirement of sharps disposal [16]. Designed with an internal channel, hollow MNs are similar to hypodermic needles, enabling the direct infusion of liquid drugs into the skin once the MNs have been inserted. Although they enable the administration of larger doses with flow control, they pose technical challenges such as risk of blockage and high manufacturing costs. Despite this, hollow MNs have been widely investigated for insulin delivery, achieving a precise and controllable dosage, as well as cosmetic applications. Although there are no hollow MNs approved for use by the FDA, interest for aesthetic means and drug delivery remains very prominent. A single-use attachable head, containing 3 silicone micro-cones of 0.6 mm in length, is produced by FillMed Laboratories to aid delivery of cosmetics to the dermis as well as a hollow microstructured transdermal system, of 12 individual needles of 1.5 mm, manufactured by 3 M, who are actively seeking partnership with pharmaceutical companies for preclinical studies [17, 18]. The latest development of hydrogel-forming MNs consists of crosslinked polymeric projections that are inserted in a dry, rigid state, and swell upon contact with interstitial fluid, forming a gel-like matrix that enables controlled drug release from an attached reservoir. The release rate in this system, as with dissolving microneedles, can be modulated by the degree of polymer crosslinking. However, its drug loading capacity is determined by the size of a separate reservoir rather than being confined to the needle volume. This distinction allows for higher payloads, overcoming a key limitation of dissolving MNs. While the aqueous hydrogel matrix limits its use with hydrophobic drugs due to poor diffusion, the system has shown effectiveness in delivering hydrophilic compounds such as metronidazole and theophylline and has been explored for diagnostic purposes through interstitial fluid extraction [15, 19].

3 Manufacture of Microneedles

The composition of MNs is subject to the design type and is integral to their intended function, influencing several key factors, such as drug loading capability, mechanical strength, dissolution rate, and biocompatibility, to name a few. The diversity of MN type and material enables a customized drug delivery approach, as depicted in Table 1, whereby the different therapeutic uses investigated are highlighted, coupled with the advantages and disadvantages of each respective category. Briefly, solid

Table 1 Overview of the different microneedle types detailing the different material components, functions, benefits, and limitations

Type of MN	Materials	Advantages	Disadvantages	Example applications	Ref.
Solid (poke and patch)	Silicon, polylactic acid, ceramic, stainless steel, titanium, nickel	Simple manufacture	Two-step application (less patient compliance)	Diphtheria toxoid, tetanus toxoid, potassium chloride, ovalbumin delivery	[20–24]
Coated (coat and poke)	Metal, silicon and polymer	Scalable manufacture (use of spray/dip-coating) Limited processing (avoid high temperature exposure) Quick delivery	Limited drug loading capacity (dose usually <1 mg) Uniformity of coating (issues with dosing variability) Limited release kinetics (immediate release profile only)	Fluorescein, vitamin B, Calcein or Sulforhodamine, paclitaxel, plasmid DNA, ovalbumin, siRNA, influenza vaccine and human delivery growth hormone	[14, 25–31]
Hollow (poke and flow)	Silicon, metal, glass, PVA	Large drug delivery amount (not limited to size of MNs) Controlled and consistent drug delivery	Complex application (driving force such as a syringe to contain formulation for delivery) Potential for blockage (delivery inaccuracy) Weaker needles amongst other types (intricate needle design and careful insertion) Precise manufacture techniques	Glucose measurement, insulin, methyl nicotine, Sulforhodamine solution, phenylephrine hydrochloride, Teriflunomide and ceftriaxone sodium.	[24, 32–38]
Dissolving (poke and dissolve)	Poly(vinylpyrrolidone) (PVP), hyaluronic acid, and Poly (methylvinylether co. maleic anhydride) (PMVE/MAH), PVA (9000–1000 MW) and PVP (K32/29), Chitosan, Maltose and hydroxypropyl methylcellulose (HMPC)	One-step application (patient friendly) Biocompatibility (deposited polymer can be safely excreted) Cost-efficient materials Tailored release profile (polymer composition will determine release rate)	Inability to sterilize (potential cross-contamination) Dose is limited to matrix inclusion	Deferasirox, simvastatin, risperidone, Etravirine, celecoxib, Triamcinolone acetonide, Itraconazole, Pramipexole, Flurbiprofen axetil, vitamin B12, influenza, Asiaticoside and near-IR photosensitizer	[39–53]

Hydrogel (poke and release)	Gantrez S-97 a co-polymer of poly(methylvinylether co. maleic acid) (PMVE/MA), (PMVE/MAH), PMVE/ MA + Na_2CO_3, methacrylate hyaluronic acid (MeHA), polyvinyl alcohol (PVA), poly 2-hydroxyethyl methacrylate (pHEMA)	High drug loading capacities and enhanced drug stability (use of external drug reservoirs) Biocompatibility (use of low-toxicity polymers) Personalized medicine (diagnostic and therapeutic response due to physiological environment and interstitial fluid extraction) Complete MN removal (no deposition of polymer in the skin)	Strict storage conditions (MNs must be protected from moisture to prevent premature swelling) Limitations of rate and degree of drug delivery (a burst release followed by a sustained release is often observed with delivery favoring hydrophilic water-soluble drugs	Delivery of fluorescein, FITC-dextran, doxorubicin, light-responsive ibuprofen conjugates, Albendazole, sildenafil citrate, rifampicin isoniazid, pyrazinamide, ethambutol delivery. Interstitial fluid testing, detection of caffeine, theophylline and glucose	[4, 54–64]

MNs are typically fabricated from metals and silicon, as such this design necessitates mechanical strength to puncture the skin and create micropores with the removal of intact needles for subsequent topical drug application. In addition to metals and silicon, polymers have also been employed to manufacture coated MNs, with the successful delivery dependent on the ability of the coating drug-containing layer to adhere to the individual MNs. The production of dissolving MNs utilizes biocompatible water-soluble materials, providing a one-step application process, to eliminate the requirement of sharps disposal while providing drug delivery in a controlled manner. As such, delivery can be specifically dictated based on the chosen polymer matrix and drug loading strategy, with release profiles dependent on the molecular weight, crosslinking, MNs design, and encapsulation of nanoparticles. Hollow MNs are tuneable devices, with variable material and fabrication methods, containing a hollow chamber to enable delivery of liquid formulations, into the epidermis and dermis, in an injectable fashion [65]. Due to the intricate geometry and requirement to deliver a precise amount of fluid, construction utilizing metal has been shown to be the most widely explored commercially, with products like Fillmed Nanosoft™ existing within the cosmetic field. Despite this, glass and polymers have also been used within literature, permitting visibility during delivery and ease of manufacture, respectively [66, 67]. Meanwhile, the manufacture of hydrogel MNs exploits the ability of hydrogels to possess sufficient strength to insert into the skin upon application, while harnessing the different cross-linking capabilities of said polymers to absorb interstitial fluid, swell, and form micro-conduits authorizing the passage of APIs beneath the skins surface, in a sustained controlled manner.

Due to the wide variety of MN types, composition materials, and intended use, it is not surprising that there is a broad range of production methods, including lithography and etching, injection molding, microemulsion, laser drilling, two-photon polymerization and 3D printing [6, 68]. Employment of laser cutting and laser ablation are two methods that have shown dominance to prepare solid, coated, and hollow metal MNs due to the scalability, simplicity, and precision of the MNs produced [69, 70]. Lithography and etching are primarily employed in the production of silicon MNs [6]. Photolithography, in particular, involves coating a silicon wafer with a photoresist material and exposing it to UV light (typically between 150 and 500 nm) through a mask to accurately transfer the desired patterns [71]. The oxide layer is then removed via etching, which can be either wet (isotropic or anisotropic) or dry (such as reactive ion etching or ion-beam milling), to control the uniformity of key parameters, such as height and sharpness, of the MNs produced [70]. Yet, for the last 10 years, the man-made polymer silicone, derived from silicon, has been extensively researched to fabricate negative master molds, to prepare both dissolving and hydrogel MNs, as opposed to the starting material of the MN itself [48, 70, 72, 73]. Following this, dissolving MNs, which show continuing promise in the drug delivery field with adaptable release centric to formulation parameters, can be produced in simple manner, utilizing micro molding with negative master molds, providing an affordable flexible means to prepare MNs of high consistency [74]. In effort to reduce the multiple processing steps which accompany techniques such as microemulsion, application of 3D-printing has been frontiered, yet limitations in ability to achieve high resolution, exclusivity of formulation materials, and high cost of the printer, at present hinders the viability of this option [75].

4 Nanocrystal-loaded Microneedles

The integration of nanocrystal (NC) technology with MN systems has emerged as a promising strategy to overcome the delivery limitations of poorly water-soluble drugs. A significant proportion of new drug candidates and marketed drugs exhibit high molecular weight and elevated Log P values, making them poorly soluble in aqueous environments [76]. NCs offer a practical and efficient means to enhance the apparent solubility and dissolution rate of these challenging compounds. When incorporated into MN patches, NCs can be directly delivered across the *stratum corneum*, bypassing gastrointestinal and hepatic first-pass barriers. This approach not only improves the bioavailability of poorly soluble drugs but also enables localized or systemic delivery with enhanced pharmacokinetic control. MNs, particularly those that dissolve or form hydrogels upon insertion, are well-suited to incorporate drug NCs due to their minimally invasive nature, ability to encapsulate solid drug forms, and capacity for controlled release. By combining the solubility-enhancing effect of NCs with the efficient skin-penetration capability of MNs, this hybrid system holds significant potential for the transdermal delivery of Biopharmaceutics Classification System (BCS) Class II and IV drugs. To date, NCs have been encapsulated within biocompatible polymers to form dissolving MNs or formulated as liquid suspensions for administration through hollow MNs [77, 78]. As such, the materials used in MNs designed for NC delivery are typically composed of various polymers and, in some cases metals, depending on the specific MN type and drug release requirements.

4.1 Characterization Techniques

The fabrication of NC-loaded MNs involves two main steps: first, the development of the NC formulation, and subsequently, their incorporation into the MNs. The comprehensive evaluation of MNs incorporating NCs typically involves a two-phase characterization approach. Initially, the NC formulation itself is subjected to thorough solid-state and physicochemical analysis to confirm its stability, morphology, and other relevant properties. These techniques have been discussed in Chap. 1. Once the NCs have been integrated into the MN arrays, the combined drug-device system is further assessed. This includes determining the drug loading, evaluating any changes in particle size following fabrication, physicochemical characterization, microscopic analysis, examining mechanical strength, insertion efficiency, ex vivo permeation, and stability under various conditions, among others. The most commonly employed techniques for the characterization of NCs-loaded MNs are discussed below.

– Particle Size and Surface Charge

The process of loading NCs into MNs may potentially impact key nanoparticle characteristics such as particle size, polydispersity index (PDI), and zeta potential (ZP), which can, in turn, influence critical properties such as the saturation solubility and dissolution rate of the active compound. Therefore, it is essential to assess these parameters after the fabrication of the NCs-loaded MNs. For this purpose, the MNs are dissolved in water to an appropriate concentration under magnetic stirring, and the resulting dispersions are analyzed for changes in hydrodynamic diameter (particle size), PDI, and ZP using dynamic light scattering (DLS). The particle size obtained via DLS corresponds to the hydrodynamic diameter and is reported as the z-average, an intensity-weighted mean diameter of the main particle population. The PDI reflects the breadth of the size distribution. A lack of significant variation in particle size after incorporation into MNs suggests that the NCs are stabilized within the dried polymeric matrix of the MNs.

– Drug Content

To evaluate the drug content, NCs-loaded MNs are dissolved in a suitable solvent (e.g., a water:ethanol mixture). The selected solvent system should allow the polymer to precipitate while keeping the drug in solution. The resulting mixture is then centrifuged at high speed, and a portion of the supernatant is collected and analyzed using an appropriate quantification technique, typically high-performance liquid chromatography (HPLC) [46].

– Crystalline State of the NCs Loaded in the MNs

To investigate whether the crystalline state of the NCs is altered during the fabrication of MNs, techniques such as differential scanning calorimetry (DSC) or X-ray diffractometry (XRD) can be employed. This involves comparing the thermal or diffraction profiles of the NC-loaded MNs with those of the NCs alone. Such analysis is important, as any change in the crystalline state may not only impact the stability of the nanoparticles but also influence their dissolution rate [78, 79].

– Mechanical and Insertion Properties

Assessing the mechanical properties of NC-loaded MNs is essential, as the success of these systems in drug delivery largely depends on their ability to penetrate biological tissue. Consequently, it is critical that MN arrays possess adequate rigidity to enable effective insertion into the skin [80]. These properties can be evaluated using a texture analyzer. In the described method, the MN array is affixed to a cylindrical probe with a cross-sectional area of 1.5 cm^2. The probe is then driven vertically downwards at a constant speed of 0.5 mm/s using the texture analyzer, compressing the MNs against a flat aluminum block. A force of 32 N is applied for 30 seconds before the probe is retracted. The height of individual MNs is measured before and after compression using a Leica EZ4W stereomicroscope equipped with an integrated camera [48, 81]. The percentage reduction in needle height is then calculated and reported as an indicator of structural integrity under applied force. To evaluate insertion properties, the same experimental set-up is used (speed: 0.5 mm/s; force: 32 N). However, instead of an aluminum surface, the MNs are compressed against a previously validated artificial skin simulant composed of eight stacked layers of Parafilm M® [82]. Following insertion, the layers are carefully separated, and the number of perforations in each layer is counted using the Leica EZ4W stereomicroscope to determine the insertion efficiency of the MNs.

– Imaging Techniques

Microscopic evaluation of NC-loaded MNs provides valuable insights into the morphology, uniformity, and structural integrity of the arrays, which are essential parameters for ensuring consistent drug loading, mechanical performance, and reproducible insertion. Low-magnification imaging can be performed using a digital microscope to assess general shape, tip sharpness, and dimensional uniformity across the MN array. These images allow rapid screening of fabrication quality and potential defects, such as bending, incomplete mound filling, or tip deformation [81]. For higher-resolution surface characterization, scanning electron microscopy (SEM) is employed. SEM enables detailed visualization of individual MNs, including surface roughness, porosity, and potential drug crystal deposition when NCs are present on the surface. Imaging is typically performed on dried samples under high vacuum and requires sputter-coating with a conductive layer (e.g., gold or platinum) to prevent charging artefacts [45]. These microscopic techniques are particularly useful for evaluating the impact of formulation components and processing parameters on the final geometry and surface properties of MNs, and they complement mechanical and insertion tests by providing structural context to the observed performance.

Additionally, optical coherence tomography has been utilized to produce cross-sectional images, detailing the penetration depth and in situ dissolution characteristics of MNs, as a result of differences in the interference of reflected light waves [83]. This real-time nondestructive technique permits axial and lateral realization of the micrometer range with imaging depths of about 2.0 mm, which is nearly 10x that of confocal microscopy [83, 84]. This technique has been successfully utilized in vivo to assess the dynamic insertion process of MNs in the skin, during and post application, as well as extensively in vitro, for quantitative assessment of penetration, comparison of MN designs, and visualization of the dissolution behavior of MNs [82, 85].

– Ex vivo Skin Permeation and Deposition

Ex vivo skin models are widely employed to evaluate the ability of NC-loaded MNs to enhance drug permeation and deposition across the skin. These studies typically use excised animal skin mounted on vertical Franz diffusion cells, which allow quantification of the drug that permeates through the skin into a receptor medium, as well as the amount retained within the tissue. Across the literature, the use of porcine skin as a model of human skin has been extensively explored due to parallels in anatomy with regard to the thickness of the epidermis and dermal layer, which is tightly bound to subcutaneous connective tissue [86]. Two types of skin preparations are commonly utilized: full-thickness skin and dermatomed skin. Full-thickness skin includes the entire epidermis and dermis, along with a variable portion of subcutaneous tissue. In contrast, dermatomed skin is processed using a dermatome to obtain a uniform thickness (typically 500–800 μm), preserving the epidermis and superficial dermis while removing most of the subcutaneous layer. During a typical experiment, MNs are applied to the skin surface manually, which is adhered to the donor compartment. The receiver compartment is then filled with phosphate-buffered saline (PBS) solution or another suitable medium, imitating the physiological environment, to provide a sink which sustains a concentration gradient and drives diffusion. As such, the medium is maintained at physiological temperature (37 ± 1 °C) and continuously stirred. Subsequently, each Franz cell is wrapped in Parafilm M® to create a sealed environment as to prevent evaporation. At specific predetermined time points, samples are typically obtained over a 24-hour period to evaluate the skin permeation. At the end of the experiment, the MNs are removed from the skin, and the skin is carefully homogenized [87] to quantify the amount of drug deposited in the skin [88]. The amount of drug permeated and deposited is quantified using suitable analytical techniques, typically HPLC.

4.2 *Applications of Nanocrystal-Loaded Microneedles in Drug Delivery*

Several research groups have demonstrated the therapeutic potential of incorporating NCs into MNs systems for both local and systemic drug delivery. In dermatology, methotrexate NCs of 678 ± 15 nm delivered via dissolving MNs resulted in sustained intradermal retention for up to 72 hours, supporting the development of on-demand psoriatic flare therapies [81]. Similarly, itraconazole NCs [46] and tacrolimus NCs [89] exhibited improved dermatokinetic profiles and enhanced antifungal or anti-inflammatory activity when administered with MNs compared to conventional creams. Metronidazole NCs of less than 200 nm incorporated in dissolving MNs demonstrated low cytotoxicity in cell line 3T3L1, and when evaluated in a 24-hour ex vivo study, drug delivery efficiency reached 95% with MNs, compared to 0% with needle-free patches [90]. Ocular delivery using MNs loaded with triamcinolone acetonide NCs (246.65 ± 8.55 nm) significantly enhanced drug permeation through the sclera, achieving better tissue penetration (4.5 times) than coarse suspensions [80]. In the context of HIV therapy, rilpivirine and cabotegravir NCs were incorporated into dissolving MNs as self-administered alternatives to injectable long-acting formulations. These systems provided extended plasma exposure with improved patient convenience [44, 91]. More recent studies focused on pediatric-friendly platforms [92], brain-targeting delivery [77], and lymphatic system targeting [16], each demonstrating sustained drug levels well above therapeutic thresholds. To address parasitic infections, NCs of ivermectin, ricobendazole, doxycycline [93], and albendazole [94] were formulated into MNs to improve local drug retention and bypass hepatic metabolism. Volpe-Zanutto et al. (2021) developed NC-loaded MNs containing artemether and lumefantrine for malaria treatment, achieving NCs sizes below 200 nm suitable for polymeric encapsulation [79]. Recently, the intradermal delivery of amphotericin B (AmB) NCs via dissolving MNs was investigated by Wang et al. (2025) for cutaneous leishmaniasis [95]. AmB-NCs, prepared by media milling with a mean particle size of 108 ± 4 nm, showed improved dissolution and potent antileishmanial activity in vitro with reduced host cell toxicity. Ex vivo studies confirmed efficient intradermal deposition (247 ± 25 μg/0.49 cm^2), while in vivo rat studies demonstrated significantly higher AmB accumulation in axillary and lumbar lymph nodes (3.2-fold and 2.8-fold, respectively) compared to the commercial intravenous formulation AmBisome® ($p < 0.05$), alongside reduced off-target organ distribution (Fig. 3).

In neurodegenerative disease management, Telmisartan NCs (<140 nm) were incorporated into MNs to enhance delivery to the brain and reduce neuroinflammation in an Alzheimer's model, resulting in significantly improved plasma and brain bioavailability compared to oral and patch controls [96]. Likewise, MNs loaded with Rotigotine NCs (274.09 ± 7.43 nm) provided therapeutically relevant levels for 48 h following a single administration, outperforming commercial transdermal patches in both efficiency and size [73]. Collectively, these studies highlight the versatility and clinical promise of NCs-loaded MNs as minimally invasive, high-efficiency drug delivery platforms for a wide range of therapeutic areas, including inflammatory, infectious, neurological, and chronic diseases.

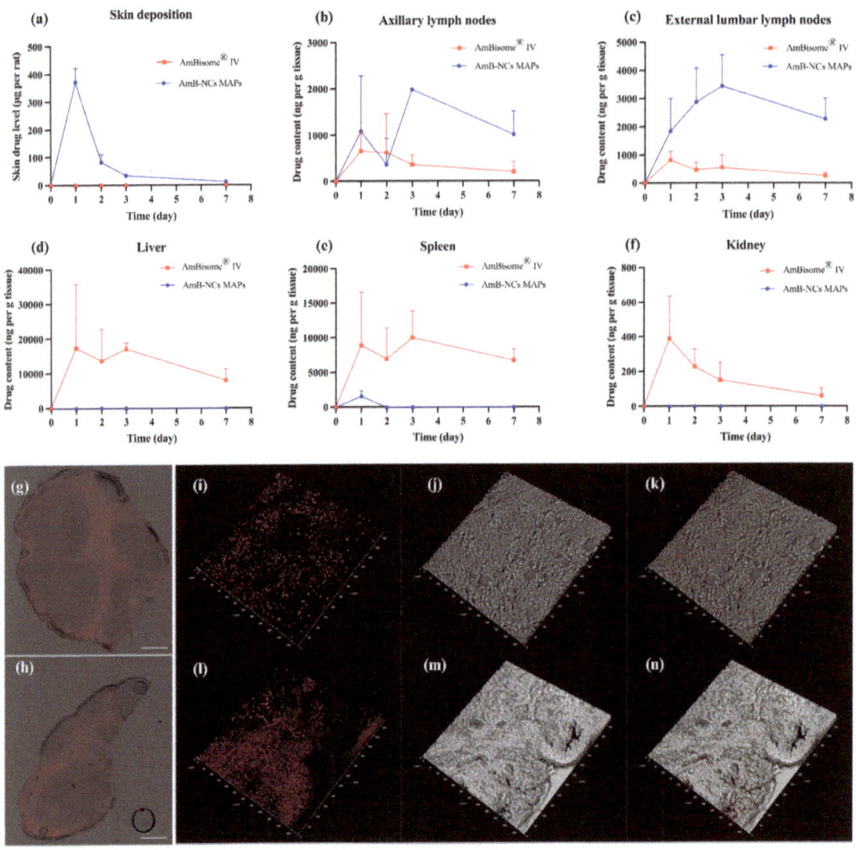

Fig. 3 In vivo (**a**) intradermal drug deposition, (**b**) axillary lymphatic concentrations, (**c**) external lumber lymphatic concentrations (**d**) liver concentrations, (**e**) spleen concentrations, and (**f**) kidney concentrations after application of AmB-NCs loaded MNs and AmBisome® IV injection (Means ± S.D., $n = 4$). Representative confocal of in vivo (**g**) axillary lymph node section and (**h**) external lumber lymph node section after application of AmB-NCs MAPs, where API was observed under mCherry channel in red (Scale bar:1 mm). Representative confocal Z-stack 3D images of in vivo axillary lymph node section (**i**) API (**j**) lymph node tissue (**k**) merge image and external lumber lymph node section (**l**) API (**m**) lymph node tissue (**n**) merge image after application of AmB-NCs loaded MNs (Z-depth: 20 µm). (Reprinted from reference [95]. Copyright© (2025), with permission from Elsevier)

5 Conclusion

The integration of NC technology with MN systems represents a highly promising advancement in transdermal drug delivery, particularly for poorly water-soluble drugs. By combining the enhanced solubility and dissolution rates afforded by NCs with the minimally invasive and efficient skin penetration capabilities of MNs, this hybrid platform offers significant potential to overcome key challenges in drug bioavailability and therapeutic efficacy.

References

1. M.S. Gerstel, V.A. Place, Drug delivery device (1971)
2. E. Larrañeta, R.E.M. Lutton, A.D. Woolfson, R.F. Donnelly, Microneedle arrays as transdermal and intradermal drug delivery systems: Materials science, manufacture and commercial development. Mater. Sci. Eng. R Rep. **104**, 1–32 (2016). https://doi.org/10.1016/J.MSER.2016.03.001
3. L. Lopez-Vidal, K. Juskaite, I.K. Ramöller, D.A. Real, P.E. McKenna, J. Priotti, R.F. Donnelly, A.J. Paredes, Advanced drug delivery systems for the management of local conditions. Ther. Deliv. (2024). https://doi.org/10.1080/20415990.2024.2437978/ASSET/363BFEEC-2364-4ECA-A354-8CADB7A1C07B/ASSETS/GRAPHIC/ITDE_A_2437978_F0004_OC.JPG
4. R.F. Donnelly, T.R.R. Singh, M.J. Garland, K. Migalska, R. Majithiya, C.M. McCrudden, P.L. Kole, T.M.T. Mahmood, H.O. McCarthy, A.D. Woolfson, Hydrogel-forming microneedle arrays for enhanced transdermal drug delivery. Adv. Funct. Mater. **22**, 4879–4890 (2012). https://doi.org/10.1002/ADFM.201200864
5. S.H. Bariya, M.C. Gohel, T.A. Mehta, O.P. Sharma, Microneedles: An emerging transdermal drug delivery system. J. Pharm. Pharmacol. **64**, 11–29 (2012). https://doi.org/10.1111/J.2042-7158.2011.01369.X
6. E. Larrañeta, M.T.C. McCrudden, A.J. Courtenay, R.F. Donnelly, Microneedles: A new frontier in nanomedicine delivery. Pharm. Res. **33**, 1055–1073 (2016). https://doi.org/10.1007/S11095-016-1885-5
7. R. Donnelly, P. McCarron, D. Morrow, A. Morrissey, D. Woolfson, Microneedles/Delivery Device and method (Patent No. WO2009040548A1), 2008
8. E.J. Catlin, L. Lopez-Vidal, R.F. Donnelly, A.J. Paredes, A dynamic duo comes of age: nanocrystals and microneedles for hydrophobic drug delivery. Expert Opin. Drug Deliv. (2025). https://doi.org/10.1080/17425247.2025.2531059
9. M.R. Prausnitz, Microneedles for transdermal drug delivery. Adv. Drug Deliv. Rev. **56**, 581–587 (2004). https://doi.org/10.1016/J.ADDR.2003.10.023
10. Y.C. Kim, J.H. Park, M.R. Prausnitz, Microneedles for drug and vaccine delivery. Adv. Drug Deliv. Rev. **64**, 1547–1568 (2012). https://doi.org/10.1016/J.ADDR.2012.04.005
11. Y.C. Kim, F.S. Quan, R.W. Compans, S.M. Kang, M.R. Prausnitz, Formulation of microneedles coated with influenza virus-like particle vaccine. AAPS PharmSciTech **11**, 1193–1201 (2010). https://doi.org/10.1208/S12249-010-9471-3/FIGURES/7
12. A. Vrdoljak, M.G. McGrath, J.B. Carey, S.J. Draper, A.V.S. Hill, C. O'Mahony, A.M. Crean, A.C. Moore, Coated microneedle arrays for transcutaneous delivery of live virus vaccines. J. Control. Release **159**, 34–42 (2012). https://doi.org/10.1016/j.jconrel.2011.12.026
13. X. He, J. Sun, J. Zhuang, H. Xu, Y. Liu, D. Wu, Microneedle system for transdermal drug and vaccine delivery: devices, safety, and prospects. Dose-Response **17**(4), 1559325819878585 (2019)
14. J. Jiang, H.S. Gill, D. Ghate, B.E. McCarey, S.R. Patel, H.F. Edelhauser, M.R. Prausnitz, Coated microneedles for drug delivery to the eye. Invest. Ophthalmol. Vis. Sci. **48**, 4038–4043 (2007). https://doi.org/10.1167/IOVS.07-0066
15. T.M. Tuan-Mahmood, M.T.C. McCrudden, B.M. Torrisi, E. McAlister, M.J. Garland, T.R.R. Singh, R.F. Donnelly, Microneedles for intradermal and transdermal drug delivery. Eur. J. Pharm. Sci. **50**, 623–637 (2013). https://doi.org/10.1016/J.EJPS.2013.05.005
16. I.K. Ramöller, F. Volpe-Zanutto, L.K. Vora, M.T.A. Abbate, A.R.J. Hutton, P.E. McKenna, K. Peng, I.A. Tekko, A. Sabri, E. McAlister, H.O. McCarthy, A.J. Paredes, R.F. Donnelly, Intradermal delivery of the antiretroviral drugs cabotegravir and rilpivirine by dissolving microarray patches: Investigation of lymphatic uptake. J. Control. Release **366**, 548–566 (2024). https://doi.org/10.1016/J.JCONREL.2024.01.010
17. R. Dell'Avanzato, A. Gualdi, E. Vescovi, M. Basso, F. Marchetti, L. Lanfranchi, A. Redaelli, M. Vitale, Facial rejuvenation using a microneedle-based device with a revitalizing solution and

free hyaluronic acid. J. Cosmet. Dermatol. **20**, 3701–3703 (2021). https://doi.org/10.1111/JOCD.14518
18. S.A. Burton, C.Y. Ng, R. Simmers, C. Moeckly, D. Brandwein, T. Gilbert, N. Johnson, K. Brown, T. Alston, G. Prochnow, K. Siebenaler, K. Hansen, Rapid intradermal delivery of liquid formulations using a hollow microstructured array. Pharm. Res. **28**, 31–40 (2011). https://doi.org/10.1007/S11095-010-0177-8/FIGURES/9
19. J.G. Turner, L.R. White, P. Estrela, H.S. Leese, Hydrogel-forming microneedles: Current advancements and future trends. Macromol. Biosci. **21**, 2000307 (2021). https://doi.org/10.1002/MABI.202000307;WGROUP:STRING:PUBLICATION
20. N. Wilke, A. Mulcahy, S.R. Ye, A. Morrissey, Process optimization and characterization of silicon microneedles fabricated by wet etch technology. Microelectron. J. **36**, 650–656 (2005). https://doi.org/10.1016/J.MEJO.2005.04.044
21. R.F. Donnelly, T.R.R. Singh, M.M. Tunney, D.I.J. Morrow, P.A. McCarron, C. O'Mahony, A.D. Woolfson, Microneedle arrays allow lower microbial penetration Than hypodermic needles in vitro. Pharm. Res. **26**, 2513 (2009). https://doi.org/10.1007/S11095-009-9967-2
22. Q.Y. Li, J.N. Zhang, B.Z. Chen, Q.L. Wang, X.D. Guo, A solid polymer microneedle patch pretreatment enhances the permeation of drug molecules into the skin. RSC Adv. **7**, 15408–15415 (2017). https://doi.org/10.1039/C6RA26759A
23. A.M. de Groot, A.C. Anouk, N. Kuijt, P.J.S. van Kooten, P.J. Vos, A.J.A.M. Sijts, K. van der Maaden, Nanoporous microneedle arrays effectively induce antibody responses against diphtheria and tetanus toxoid. Front. Immunol. **8**, 309185 (2017). https://doi.org/10.3389/FIMMU.2017.01789/BIBTEX
24. D.V. McAllister, P.M. Wang, S.P. Davis, J.H. Park, P.J. Canatella, M.G. Allen, M.R. Prausnitz, Microfabricated needles for transdermal delivery of macromolecules and nanoparticles: Fabrication methods and transport studies. Proc. Natl. Acad. Sci. **100**, 13755–13760 (2003). https://doi.org/10.1073/PNAS.2331316100
25. H.S. Gill, M.R. Prausnitz, Coated microneedles for transdermal delivery. J. Control. Release **117**, 227 (2006). https://doi.org/10.1016/J.JCONREL.2006.10.017
26. K.J. Lee, S.H. Park, J.Y. Lee, H.C. Joo, E.H. Jang, Y.N. Youn, W. Ryu, Perivascular biodegradable microneedle cuff for reduction of neointima formation after vascular injury. J. Control. Release **192**, 174–181 (2014). https://doi.org/10.1016/j.jconrel.2014.07.007
27. G. Widera, J. Johnson, L. Kim, L. Libiran, K. Nyam, P.E. Daddona, M. Cormier, Effect of delivery parameters on immunization to ovalbumin following intracutaneous administration by a coated microneedle array patch system. Vaccine **24**, 1653–1664 (2006). https://doi.org/10.1016/J.VACCINE.2005.09.049
28. R.H.E. Chong, E. Gonzalez-Gonzalez, M.F. Lara, T.J. Speaker, C.H. Contag, R.L. Kaspar, S.A. Coulman, R. Hargest, J.C. Birchall, Gene silencing following siRNA delivery to skin via coated steel microneedles: In vitro and in vivo proof-of-concept. J. Control. Release **166**, 211–219 (2013). https://doi.org/10.1016/J.JCONREL.2012.12.030
29. Y.C. Kim, F.S. Quan, D.G. Yoo, R.W. Compans, S.M. Kang, M.R. Prausnitz, Enhanced memory responses to seasonal H1N1 influenza vaccination of the skin with the use of vaccine-coated microneedles. J. Infect. Dis. **201**, 190–198 (2010). https://doi.org/10.1086/649228
30. M. Pearton, V. Saller, S.A. Coulman, C. Gateley, A.V. Anstey, V. Zarnitsyn, J.C. Birchall, Microneedle delivery of plasmid DNA to living human skin: Formulation coating, skin insertion and gene expression. J. Control. Release **160**, 561–569 (2012). https://doi.org/10.1016/J.JCONREL.2012.04.005
31. M. Ameri, M. Kadkhodayan, J. Nguyen, J.A. Bravo, R. Su, K. Chan, A. Samiee, P.E. Daddona, Human growth hormone delivery with a microneedle transdermal system: Preclinical formulation, stability, delivery and PK of therapeutically relevant doses. Pharmaceutics **6**, 220–234 (2014). https://doi.org/10.3390/PHARMACEUTICS6020220

32. J. Gupta, E.I. Felner, M.R. Prausnitz, Minimally invasive insulin delivery in subjects with type 1 diabetes using hollow microneedles. Diabetes Technol. Ther. **11**, 329 (2009). https://doi.org/10.1089/DIA.2008.0103
33. R.K. Sivamani, B. Stoeber, G.C. Wu, H. Zhai, D. Liepmann, H. Maibach, Clinical microneedle injection of methyl nicotinate: Stratum corneum penetration. Skin Res. Technol. **11**, 152–156 (2005). https://doi.org/10.1111/J.1600-0846.2005.00107.X
34. S.P. Davis, W. Martanto, M.G. Allen, M.R. Prausnitz, Hollow metal microneedles for insulin delivery to diabetic rats. IEEE Trans. Biomed. Eng. **52**, 909–915 (2005). https://doi.org/10.1109/TBME.2005.845240
35. J. Jiang, J.S. Moore, H.F. Edelhauser, M.R. Prausnitz, Intrascleral drug delivery to the eye using hollow microneedles. Pharm. Res. **26**, 395–403 (2009). https://doi.org/10.1007/S11095-008-9756-3/FIGURES/7
36. C.Y. Lee, K. Lee, Y.S. You, S.H. Lee, H. Jung, Tower microneedle via reverse drawing lithography for innocuous Intravitreal drug delivery. Adv. Healthc. Mater. **2**, 812–816 (2013). https://doi.org/10.1002/ADHM.201200239
37. H. Abd-El-Azim, H. Abbas, N.S. El Sayed, A.M. Fayez, M. Zewail, Non-invasive management of rheumatoid arthritis using hollow microneedles as a tool for transdermal delivery of teriflunomide loaded solid lipid nanoparticles. Int. J. Pharm. **644**, 123334 (2023). https://doi.org/10.1016/J.IJPHARM.2023.123334
38. S. Kawre, P. Suryavanshi, D.S. Lalchandani, M.K. Deka, P. Kumar Porwal, S. Kaity, S. Roy, S. Banerjee, Bioinspired labrum-shaped stereolithography (SLA) assisted 3D printed hollow microneedles (HMNs) for effectual delivery of ceftriaxone sodium. Eur. Polym. J. **204**, 112702 (2024). https://doi.org/10.1016/J.EURPOLYMJ.2023.112702
39. L.E. Moore, S. Vucen, A.C. Moore, Trends in drug- and vaccine-based dissolvable microneedle materials and methods of fabrication. Eur. J. Pharm. Biopharm. **173**, 54–72 (2022). https://doi.org/10.1016/J.EJPB.2022.02.013
40. M.B. McGuckin, A.R.J. Hutton, E.R. Davis, A.H.B. Sabri, A. Ripolin, A. Himawan, Y.A. Naser, R. Ghanma, B. Greer, H.O. McCarthy, A.J. Paredes, E. Larrañeta, R.F. Donnelly, Transdermal delivery of pramipexole using microneedle technology for the potential treatment of Parkinson's disease. Mol. Pharm. **21**, 2512–2533 (2024). https://doi.org/10.1021/acs.molpharmaceut.4c00065
41. X. Yang, W. Cao, X. Gu, L. Zheng, Q. Wang, Y. Li, F. Wei, T. Ma, L. Zhang, Q. Wang, Simvastatin nanocrystals-based dissolving microneedles for wound healing. Int. J. Pharm. **647** (2023). https://doi.org/10.1016/j.ijpharm.2023.123543
42. H.S. Faizi, L.K. Vora, M.I. Nasiri, Y. Wu, D. Mishra, Q.K. Anjani, A.J. Paredes, R.R.S. Thakur, M.U. Minhas, R.F. Donnelly, Deferasirox nanosuspension loaded dissolving microneedles for intradermal delivery. Pharmaceutics **14** (2022). https://doi.org/10.3390/pharmaceutics14122817
43. R. Ghanma, Y.A. Naser, Q. Kurnia Anjani, et al., Dissolving microarray patches for transdermal delivery of risperidone for schizophrenia management. Int. J. Pharm. **660** (2024). https://doi.org/10.1016/j.ijpharm.2024.124342
44. S. Rojekar, L.K. Vora, I.A. Tekko, F. Volpe-Zanutto, H.O. McCarthy, P.R. Vavia, R.F. Ryan, Etravirine-loaded dissolving microneedle arrays for long-acting delivery. Eur. J. Pharm. Biopharm. **165**, 41–51 (2021). https://doi.org/10.1016/j.ejpb.2021.04.024
45. Q. Wang, X. Yang, X. Gu, F. Wei, W. Cao, L. Zheng, Y. Li, T. Ma, C. Wu, Q. Wang, Celecoxib nanocrystal-loaded dissolving microneedles with highly efficient for osteoarthritis treatment. Int. J. Pharm. **625** (2022). https://doi.org/10.1016/j.ijpharm.2022.122108
46. A.D. Permana, A.J. Paredes, F. Volpe-Zanutto, Q.K. Anjani, E. Utomo, R.F. Donnelly, Dissolving microneedle-mediated dermal delivery of itraconazole nanocrystals for improved treatment of cutaneous candidiasis. Eur. J. Pharm. Biopharm. **154**, 50–61 (2020). https://doi.org/10.1016/j.ejpb.2020.06.025

47. J. Xie, X. Zhu, M. Wang, C. Liu, G. Ling, P. Zhang, Dissolving microneedle-mediated transdermal delivery of flurbiprofen axetil-loaded pH-responsive liposomes for arthritis treatment. Chem. Eng. J. **482**, 148840 (2024). https://doi.org/10.1016/J.CEJ.2024.148840
48. I.K. Ramöller, I.A. Tekko, H.O. McCarthy, R.F. Donnelly, Rapidly dissolving bilayer microneedle arrays – a minimally invasive transdermal drug delivery system for vitamin B12. Int. J. Pharm. **566**, 299–306 (2019). https://doi.org/10.1016/J.IJPHARM.2019.05.066
49. W. Disphanurat, N. Sivapornpan, B. Srisantithum, J. Leelawattanachai, Efficacy of a triamcinolone acetonide-loaded dissolving microneedle patch for the treatment of hypertrophic scars and keloids: a randomized, double-blinded, placebo-controlled split-scar study. Arch. Dermatol. Res. **315**, 989–997 (2023). https://doi.org/10.1007/S00403-022-02473-6/FIGURES/3
50. H. Huang, Y. Shen, X. Yang, C. Hou, X. Ke, R. Yang, Dissolvable microneedles loaded with asiaticoside nanocrystals stabilized by Panax Notoginseng saponins for hypertrophic scar treatment. J. Drug Deliv. Sci. Technol. **98** (2024). https://doi.org/10.1016/j.jddst.2024.105854
51. I.M.N. Hamdan, I.A. Tekko, K.B. Matchett, L.G. Arnaut, C.S. Silva, H.O. McCarthy, R.F. Donnelly, Intradermal delivery of a near-infrared photosensitizer using dissolving microneedle arrays. J. Pharm. Sci. **107**, 2439–2450 (2018). https://doi.org/10.1016/J.XPHS.2018.05.017/ASSET/80C33D8B-D056-41FD-AA20-0C14C11D1C6D/MAIN.ASSETS/GR10.JPG
52. M.C. Chen, M.H. Ling, K.Y. Lai, E. Pramudityo, Chitosan microneedle patches for sustained transdermal delivery of macromolecules. Biomacromolecules **13**, 4022–4031 (2012). https://doi.org/10.1021/BM301293D/ASSET/IMAGES/MEDIUM/BM-2012-01293D_0013.GIF
53. B. Chanabodeechalermrung, T. Chaiwarit, S. Chaichit, S. Udomsom, P. Baipaywad, P. Worajittiphon, P. Jantrawut, HPMC/PVP K90 dissolving microneedles fabricated from 3D-printed master molds: Impact on microneedle morphology, mechanical strength, and topical dissolving property. Polymers (Basel) **16**, 452 (2024). https://doi.org/10.3390/POLYM16040452/S1
54. J.G. Hardy, E. Larrañeta, R.F. Donnelly, N. McGoldrick, K. Migalska, M.T.C. McCrudden, N.J. Irwin, L. Donnelly, C.P. McCoy, Hydrogel-forming microneedle arrays made from light-responsive materials for on-demand transdermal drug delivery. Mol. Pharm. **13**, 907–914 (2016). https://doi.org/10.1021/ACS.MOLPHARMACEUT.5B00807/ASSET/IMAGES/LARGE/MP-2015-00807W_0007.JPEG
55. R. He, Y. Niu, Z. Li, A. Li, H. Yang, F. Xu, F. Li, A hydrogel microneedle patch for point-of-care testing based on skin interstitial fluid. Adv. Healthc. Mater. **9**, 1901201 (2020). https://doi.org/10.1002/ADHM.201901201
56. M. Han, D.K. Kim, S.H. Kang, H.R. Yoon, B.Y. Kim, S.S. Lee, K.D. Kim, H.G. Lee, Improvement in antigen-delivery using fabrication of a grooves-embedded microneedle array. Sens Actuators B Chem **137**, 274–280 (2009). https://doi.org/10.1016/J.SNB.2008.11.017
57. R.F. Donnelly, M.T.C. McCrudden, A.Z. Alkilani, E. Larrañeta, E. McAlister, A.J. Courtenay, M.C. Kearney, T.R. Raj Singh, H.O. McCarthy, V.L. Kett, E. Caffarel-Salvador, S. Al-Zahrani, A.D. Woolfson, Hydrogel-forming microneedles prepared from "super swelling" polymers combined with lyophilised wafers for transdermal drug delivery. PLoS One **9** (2014). https://doi.org/10.1371/JOURNAL.PONE.0111547
58. H. Chang, M. Zheng, X. Yu, A. Than, R.Z. Seeni, R. Kang, J. Tian, D. Phan Khanh, L. Liu, P. Chen, C. Xu, H. Chang, M. Zheng, A. Than, R.Z. Seeni, R. Kang, J. Tian, D.P. Khanh, P. Chen, C. Xu, X. Yu, L. Liu, A Swellable microneedle patch to rapidly extract skin interstitial fluid for timely metabolic analysis. Adv. Mater. **29**, 1702243 (2017). https://doi.org/10.1002/ADMA.201702243
59. E.M. Migdadi, A.J. Courtenay, I.A. Tekko, M.T.C. McCrudden, M.C. Kearney, E. McAlister, H.O. McCarthy, R.F. Donnelly, Hydrogel-forming microneedles enhance transdermal delivery of metformin hydrochloride. J. Control. Release **285**, 142–151 (2018). https://doi.org/10.1016/j.jconrel.2018.07.009

60. E. Caffarel-Salvador, A.J. Brady, E. Eltayib, T. Meng, A. Alonso-Vicente, P. Gonzalez-Vazquez, B.M. Torrisi, E.M. Vicente-Perez, K. Mooney, D.S. Jones, S.E.J. Bell, C.P. McCoy, H.O. McCarthy, J.C. McElnay, R.F. Donnelly, Hydrogel-forming microneedle arrays allow detection of drugs and glucose in vivo: Potential for use in diagnosis and therapeutic drug monitoring. PLoS One **10**, e0145644 (2015). https://doi.org/10.1371/JOURNAL.PONE.0145644
61. N.G. Oh, S.Y. Hwang, Y.H. Na, Fabrication of a PVA-based hydrogel microneedle patch. ACS Omega **7**, 25179–25185 (2022). https://doi.org/10.1021/ACSOMEGA.2C01993
62. A.Y.R. Aziz, N.A. Hasir, N.B.P. Imran, M.F. Hamdan, U. Mahfufah, N. Wafiah, A. Arjuna, R.N. Utami, A.D. Permana, Development of hydrogel-forming microneedles for transdermal delivery of albendazole from liquid reservoir. J. Biomater. Sci. Polym. Ed. **34**, 1101–1120 (2023). https://doi.org/10.1080/09205063.2022.2157671
63. D. Elim, A.M.N. Fitri, et al., Hydrogel forming microneedle-mediated transdermal delivery of sildenafil citrate from polyethylene glycol reservoir: An ex vivo proof of concept study. Colloids Surf. B Biointerfaces **222**, 113018 (2023). https://doi.org/10.1016/J.COLSURFB.2022.113018
64. Q.K. Anjani, A.D. Permana, Á. Cárcamo-Martínez, J. Domínguez-Robles, I.A. Tekko, E. Larrañeta, L.K. Vora, D. Ramadon, R.F. Donnelly, Versatility of hydrogel-forming microneedles in in vitro transdermal delivery of tuberculosis drugs. Eur. J. Pharm. Biopharm. **158**, 294–312 (2021). https://doi.org/10.1016/J.EJPB.2020.12.003
65. F.K. Aldawood, A. Andar, S. Desai, A comprehensive review of microneedles: Types, materials, processes, characterizations and applications. Polymers (Basel) **13**, 2815 (2021). https://doi.org/10.3390/POLYM13162815
66. P.M. Wang, M. Cornwell, J. Hill, M.R. Prausnitz, Precise microinjection into skin using hollow microneedles. J. Invest. Dermatol. **126**, 1080–1087 (2006). https://doi.org/10.1038/SJ.JID.5700150
67. P. Dardano, S. De Martino, M. Battisti, B. Miranda, I. Rea, L. De Stefano, One-shot fabrication of polymeric hollow microneedles by standard photolithography. Polymers **13**, 520 (2021). https://doi.org/10.3390/POLYM13040520
68. Z.F. Rad, P.D. Prewett, G.J. Davies, An overview of microneedle applications, materials, and fabrication methods. Beilstein J. Nanotechnol. **12**, 1034–1046 (2021). https://doi.org/10.3762/BJNANO.12.77
69. P.M. Santos, C.O. Rangel-Yagui, A.V. Janorkar, L. Luo, N. Sargioti, T.J. Levingstone, et al., Metallic microneedles for transdermal drug delivery: Applications, fabrication techniques and the effect of geometrical characteristics. Bioengineering **10**, 24 (2022). https://doi.org/10.3390/BIOENGINEERING10010024
70. J.H. Jung, S.G. Jin, Microneedle for transdermal drug delivery: current trends and fabrication. J. Pharm. Investig. **51**, 503–517 (2021). https://doi.org/10.1007/s40005-021-00512-4
71. P. Dardano, A. Caliò, V. Di Palma, M.F. Bevilacqua, A. Di Matteo, L. De Stefano, A photolithographic approach to polymeric microneedles array fabrication. Materials **8**(12), 8661–8673 (2015)
72. R.E.M. Lutton, E. Larrañeta, M.C. Kearney, P. Boyd, A.D. Woolfson, R.F. Donnelly, A novel scalable manufacturing process for the production of hydrogel-forming microneedle arrays. Int. J. Pharm. **494**, 417–429 (2015). https://doi.org/10.1016/J.IJPHARM.2015.08.049
73. Y. Li, J. Wang, L.K. Vora, A.H. Bin Sabri, M.B. McGuckin, A.J. Paredes, R.F. Donnelly, Dissolving microarray patches loaded with a rotigotine nanosuspension: A potential alternative to Neupro® patch. J. Control. Release **372**, 304–317 (2024). https://doi.org/10.1016/j.jconrel.2024.06.039
74. T.N. Tarbox, A.B. Watts, Z. Cui, R.O. Williams, An update on coating/manufacturing techniques of microneedles. Drug Deliv. Transl. Res. **8**, 1828–1843 (2017). https://doi.org/10.1007/S13346-017-0466-4

75. M. Olowe, S.K. Parupelli, S. Desai, A review of 3D-printing of microneedles. Pharmaceutics **14**, 2693 (2022). https://doi.org/10.3390/PHARMACEUTICS14122693
76. B.E. Rabinow, Nanosuspensions in drug delivery. Nat. Rev. Drug Discov. **3**, 785–796 (2004)
77. M.T.A. Abbate, I.K. Ramöller, A.H. Sabri, A.J. Paredes, A.J. Hutton, P.E. McKenna, K. Peng, J.A. Hollett, H.O. McCarthy, R.F. Donnelly, Formulation of antiretroviral nanocrystals and development into a microneedle delivery system for potential treatment of HIV-associated neurocognitive disorder (HAND). Int. J. Pharm. **640**, 123005 (2023). https://doi.org/10.1016/J.IJPHARM.2023.123005
78. L.K. Vora, P.R. Vavia, E. Larrañeta, S.E.J. Bell, R.F. Donnelly, Novel nanosuspension-based dissolving microneedle arrays for transdermal delivery of a hydrophobic drug. J Interdiscip Nanomed **3**, 89–101 (2018). https://doi.org/10.1002/jin2.41
79. F. Volpe-Zanutto, L.T. Ferreira, A.D. Permana, M. Kirkby, A.J. Paredes, L.K. Vora, A.P. Bonfanti, I. Charlie-Silva, C. Raposo, M.C. Figueiredo, I.M.O. Sousa, A. Brisibe, F.T.M. Costa, R.F. Donnelly, M.A. Foglio, Artemether and lumefantrine dissolving microneedle patches with improved pharmacokinetic performance and antimalarial efficacy in mice infected with plasmodium yoelii. J. Control. Release **333**, 298–315 (2021). https://doi.org/10.1016/J.JCONREL.2021.03.036
80. Y. Wu, L.K. Vora, D. Mishra, M.F. Adrianto, S. Gade, A.J. Paredes, R.F. Donnelly, T.R.R. Singh, Nanosuspension-loaded dissolving bilayer microneedles for hydrophobic drug delivery to the posterior segment of the eye. Biomater. Adva. **137**, 212767 (2022). https://doi.org/10.1016/J.BIOADV.2022.212767
81. I.A. Tekko, A.D. Permana, L. Vora, T. Hatahet, H.O. McCarthy, R.F. Donnelly, Localised and sustained intradermal delivery of methotrexate using nanocrystal-loaded microneedle arrays: Potential for enhanced treatment of psoriasis. Eur. J. Pharm. Sci. **152**, 105469 (2020). https://doi.org/10.1016/J.EJPS.2020.105469
82. E. Larrañeta, J. Moore, E.M. Vicente-Pérez, P. González-Vázquez, R. Lutton, A.D. Woolfson, R.F. Donnelly, A proposed model membrane and test method for microneedle insertion studies. Int. J. Pharm. **472**, 65–73 (2014). https://doi.org/10.1016/J.IJPHARM.2014.05.042
83. R.F. Donnelly, M.J. Garland, D.I.J. Morrow, K. Migalska, T.R.R. Singh, R. Majithiya, A.D. Woolfson, Optical coherence tomography is a valuable tool in the study of the effects of microneedle geometry on skin penetration characteristics and in-skin dissolution. J. Control. Release **147**, 333–341 (2010). https://doi.org/10.1016/J.JCONREL.2010.08.008
84. S. Aumann, S. Donner, J. Fischer, F. Müller, Optical coherence tomography (OCT): Principle and technical realization, in *High Resolution Imaging in Microscopy and Ophthalmology*, (2019), pp. 59–85. https://doi.org/10.1007/978-3-030-16638-0_3
85. R.R.S. Thakur, I.A. Tekko, F. Al-Shammari, A.A. Ali, H. McCarthy, R.F. Donnelly, Rapidly dissolving polymeric microneedles for minimally invasive intraocular drug delivery. Drug Deliv. Transl. Res. **6**, 800–815 (2016). https://doi.org/10.1007/S13346-016-0332-9/FIGURES/12
86. A. Summerfield, F. Meurens, M.E. Ricklin, The immunology of the porcine skin and its value as a model for human skin. Mol. Immunol. **66**, 14–21 (2015). https://doi.org/10.1016/J.MOLIMM.2014.10.023
87. A.J. Paredes, P.E. McKenna, I.K. Ramöller, Y.A. Naser, F. Volpe-Zanutto, M. Li, M.T.A. Abbate, L. Zhao, C. Zhang, J.M. Abu-Ershaid, X. Dai, R.F. Donnelly, Microarray patches: Poking a hole in the challenges faced when delivering poorly soluble drugs. Adv. Funct. Mater. **31**, 2005792 (2021). https://doi.org/10.1002/ADFM.202005792;SUBPAGE:STRING:FULL
88. N. Qin, M. Li, L.K. Vora, K. Peng, A.H. Bin Sabri, Y. Tao, A.J. Paredes, H.O. McCarthy, R.F. Donnelly, Enhanced long-acting simvastatin delivery via effervescent powder-carrying hollow microneedles and nanocrystal-loaded microneedles. Int. J. Pharm. **665**, 124691 (2024). https://doi.org/10.1016/J.IJPHARM.2024.124691

89. Z. Men, T. Su, Z. Tang, J. Liang, T. Shen, Tacrolimus nanocrystals microneedle patch for plaque psoriasis. Int. J. Pharm. **627**, 122207 (2022). https://doi.org/10.1016/J.IJPHARM.2022.122207
90. Q.K. Anjani, A.H. Bin Sabri, J. Domínguez-Robles, N. Moreno-Castellanos, E. Utomo, L.A.H. Wardoyo, E. Larrañeta, R.F. Donnelly, Metronidazole nanosuspension loaded dissolving microarray patches: An engineered composite pharmaceutical system for the treatment of skin and soft tissue infection. Biomater. Adv. **140**, 213073 (2022). https://doi.org/10.1016/J.BIOADV.2022.213073
91. M.T.C. Mc Crudden, E. Larrañeta, A. Clark, C. Jarrahian, A. Rein-Weston, S. Lachau-Durand, N. Niemeijer, P. Williams, C. Haeck, H.O. McCarthy, D. Zehrung, R.F. Donnelly, Design, formulation and evaluation of novel dissolving microarray patches containing a long-acting rilpivirine nanosuspension. J. Control. Release **292**, 119–129 (2018). https://doi.org/10.1016/J.JCONREL.2018.11.002
92. K. Moffatt, I.A. Tekko, L. Vora, F. Volpe-Zanutto, A.R.J. Hutton, J. Mistilis, C. Jarrahian, N. Akhavein, A.D. Weber, H.O. McCarthy, R.F. Donnelly, Development and evaluation of dissolving microarray patches for co-administered and repeated intradermal delivery of long-acting Rilpivirine and Cabotegravir nanosuspensions for Paediatric HIV antiretroviral therapy. Pharm. Res. **40**, 1673–1696 (2023). https://doi.org/10.1007/S11095-022-03408-6
93. A.D. Permana, M.T.C. McCrudden, R.F. Donnelly, Enhanced intradermal delivery of nanosuspensions of Antifilariasis drugs using dissolving microneedles: A proof of concept study. Pharmaceutics **11**, 346 (2019). https://doi.org/10.3390/PHARMACEUTICS11070346
94. A.D. Permana, A.J. Paredes, F.V. Zanutto, M.N. Amir, I. Ismail, M.A. Bahar, Sumarheni, S.D. Palma, R.F. Donnelly, Albendazole nanocrystal-based dissolving microneedles with improved pharmacokinetic performance for enhanced treatment of cystic echinococcosis. ACS Appl. Mater. Interfaces **13**, 38745–38760 (2021). https://doi.org/10.1021/ACSAMI.1C11179/ASSET/IMAGES/MEDIUM/AM1C11179_0012.GIF
95. J. Wang, Y. Li, R.M. Reguera, A. Sabri, N. García-Fernández, M. Álvarez-Bardón, L.K. Vora, L. Lopez-Vidal, M.G. Fuster, Y. Luo, H.O. McCarthy, R.F. Donnelly, A.J. Paredes, Lymphatic-targeted amphotericin B nanocrystals delivered using microarray patches applied to cutaneous leishmaniasis. J. Control. Release **386**, 114115 (2025). https://doi.org/10.1016/J.JCONREL.2025.114115
96. A.F. Madani, M.A. Syauqi, J.A. Permatasari, A.A. Putri, M. Fadel, A.D. Permana, Development of Telmisartan nanocrystal-based dissolving microneedle for brain targeting via trigeminal pathway: A potentially promising treatment for Alzheimer's with an improved pharmacokinetic profile. ACS Appl. Bio. Mater. **7**, 2582–2593 (2024)

Drug Delivery to the Central Nervous System Using Drug Nanocrystals

Yushi Tao, Marta G. Fuster, Martina Sangalli, and Alejandro J. Paredes

Abstract Nanocrystals (NCs) have emerged as a platform with significant potential for drug delivery to the central nervous system (CNS). They offer considerable advantages, particularly in improving the solubility, bioavailability, and targeting of hydrophobic therapeutics. Composed almost entirely of active pharmaceutical ingredients, NCs provide high drug-loading capacity without the need for additional carriers. Their small size and surface modifications, such as PEGylation and coating targeting ligands, enhance penetration across various physiological barriers like the blood–brain barrier (BBB), making them an attractive option for CNS-targeted therapies. In addition, NCs have the potential for controlled and sustained drug release, which is crucial for managing chronic CNS conditions. This chapter summarises current administration routes for CNS-targeted NCs and examines how biological barriers and disease-mediated alterations affect the precision delivery of drug NCs to the CNS. Additionally, diverse surface modification strategies aimed at enhancing CNS targeting are highlighted. Last but not least, key challenges are outlined, and future perspectives are offered to advance NCs-based therapies in managing CNS disorders.

Keywords Nanocrystals · Central nervous system targeting · Blood–brain barrier · Targeted nanocrystals · Nanosuspensions · Solid drug nanoparticles

1 Introduction

The central nervous system (CNS), comprising the brain and spinal cord, serves as the body's primary control hub, orchestrating sensory, motor, and cognitive functions, essential to enable complex behavioural responses and maintain homeostasis. The brain interprets sensory inputs, generates thoughts and emotions, and directs motor actions, whereas the spinal cord acts as a communication highway, relaying

Y. Tao · M. G. Fuster · M. Sangalli · A. J. Paredes (✉)
School of Pharmacy, Queen's University Belfast, Belfast, UK
e-mail: a.paredes@qub.ac.uk

A. J. Paredes (ed.), *Nanocrystals*, https://doi.org/10.1007/978-3-032-10238-6_7

signals between the brain and the body [1]. However, the CNS is vulnerable to a series of disorders or injuries, severely impacting its functions and resulting in various CNS diseases [2]. Neurodegenerative diseases such as Alzheimer's and Parkinson's lead to progressive cognitive decline, motor dysfunction, and memory deficits [3], while demyelinating conditions like multiple sclerosis disrupt nerve signal transmission by damaging the myelin sheath [4]. Cerebrovascular incidents, like strokes, cause sudden and debilitating neurological impairments due to impaired blood flow [5]. Infectious diseases, including meningitis and encephalitis, trigger life-threatening inflammation in the brain or spinal cord [6]. Moreover, traumatic injuries, autoimmune conditions, and parasitic infections (like cerebral malaria) further illustrate the complexity of CNS pathologies, often requiring specialised medical interventions with a multidisciplinary approach [7]. For patients with primary brain cancers or cerebral metastases, treatment options remain limited, including systemic chemotherapy, radiation therapy, and surgical resection [8, 9]. Despite the rising number of therapeutic candidates that have been developed, CNS drugs still face significant challenges and exhibit more failures in clinical trials compared with those aimed at non-CNS conditions, largely attributed to inadequate therapeutic efficacy and persistent side effects in the brain [10]. Besides, the burden of CNS disorders is escalating globally, posing a significant and costly challenge for healthcare systems.

Various administration pathways are applied for drug delivery to the brain, including topical applications such as intranasal (IN) drug delivery, systemic injections, and oral administration [11]. However, a substantial obstacle to systemic drug delivery is passing through the biological barriers, such as the blood–brain barrier (BBB) and the blood–cerebrospinal fluid (CSF) barrier. These highly restrictive barriers impede most therapeutics, especially macromolecules, separating the target site in the CNS from the circulating blood [12]. The BBB (Fig. 1) is a highly selective semipermeable membrane of the epithelium, composed of tight junctions, transport proteins, and an enzymatic barrier. The membrane surface provides anionic sites and is negatively charged due to the carboxylate groups of sialic acid residues on the glycoproteins [13]. The BBB allows the passive diffusion of hydrophobic or some small non-polar molecules. At the same time, it restricts the passage of solutes, ions, and hydrophilic macromolecules into the brain, unless they possess specialised transporters (such as glucose transporters or amino acid transporters) that can be actively transported via receptor- or carrier-mediated transport [14, 15]. Briefly, it effectively protects the brain from potentially harmful substances in the bloodstream, but also impedes drugs delivery to the brain due to its physicochemical properties, thus limiting the therapeutic efficacy in managing CNS diseases [14]. For effective BBB penetration, based on Lipinski's Rule of Five, molecules typically have a molecular weight (MW) closer to 400 Da, a log P (lipophilicity) value of 1–3, and limited hydrogen bonding to ensure better diffusion through the lipid-rich BBB [12]. Besides, factors like P-glycoprotein efflux and active transport mechanisms further restrict drug accumulation in target areas.

Under numerous CNS pathological conditions, BBB disruptions commonly happen and are characterised by structural changes, such as endothelial

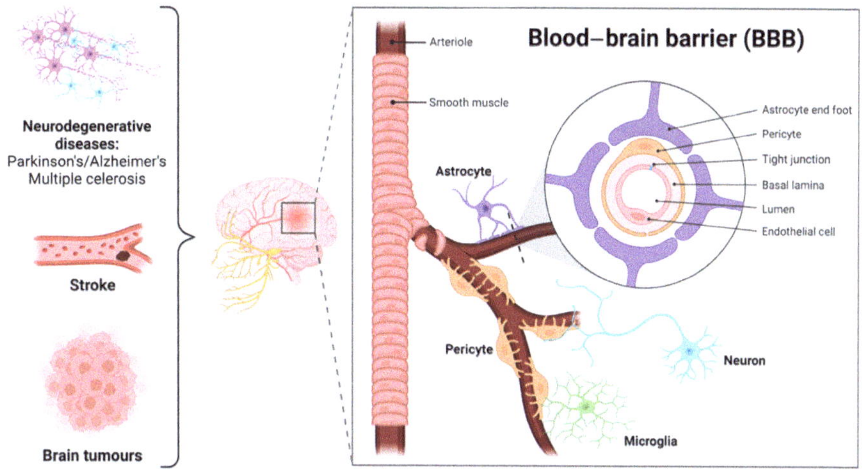

Fig. 1 Schematic representation of major CNS diseases and the structure of the BBB. The BBB is composed of brain vascular specialised endothelial cells, pericytes, astrocytes, and tight junctions that restrict the passage of most drugs. Microglia and neurons surround the blood vessels. (Image was created using BioRender 2025)

degeneration, reduced tight junction expression, thickening or disintegration of the basement membrane, pericyte dissociation, and detachment of astrocytic end feet [16]. These alterations result in elevated BBB permeability, potentially enhancing drug accumulation in the brain. However, they also present several complexities, like inducing the formation of fibrotic scars through extracellular matrix deposition, changing the BBB physiological composition, and potentially limiting drug access to the brain parenchyma [17]. Moreover, the extent and duration of BBB disruption vary among diseases, whereas the BBB often undergoes recovery within 2 weeks post-injury. It influences therapeutic windows and dosing regimens, requiring precise coordination to optimise CNS drug delivery [18]. Beyond the BBB, interstitial barriers such as the extracellular matrix and neurotoxic debris, as well as variations in interstitial flow rate and pressure, can affect drug diffusion, presenting challenges to drug transport efficiency [19]. In addition, cellular barriers, consisting of the cell membrane and endosome, are able to further impede drug absorption [20].

Apart from biological barriers, the complexity of brain targeting, managing drug toxicity, and addressing drug resistance must be taken into consideration. Current therapies often lack precision, leading to widespread drug distribution throughout the brain or even the whole body, resulting in off-target effects and limited therapeutic efficacy [21]. The brain's intricate architecture (Fig. 2), with specific regions responsible for distinct functions, makes certain areas particularly susceptible to various CNS diseases. For example, the prefrontal cortex, located in the frontal lobe, is closely linked to cognitive decline, a hallmark of neurodegenerative disorders like Alzheimer's disease [22]. This selective vulnerability extends beyond regional differences to the cellular level, particularly in the various cells comprising the

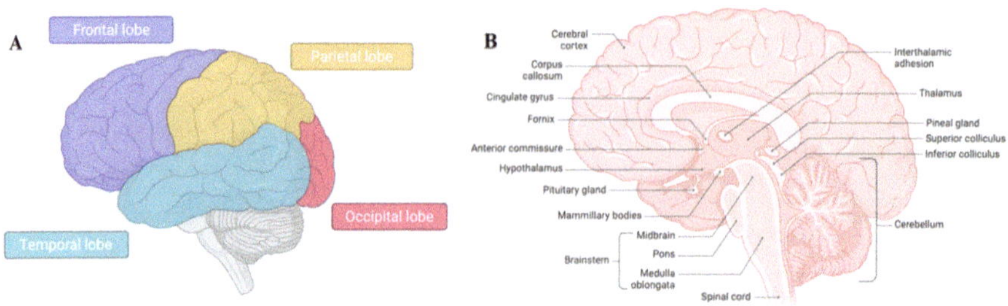

Fig. 2 (**a**) Different regions and (**b**) anatomy of the human brain. (Image was created using BioRender 2025)

BBB. In conditions like Parkinson's disease and other disorders involving cerebellar damage, Purkinje neurons are notably affected, often exhibiting axonal swelling [23]. These region- and cell-specific susceptibilities underscore the importance of precision-targeted therapies in addressing the unique pathological features of CNS diseases.

CNS drugs can pose a high risk of neurotoxicity, which can lead to cognitive impairment, neuroinflammation, and other adverse effects, particularly in long-term treatments [24]. Drug resistance, such as the overexpression of efflux transporters, further complicates treatment, especially in epilepsy. Over time, some CNS drugs may lose their effectiveness as tolerance develops, necessitating higher doses, which increases the risk of side effects [25]. In addition, rapid clearance from the brain, driven by factors such as CSF dynamics, restricts the ability to maintain therapeutic levels, whereas pharmacokinetic and pharmacodynamic variability leads to unpredictable drug responses between individuals [26]. To address these multifaceted challenges, innovative strategies like nanocrystal-based therapies are being explored to provide precisely targeted, controlled drug delivery across or bypassing the BBB to CNS disease sites, reducing side effects, settling problems on drug resistance, and improving patient adherence [27].

Nanoparticles (NPs), encompassing three main categories, organic, inorganic, and carbon-based, have shown both protective and degenerative effects in CNS disorders [28], as illustrated in Fig. 3. In detail, NPs offer a promising approach to mitigate neurotoxicity and inflammation, where their dual action on neurons and microglia is critical [29, 30]. By offering neuroprotection, NPs can reduce oxidative stress, modulate neurotransmitter release, and deliver neurotrophic factors, which aid in preventing neuronal damage from neurodegenerative conditions such as Alzheimer's and Parkinson's disease [31]. Their primary advantages include enhancing drug solubility and stability, reducing side effects while improving bioavailability via targeting specific tissues, and enabling controlled drug release [32]. NPs can cross the BBB to deliver therapeutic agents such as antioxidants and anti-inflammatory drugs directly to the brain. Despite these benefits, NP-based therapies face challenges such as low drug loading (typically 5–15% w/w), restricted long-term stability, and often require organic solvents during manufacturing

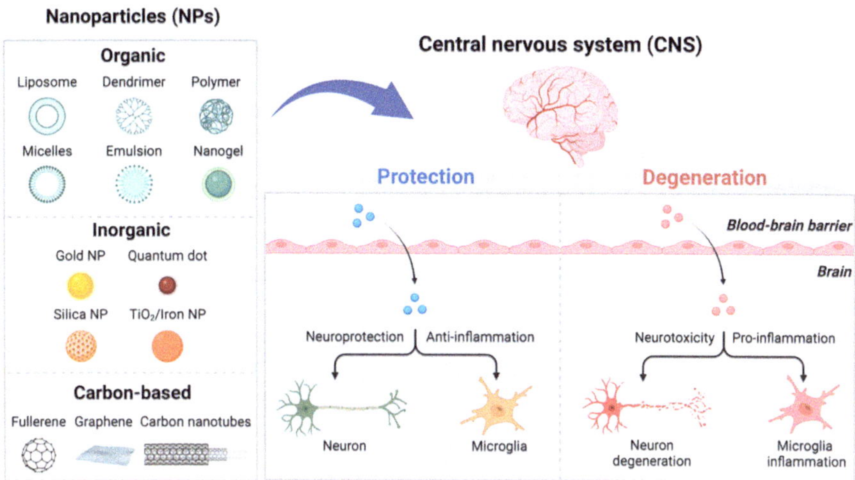

Fig. 3 Classification of NPs and their roles in CNS disorders. NPs cross the BBB into the brain cells, consisting of neurons, microglia, and astrocytes, to exert protective and degenerative effects. (Image was created using BioRender 2025)

[33]. Some NPs may inadvertently induce neurotoxicity, stimulate immune responses, or accumulate in brain tissues, leading to potential chronic toxicity. Under this circumstance, nanocrystals (NCs) arise as an outstanding subfamily of NPs, which can address these limitations more effectively, providing an optimised platform for CNS-targeted drug delivery with enhanced efficiency and safety [34].

In essence, NCs are crystalline drug particles with a mean size in the nanometre range (diameter < 1 μm), composed of pure drugs surrounded by a thin layer of surfactant or polymeric material, working as highly loaded drug vectors [11, 34]. NCs have been reported to potentially improve the solubility and efficacy of Biopharmaceutical Classification System (BCS) class II and IV drugs [35]. The Food and Drug Administration (FDA) approved five NCs-based formulations from 2000 to 2005, and more than 20 NCs-based hydrophobic drugs have entered the market in recent years, used to increase poorly soluble oral absorption or enable long-acting release from intramuscular depots [11, 36]. Unlike other NPs-based drug delivery systems, NCs possess unique advantages including high drug loading capacity, reduced toxicity or side effects, improved long-term stability, extended shelf life of drugs prone to degradation, activation of specific cell populations, and various fabrication methods [34]. NCs-based formulations are universally applicable to all drugs, even those insoluble in aqueous and organic media, and capable of being manufactured in both solvent-free and neutral pH conditions [37]. Furthermore, according to the Noyes–Whitney equation, NCs significantly enlarge the surface area-to-volume ratio and improve dissolution rate [11]. They consequently enhance the saturation solubility and bioadhesion, as well as facilitate better penetration through biological barriers compared with microparticles [34, 37]. NCs offer a potential solution to facilitate targeted delivery and increase drug bioavailability as drug NCs can be engineered to possess special surface properties or functionalised with targeting molecules. Tailored NCs are capable of crossing the BBB directly or through receptor-mediated endocytosis [27]. Once passed, they can achieve controlled release in the brain, consequently improving drug efficacy and reducing side effects. This approach is widely explored for treating CNS diseases, making it a cutting-edge area of research at the intersection of nanotechnology and neuroscience.

In this chapter, a comprehensive overview of current state-of-the-art NCs' delivery pathways is presented, along with an evaluation of biological barriers limiting CNS access. Then, key surface modification strategies to enhance NCs' selectivity for brain targeting are highlighted, along with additional applications such as long-acting drug delivery and diagnostic advancements [38]. Finally, the chapter concludes with perspectives on the challenges, anticipated clinical impacts, and future directions for NCs-based CNS delivery systems.

2 Drug Nanocrystals for Central Nervous System Delivery

NCs represent a transformative advancement in drug delivery, particularly for overcoming the challenges of transporting therapeutics across the BBB to treat neurological disorders [11, 34]. Their fabrication through top-down and bottom-up approaches enables precise control over particle size, zeta potential (ζ-potential), and surface modifications to optimise bioavailability and targeted delivery. Typically prepared as nanosuspensions (NSs) stabilised to prevent aggregation, NCs exhibit the capacity to traverse biological barriers while remaining intact, as confirmed by Förster Resonance Energy Transfer (FRET), which tracks their integrity during cellular uptake and transport. This technique is instrumental in determining the stage of NCs dissolution, which can occur either immediately after internalisation or later, closer to the target site [39]. In cases where intact NCs are not explicitly demonstrated to cross the BBB, brain accumulation is likely facilitated by systemic circulation and subsequent diffusion [40]. Studies using in vitro and ex vivo models have displayed the internalisation of NCs by various barrier cells, such as nasal and gastrointestinal (GI) epithelium, attributed to the small size of NCs and the use of surfactants [39, 41, 42]. Surface modifications like PEGylation and stimuli-responsive gels further enhance drug NCs' permeability, reduce undesired interactions (e.g., with mucins), and tailor drug release profiles, improving therapeutic efficacy, especially in the nose-to-brain (N2B) pathway [43, 44]. This route bypasses systemic metabolism, leveraging nasal mucosa and olfactory pathways to directly access the CNS and achieve higher brain drug concentrations while avoiding enzymatic degradation and hepatic first-pass metabolism (FPM). Additionally, innovative approaches like the facial intradermal microneedle array patch (MAP) facilitate direct NCs transport into microvasculature, improving pharmacokinetics and bioavailability [45]. Collectively, these strategies highlight the versatility of NCs to deliver hydrophobic drugs across physiological barriers, reduce toxicity, and improve therapeutic outcomes, positioning them as a promising platform for CNS-targeted therapies. Table 1 summarises various examples of NCs, along with their preparation methods and administration routes, illustrating their diverse applications and remarkable efficacy in treating different neurological conditions.

2.1 Administration Routes to Deliver Drug NCs to the CNS

As a transformative technology to enhance brain delivery, NCs leverage various minimally or non-invasive administration pathways (Fig. 4) to overcome the obstacles posed by the BBB [12, 48]. Each administration route offers unique benefits, enhancing efficacy while diminishing off-target effects and systemic toxicity. Through intramuscular (IM) injection, via retrograde axonal transport, NCs can be taken up by nerve terminals at the injection site and transported to the CNS, potentially overcoming the BBB limitations, an approach always beneficial for prolonged treatment of Alzheimer's, Parkinson's, or amyotrophic lateral sclerosis (ALS) [11, 49]. Intravenous (IV) injection is widely applied and provides rapid, direct access to the bloodstream, allowing drug NCs to reach immediate therapeutic concentrations, which is crucial for acute conditions such as severe cerebral malaria [50]. It not only ensures high bioavailability via bypassing FPM but also permits drug NCs to circulate systemically and penetrate the BBB by passive diffusion or transporter-mediated mechanisms [27]. MAP is emerging as a promising alternative to both IV and IM injections for CNS-targeted therapies, offering enhanced brain penetration through transdermal drug delivery (TDD) [51]. It includes strategic facial applications near the trigeminal nerve, facilitating more efficient transport of NCs via bypassing the BBB [45]. As a minimally invasive approach, MAPs cause little to no pain and can be self-administered without medical expertise, making them accessible and convenient for patients, especially in resource-limited settings [52]. Furthermore, MAPs mitigate infection risks associated with repeated needle use, as well as bypass GI degradation and hepatic FPM, significantly enhancing the bioavailability of drug NCs [53]. They are engineered to provide controlled and sustained drug release, which maintains steady therapeutic levels and reduces dosing frequency [52]. This innovative delivery technology can decrease systemic toxicity associated with the high peak concentrations of IV and IM injections, offering a patient-friendly, efficient, and safe option for therapeutic administration [51].

Oral administration acts as a convenient and non-invasive route, particularly for mild cases or preventive treatments [38]. The use of NCs enhances the bioavailability of poorly soluble drugs, improving absorption in the GI tract [11]. However, in severe CNS conditions, the slow onset and impact of FPM can limit its efficacy. Although the intestinal epithelium cannot directly affect CNS drug delivery, it plays an essential role in the initial absorption phase, after which the drug NCs still need to cross the BBB to exert their therapeutic effect [27]. Intranasal administration is a non-invasive and efficient approach for direct drug delivery to the brain, circumventing the BBB and facilitating targeted CNS delivery via the N2B pathway

Table 1 Drug NCs examples were used to treat CNS disorders by crossing physiological barriers

CNS disease	Drug NCs name with particle size	NCs preparation method	Biological barrier crossed	Administration	Key findings from the reports	Animal studies? (Y/N)
General CNS targeting	Curcumin NCs (CUR-NCs): 328.7 ± 16.9 nm [42]	Solvent–antisolvent Sono precipitation	Nasal epithelial barrier, BBB	N2B delivery in vitro model	1. CUR-NCs displayed strong intracellular fluorescence and indicated significant uptake by olfactory Ensheathing cells (OECs)	No
	CUR-NCs: 200.0 ± 5.0 nm [44]	Wet bead milling	Nasal mucosa, BBB	N2B delivery in vitro model	1. Both high cross-linking surface (HCS) and low cross-linking surface (LCS) NCs demonstrated enhanced mucosal barrier penetration 2. CUR-NCs maintained high drug content (>80%) and stability 3. Effective in crossing epithelial monolayers (~40% penetration over 6 h) 4. Surface PEGylation did not alter CUR bioactivity	No
	Resveratrol nanosuspensions (RES-NSs) in stimuli-responsive in situ gel [43]	Antisolvent precipitation	Nasal mucosa, BBB	Intranasal	1. RES-NSs effectively penetrated nasal mucosa and bypassed the BBB 2. Increased retention time in the nasal cavity and enhanced bioavailability in the brain compared to IV administration 3. NSs were directly delivered via the olfactory and trigeminal nerve pathways, allowing efficient brain targeting	Yes

Parkinson's disease	Paeoniflorin NCs (PA-NCs): 139.6 ± 1.3 nm [39]	Antisolvent precipitation	Nasal mucosa, BBB	Intranasal	1. In ex vivo studies, PA-NCs showed significantly higher permeation (87.14% ± 5.34% at 24 h) in the nasal mucosa and epithelial cells compared to free PA 2. FRET imaging confirmed that the intact NCs were internalised into nasal epithelial cells 3. In vivo studies demonstrated that intranasal (IN) administration achieved a significantly higher maximum concentration (C_{max}) in the brain (108.47 ± 9.51 ng/g) compared to intravenous (IV) administration (47.19 ± 5.69 ng/g), highlighting the superior efficiency of the IN route in delivering therapeutics directly to the CNS	Yes
	Puerarin NCs (PU-NCs): 83.1 ± 2.0 nm [41]	Antisolvent precipitation	GI barrier, BBB	Oral	1. In vitro studies illustrated that PU-NCs exhibited enhanced permeability and cellular uptake (6.1- and 5.1-fold increases rather than PU crudes, respectively) 2. Intact NCs were detected in the intestine and brain of zebrafish after oral gavage 3. Enhanced brain and plasma pharmacokinetics of PU-NCs (3.79- and 7.14-fold increase in brain and plasma C_{max})	Yes

(continued)

Table 1 (continued)

CNS disease	Drug NCs name with particle size	NCs preparation method	Biological barrier crossed	Administration	Key findings from the reports	Animal studies? (Y/N)
	Resveratrol NCs (RES-NCs): 222.5 ± 1.7 nm [46]	Antisolvent precipitation	GI barrier, BBB	Oral	1. Improved dissolution rate and increased cellular uptake (2.26 ± 0.04 μg/mg) of RES-NCs in MDCK cells confirmed enhanced absorption through the GI tract 2. Pharmacokinetics in rats demonstrated higher plasma C_{max} and $AUC_{0\text{-}t}$ (3.09- and 3.52-fold increases, respectively).	Yes
	Schisantherin A NCs (SA-NCs): ~160 nm [40]	Antisolvent precipitation	Intestinal epithelium, BBB	Oral	1. FRET demonstrated that intact SA-NCs were internalised by MDCK cells and dissolved intracellularly 2. Enhanced brain concentrations (C_{max} ~ 5.47-fold increase) after oral administration of SA-NCs indicated successful delivery across the BBB	Yes
HIV-associated neurocognitive disorder (HAND)	Rilpivirine NCs and cabotegravir NCs (RPV-NCs and CAB-NCs): 163.1 ± 5.5 nm and 194.8 ± 2.7 nm, respectively [45]	Wet bead milling	Skin barrier, BBB	Facial intradermal injection via MAPs	1. RPV-NCs reached brain C_{max} at 21 days (619.17 ± 73.32 ng/g), exceeding plasma IC_{90} levels, with potential therapeutic levels for 28 days 2. CAB-NCs achieved brain C_{max} at 28 days (478.31 ± 320.86 ng/g), below IC_{90} but showed potential for optimisation. Lymph and plasma levels exceeded therapeutic thresholds	Yes

Schizophrenia	Clozapine nanosuspensions (CLZ-NSs): 281.0 ± 7.5 nm [47]	High-speed homogenisation	Nasal mucosa, BBB	Intranasal	1. Increased drug permeability of CLZ-NSs through nasal mucosa after 24 h (96.15%) compared to conventional suspension (41.12%) 2. 3.56-fold brain drug concentration compared to the oral route, even with a 528-fold lower administered dose	Yes

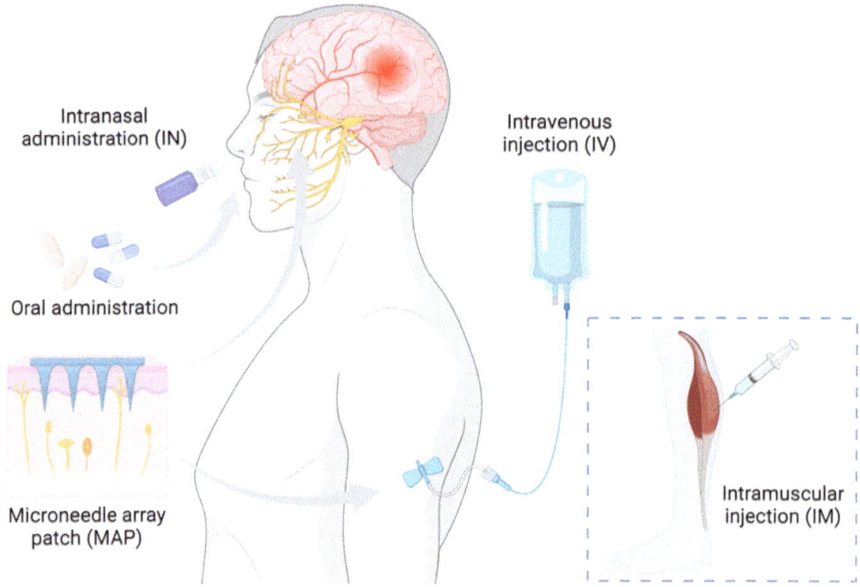

Fig. 4 Various NCs-based therapeutics administration routes for the treatment of CNS disorders. (Image was created using BioRender 2025)

[54]. Drug NCs must penetrate the nasal mucus layer and epithelium while evading mucociliary clearance [39]. This route minimises systemic exposure and side effects, in order to achieve a faster onset of action [55]. Although promising, future studies will focus on obtaining adequate bloodstream concentrations of drugs for effective brain targeting. Therefore, optimising drug NCs is a key strategy to navigate these biological barriers, especially the BBB, maximising therapeutic potential and enhancing patient outcomes in the management of CNS disorders [56].

2.2 Pharmaceutical Technology Strategies to Target the CNS

Considering the points discussed in the previous section, multiple strategies (Fig. 5) have been explored to improve the performance of NCs in overcoming physiological barriers that are pivotal to achieving CNS targeting. One key technique is chemical modification, such as PEGylation of NCs, which is achieved through either physical adsorption or covalent bonding of functionalised polyethylene glycol (PEG) molecules to the NCs surface [57]. Physical adsorption leverages hydrophobic or electrostatic interactions, whereas covalent attachment involves specific chemistries, such as thiol–maleimide reactions, to create stable bonds with functional groups like amines, thiols, or carboxyls [58]. It significantly enhances the solubility, bioavailability, and pharmacokinetics of drug NCs to cross biological barriers by

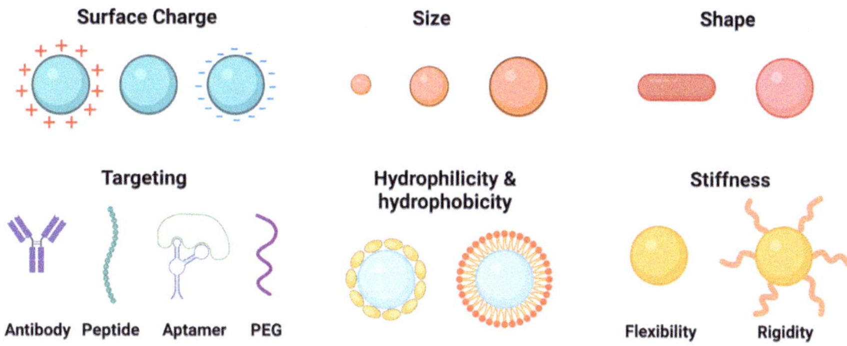

Fig. 5 Characteristics of NCs significantly affect their performance (e.g. adsorption, distribution, across the physiological barriers, and clearance) to target the CNS. (Image was created using BioRender 2025)

prolonging circulation time and reducing immune clearance by the mononuclear phagocyte system (MPS), primarily due to its hydrophilic nature and 'stealth' effect [59]. These attributes make PEGylated NCs a highly effective platform for improving the therapeutic performance of poorly water-soluble drugs, particularly in advanced drug delivery applications such as brain targeting and sustained release to combat drug-resistant conditions [60]. Besides, PEGylation prevents particle agglomeration by steric stabilisation, ensuring NCs remain stable and discrete, thereby protecting them from enzymatic degradation and premature drug release. It not only preserves the integrity of therapeutic agents but also minimises accumulation on the endothelial cell membrane, which is ideally used in long-acting drug delivery and advantageous in diagnostic imaging [61]. Furthermore, their biocompatibility and non-immunogenic nature can reduce drug toxicity, making PEGylated NCs highly suitable to be functionalised with targeting ligands for increasing the specificity, consequently treating brain disorders and cancers [62].

Apart from surface chemistry, surface charge also plays a vital role, where positively charged NCs promote adsorptive-mediated transcytosis across the BBB, whereas uncoated or excessively positively charged particles may result in rapid immune clearance by macrophages. In contrast, neutral or slightly anionic NCs exhibit longer circulation times because of reduced immune recognition [63]. Size and shape also impact NCs behaviours in the bloodstream and interactions with cells, where small NCs penetrate tissues more efficiently, exhibiting enhanced retention [63]. Spherical NCs tend to marginate more easily during circulation, leading to more efficient cellular internalisation. Rod-shaped NCs have longer circulation times and tend to extravasate; however, their irregular shape can reduce internalisation efficiency and increase protein corona formation [64, 65]. Moreover, non-spherical NCs display accessible adherence to the brain microvascular walls, further complicating drug delivery [65, 66]. Both biocompatibility and biodegradability allow for controlled drug release, with materials such as chitosan and cellulose breaking down over time, whereas non-degradable materials may

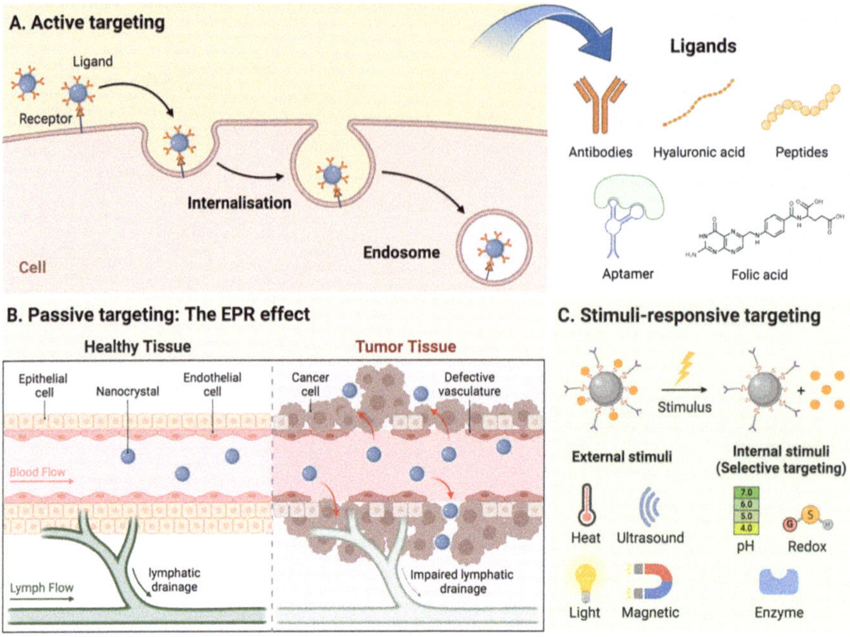

Fig. 6 Three principal NCs targeting strategies, namely (**a**) active targeting (with several ligand examples), (**b**) passive targeting, and (**c**) stimuli-responsive targeting. EPR effect—enhanced permeability and retention effect. (Image was created using BioRender 2025)

accumulate in several organs [67, 68]. Finally, targeting strategies are essential to improve the precision, efficiency, and safety of NC-based therapeutics in managing CNS disorders, and are broadly divided into active, passive, and stimuli-responsive targeting [56, 69], as indicated in Fig. 6. The choice of strategies is based on disease characteristics, NCs formulations, and desired therapeutic outcomes.

Active targeting, or specific targeting, is a precision-focused approach in which drug NCs are directed to specific cells or tissues by leveraging recognition mechanisms on the surface of target cells [56, 70]. It is typically achieved by attaching targeting ligands (such as antibodies, cell-penetrating peptides, and aptamers) to receptors overexpressed or uniquely presented on diseased cells, consequently enhancing the precision of drug delivery [71]. Once bound, drug NCs are internalised into cells via receptor-mediated endocytosis. Despite its benefits, specific targeting comes with challenges, such as the requirement for overexpressed receptors on target cells, which may not be present in all patients [72]. Developing and rationalising targeting ligands increases the complexity and cost of NCs formulations. Receptor saturation is another limitation, as target cells cannot take up drugs infinitely, restricting the overall therapeutic dose [73]. Chai et al. [74] introduced an innovative targeted drug delivery system leveraging red blood cell (RBC) membrane-coated drug NCs (RBC-NCs) for glioma treatment. The chemotherapeutic agent docetaxel (DTX), a poorly soluble drug associated with severe toxicity, was commonly employed in cancer therapy. To enhance targeting capability, the RBC membrane was functionalised with the tumour-targeting peptide c(RGDyK), which specifically binds to integrins overexpressed on tumour cells and neovasculature, enabling active targeting. The RBC-NCs formulations demonstrated excellent biocompatibility, prolonged stability, and a marked reduction in systemic toxicity compared to uncoated drug NCs. The resulting RGD-RBC-NCs exhibited irregular shapes with an average size of approximately 70 nm. In vivo, they achieved enhanced therapeutic efficacy in both subcutaneous tumour and orthotopic glioma models, underscoring their potential as biomimetic nanocarriers for active targeting in cancer therapy.

In contrast, passive targeting (non-specific targeting) relies on natural physiological processes such as blood circulation and tissue permeability to accumulate drug NCs in specific tissues, without utilising targeting ligands [75]. The enhanced permeability and retention (EPR) effect is a well-known mechanism in non-specific targeting, where the leaky vasculature and poor lymphatic drainage in tumours or inflamed tissues allow NCs to passively accumulate [69]. Smaller particles can penetrate biological barriers more easily, whereas the larger ones circulate longer and accumulate in areas of compromised vasculature [76]. Non-specific targeting is easier and cost-effective to develop; however, it lacks precision, resulting in potential off-target effects and systemic toxicity, because NCs can accumulate indiscriminately in both healthy and diseased tissues [77]. Xiong et al. [46] developed NCs formulations of resveratrol, a non-flavonoid polyphenol known for its neuroprotective effects, particularly in the treatment of Parkinson's disease. The aim of this research was to enhance the oral bioavailability and brain delivery of resveratrol, which is limited due to its poor water solubility. Drug NCs were fabricated using an antisolvent precipitation method with hydroxypropyl methylcellulose (HPMC) as a stabiliser. This approach allowed the production of resveratrol NCs (RES-NCs) with a particle size of 222.5 ± 1.7 nm and a polydispersity index (PDI) of 0.125 ± 0.035, indicating a uniform particle size distribution. The ζ-potential was measured at -9.41 ± 0.37 mV, suggesting good stability. This treatment employed a passive targeting strategy to enhance the delivery of resveratrol to the brain. This was achieved by utilising the small size of the NCs, which facilitated their accumulation in the brain through the EPR effect. The effectiveness of targeting was evaluated through in vitro dissolution studies and in vivo pharmacokinetic assessments, which demonstrated a significant increase in the bioavailability of resveratrol compared to conventional formulations. This study presented a promising NCs formulation of resveratrol that not only improved its solubility and absorption but also enhanced its therapeutic potential for treating Parkinson's disease via passive targeting mechanisms.

Furthermore, stimuli-responsive targeting consists of selective targeting and physical targeting, which leverage internal and external stimuli, respectively [78, 79]. Selective targeting improves NCs' specificity by exploiting the unique biochemical or physiological microenvironments of diseased cells, such as pH, redox status, or specific enzymes, to distinguish them from healthy cells, rather than relying solely on receptor binding [80]. In contrast, physical targeting uses magnetic fields, ultrasound, or light to direct NCs to specific tissues or to control drug release [81]. It is non-invasive and can be combined with other targeting strategies. These multifunctional methods maximise drug biodistribution and cellular uptake, consequently enhancing therapeutic efficacy in the CNS. Kamba et al. [82] developed calcium carbonate NCs as a drug delivery system for the anticancer drug doxorubicin (DOX), aimed at enhancing targeted therapy for breast cancer. Drug NCs were synthesised using an oil-in-water microemulsion technique, which involved high-pressure homogenisation to achieve a rod-shaped morphology with an average particle size of 35–60 nm. The stimuli-responsive targeting mechanism was achieved through the EPR effect, allowing for selective accumulation of NCs in tumour tissues, as confirmed by in vitro assays showing a 96% encapsulation efficiency and significant uptake by MDA MB231 breast cancer cells. The study demonstrated that the CaCO3/DOX nanocrystals exhibited stability at neutral pH (7.4) with slow drug release, whereas a faster release was observed under acidic conditions (pH 4.8), mimicking the tumour microenvironment. In vitro experiments indicated effective cancer cell death with minimal non-specific toxicity, highlighting the potential of these biobased NCs for targeted cancer therapy. The reported NCs-based therapies modified by three principal targeting methods in managing CNS indications are summarised in Table 2.

Case Study Xiong et al. developed puerarin NCs (PU-NCs) to enhance their ability to cross the BBB for the effective treatment of Parkinson's disease, consequently improving oral bioavailability [41]. This study employed an antisolvent precipitation method, dissolving puerarin in acetone, followed by rapid injection into water containing the stabiliser F68 with stirring at 1000 rpm. This rapid mixing not only facilitated the immediate precipitation of PU-NCs but also ensured that the stabiliser effectively coated the particles, preventing aggregation and maintaining a narrow size distribution. It was pivotal to use F68 that enhanced the solubility and bioavailability of PU by improving its interaction with biological membranes, thereby promoting BBB penetration. A key aspect of this research was the implementation of a passive targeting strategy, relying on the EPR effect, which allowed the drug NCs to accumulate in the brain's pathological tissues. Dynamic light scattering (DLS) and transmission electron microscopy (TEM) confirmed the small particle size of 83.1 ± 2.0 nm (Fig. 7a) and a ζ-potential of -22.41 ± 1.19 mV, with a nearly spherical morphology of the PU-NCs, indicating sufficient electrostatic repulsion to prevent aggregation and enhance stability under physiological conditions.

In vitro dissolution studies demonstrated a significantly faster release profile (Fig. 7b), with 93.1% of the drug released within 5 min, compared to less than 40% from conventional formulations after 30 min. It is crucial for quickly achieving therapeutic concentrations, particularly in acute treatment scenarios. In vivo studies further revealed plasma and brain levels of PU-NCs, showing that the maximum drug concentration (C_{max}: 2186.75 ± 161.92 ng/mL) was achieved in just 1.83 h in plasma (Fig. 7c), with brain accumulation reaching 322.13 ± 28.35 ng/g in 4.27 h in

Table 2 Various NCs-based therapies apply three primary targeting strategies in the management of CNS disorders

NC-based therapies	Targeting methods	Mechanisms	CNS indications	Main findings
RBC membrane-coated NCs (RBC-NCs) [74]	Active targeting	The RBC membrane-coated NCs employ a modified RBC membrane with RGD peptides for tumour targeting via integrin binding, enhancing drug accumulation in tumour tissues	Glioma	This system improves drug stability, prolongs circulation time, and ensures targeted delivery with reduced side effects
Paclitaxel-disulfiram NCs (PTX-DSF NCs) [83]		Reverses multidrug resistance (MDR) via P-glycoprotein (MDR-1) inactivation, increasing apoptosis, and inducing cell cycle arrest in G2/M phase	Cancer (Taxol-resistant A549 cells)	Effective targeting through the BBB and overcoming drug resistance
Harmine NCs in situ gel (HAR-NC-IG) [84]		The HAR-NC-IG utilises the ionic gelation properties of DGG to form a bioadhesive matrix that enhances retention time in the nasal cavity	Cognitive impairment (Alzheimer's disease)	Ensuring stability and controlled release of poorly soluble drugs across the BBB
PPD NCs (20 (S)-protopanaxadiol NCs) [85]	Passive targeting	The PPD NCs were fabricated using an antisolvent precipitation approach, using D-a-tocopheryl polyethylene glycol 1000 succinate (TPGS) as a stabiliser	Brain cancer	Enhanced bioavailability and drug absorption into plasma, leading to increased drug concentration in the brain
Paeoniflorin NCs (PA-NCs) [39]		Enhanced nasal mucosa permeation, increased cellular uptake via van der Waals forces between PA and TPGS, and improved transcellular transport through Calu-3 cells	Parkinson's disease	Low natural bioavailability and limited brain delivery of free PA
Resveratrol NCs (RES-NCs) [46]		The RES-NCs effectively cross the BBB, resulting in enhanced brain accumulation, neuroprotection against MPP^{+}-induced cytotoxicity, and	Parkinson's disease	Ensuring sustained release and stability of RES-NCs

(continued)

Table 2 (continued)

NC-based therapies	Targeting methods	Mechanisms	CNS indications	Main findings
		modulation of the Akt/Gsk3β signalling pathway, which supports dopaminergic neuron survival		
Donepezil NCs (D-NCs) [86]		The hydrophobic ion pair technique is used for improving hydrophobicity	Alzheimer's disease	Ensuring sustained release with prolonged plasma levels and maintaining drug stability over time
PX-18 NCs [87]		Upon administration, PX-18 inhibits sPLA2-IIA activity, reducing pro-inflammatory mediators, which decreases neuronal death, DNA damage, and glial cell activation in the hippocampal CA1 region following ischemia/reperfusion injury	Cerebral ischemia	The low solubility of PX-18 limits its efficacy to be a neuroprotective agent
Calcium carbonate NCs ($CaCO_3$/DOX) [82]	Stimuli-response targeting	pH-sensitive release of doxorubicin (DOX): slow release at neutral pH (7.4), while accelerated release at acidic pH (4.8) in the tumour microenvironment	Cancer (MDA MB231 breast cancer cells)	Ensuring selective and effective drug delivery to tumour cells without non-specific toxicity

the rat brain (Fig. 7d). These results illustrated a quick absorption of PU-NCs, likely due to their fast dissolution rate and enhanced oral absorption. Following this, a rapid elimination of PU (control group) was observed in both plasma and brain, whereas PU-NCs exhibited a slower elimination rate, suggesting a prolonged retention and higher oral bioavailability. Overall, this research underscored the capacity of PU-NCs to significantly improve the hydrophobic drug accumulation in the CNS via oral administration, which was attributed to passive targeting via the EPR effect, making them a promising approach for neurodegenerative disorder therapies.

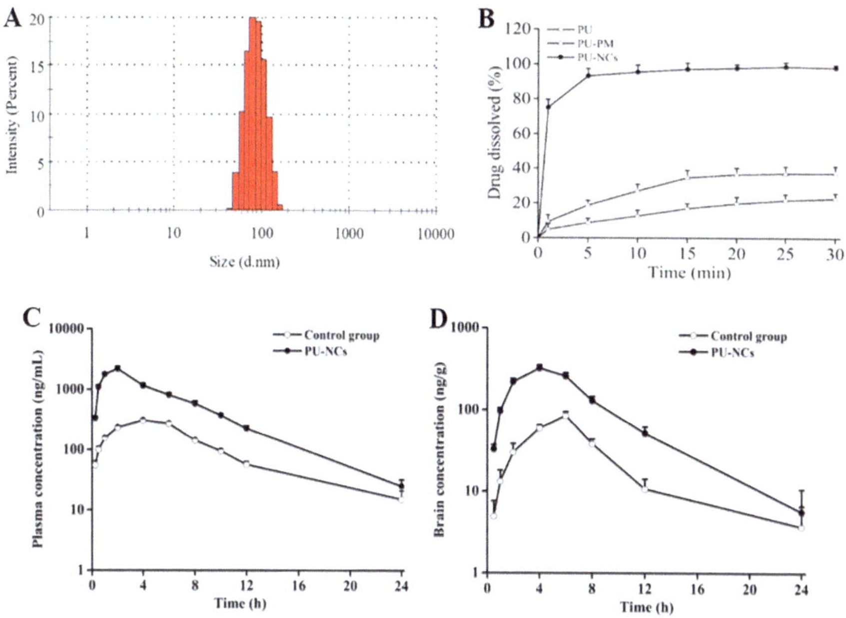

Fig. 7 (**a**) Particle size distribution of PU-NCs. (**b**) In vitro dissolution profiles of PU crudes, the physical mixture of PU and excipients, and PU-NCs. (**c**) Plasma concentration-time profiles of PU-NCs with the control group. Data were represented as means ± SD ($n = 6$). (**d**) Brain concentration-time profiles of PU-NCs with the control group. Data were presented as means ± SD ($n = 4$). (Obtained from Xiong et al. [41])

2.3 *Drug Nanocrystal Transport Approaches Across the BBB*

The transport of drug NCs from the bloodstream into the brain crossing the BBB is a highly selective and complex process. Several transcellular mechanisms have been identified to facilitate NCs' penetration, as displayed in Fig. 8. Receptor-mediated transcytosis (RMT), the most promising pathway, involves specific receptors on the endothelial cells of the BBB, which bind to ligands and initiate vesicular transport, effectively carrying NCs into the brain [88]. Among the key receptors, the transferrin receptor (TfR) plays a critical role in transporting Tf-bound NCs via endocytosis and exocytosis into the brain parenchyma [89, 90]. Besides, the insulin receptor (IR) can be exploited by conjugating NCs to insulin or insulin-like molecules, allowing brain delivery through insulin receptor-mediated transcytosis [91]. Similarly, the low-density lipoprotein receptor (LDLR) can be targeted by conjugating NCs to LDL or mimicking peptides, via an LDLR-mediated pathway [92]. For instance, surface modification of baicalin NCs with a combination of Tween 80 and TPGS enhanced brain delivery by initiating RMT across the BBB, facilitated through the adsorption of apolipoprotein E (ApoE) as a ligand for LDLR, which is abundantly

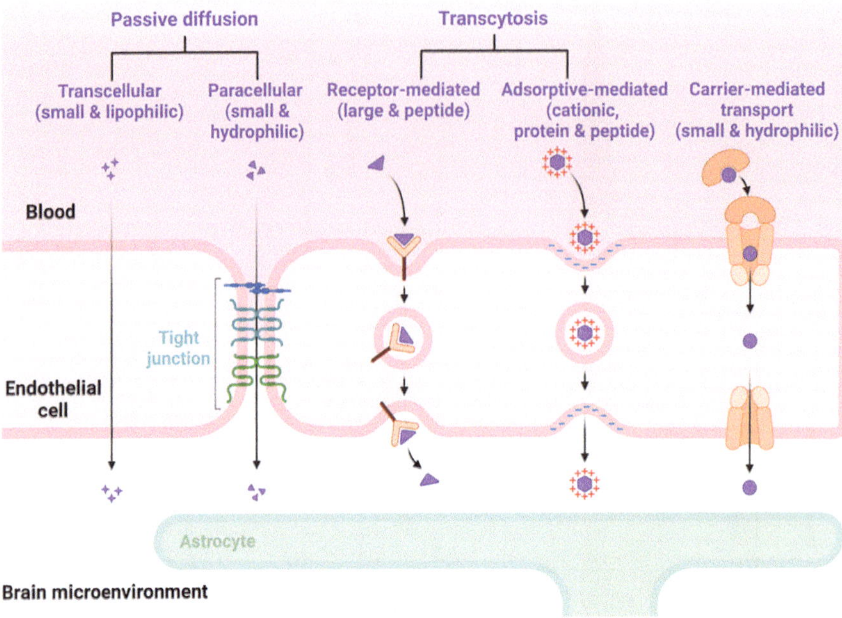

Fig. 8 Transport pathways of drug NCs from blood to the brain across the BBB. (Image was created using BioRender 2025)

expressed on the BBB [93, 94]. Moreover, carrier-mediated transcytosis (CMT) leverages specific membrane transport proteins to shuttle small molecules across the BBB. Although traditionally reserved for nutrients or amino acids, drug NCs can be engineered to mimic these natural substrates [65]. Adsorptive-mediated transcytosis (AMT) is another potential route for BBB penetration, relying on the electrostatic interaction between the positively charged NCs and negatively charged endothelial membranes, resulting in drug internalisation and transcytosis [95, 96]. This non-specific pathway requires careful optimisation to equilibrate transport efficiency with potential side effects [97]. In terms of paracellular transport, drug NCs' movement through the endothelial cells is generally restricted by the tight junctions of BBB. However, it can happen under certain pathological conditions, like inflammation, or with the use of BBB-disrupting agents [98]. This non-specific pathway typically occurs for low MW hydrophilic drugs; nonetheless, it may be exploited in combination with other strategies to enhance CNS drug access [99]. Additionally, NCs-mediated permeability enhancement can be employed to temporarily increase the BBB permeability, facilitating the entry of drug NCs [100]. For instance, NCs can be combined with osmotic agents like mannitol, which transiently disrupt the tight junctions between endothelial cells. However, it is non-selective and carries inherent risks, including increased exposure to neurotoxins [101]. Collectively, researchers are able to exploit these pathways to overcome the formidable challenges of crossing the BBB, opening new avenues for the effective management of CNS disorders.

3 Author Perspective

NCs offer significant benefits in drug delivery to the CNS due to their ability to improve the bioavailability of poorly water-soluble drugs and their ease of surface modification. They increase the surface area for drug absorption and possess high drug loading, leading to faster dissolution rates and higher drug accumulation in the brain [11]. However, some challenges remain that must be solved to fully realise the potential of NCs-based therapies. NCs are best suited for delivering small-molecule drugs, whereas they are less applicable to larger biomolecules such as proteins and nucleic acids, which are preferably delivered by other nanocarriers like liposomes or polymeric NPs [102]. On the other hand, NCs are prone to forming a protein corona, a layer of adsorbed proteins from the bloodstream, which poses serious detriments to active targeting in brain drug delivery by obscuring ligands on the NCs surface and reducing the specificity and effectiveness [103]. In addition, interactions between NCs and macrophages can trigger immune system activation, consequently inducing inflammation and tissue damage. In this case, PEGylation is commonly employed for 'stealth' coating of drug NCs, which helps them evade immune detection and opsonisation [57, 104]. Drug resistance represents an additional concern, particularly in treating brain tumours and infections such as cerebral malaria, in which inadequate drug penetration into the CNS not only diminishes therapeutic efficacy but also contributes to treatment failure [25]. Therefore, there is a pressing need for advanced delivery systems to enhance CNS drug NCs accumulation. Patient acceptance is another hurdle, where complex treatment regimens and invasive delivery methods can reduce compliance [36], especially in chronic CNS disorders such as Alzheimer's and Parkinson's, requiring clear communication and education to build trust and understanding in this innovative technology [105].

Furthermore, numerous CNS-targeted drugs still face challenges in crossing biological barriers, especially the BBB, whereas NCs excel at overcoming these formidable obstacles in CNS drug delivery [35]. The small particle size, coupled with surface modifications to possess desirable properties such as brain targeting, facilitates more efficient transport across or bypassing the BBB, making them adaptable for various administration pathways [11]. Beyond crossing the BBB, targeting specific CNS regions also poses a challenge, as treatments must be delivered precisely to the affected areas without harming healthy tissues [18]. It is crucial for treating brain tumours or neuroinflammatory conditions that require localised drug action. NCs stability at the target site is equally important to avoid degradation or premature release [106]. Besides, maintaining biocompatibility while minimising toxicity is crucial to prevent immune responses or unwanted NCs accumulation in non-target organs. In order to mitigate this, the materials used for NCs' coating must be carefully selected for safe degradation within the body, promoting a prolonged circulation for effective CNS targeting, while avoiding rapid elimination [35]. Furthermore, achieving consistent industrial-scale production (uniform size, shape, and functional properties) presents significant challenges, as potential variations during NCs manufacturing can impact their performance [36]. Environmental concerns arise from their production and disposal, having unintended consequences [107]. At the same time, there remains a considerable gap between industrial and clinical translation [108]. Finally, the regulatory approval for NCs-based therapies remains a challenge, as the regulatory landscape for nanomedicines continues to evolve, requiring standardised methods to evaluate their safety and efficacy, posing additional obstacles to bringing these advanced therapies to market [109].

Future developments in NCs delivery will aim to enhance the precision of drug targeting within the CNS. This involves the application of more sophisticated surface modifications, such as conjugating NCs with ligands that selectively bind to specific receptors on neurons, glial cells, or endothelial cells in the brain [74]. These innovations could enable region-specific or cell-specific drug delivery, thereby improving therapeutic outcomes while minimising off-target effects. Emerging fields like precision medicine and nanobiotechnology will drive the design of more intelligent NCs capable of distinguishing between healthy and diseased cells or tissues, leading to more effective and targeted treatments for CNS disorders. On the other hand, future research will focus on designing drug NCs that can evade drug efflux pumps or counteract resistance mechanisms at the BBB [110]. By integrating multi-drug delivery systems, NCs are able to offer a synergistic effect via combination therapies that reduce the required dosage of each agent, thereby minimising side effects while maintaining or enhancing efficacy [111]. For example, NCs co-loaded with anti-inflammatory and antioxidant drugs, such as curcumin and quercetin, have demonstrated synergistic effects in neurodegenerative disease treatments, improving both therapeutic outcomes and patient adherence [112]. Furthermore, innovative administration platforms to improve BBB penetration, such as utilising NCs to trigger receptor-mediated transcytosis or exploring non-invasive routes like intranasal delivery, will be crucial for enhancing the efficacy of CNS-targeted therapies [45, 113, 114].

Increasing emphasis is being placed on developing the stimuli-responsive NCs, which release drug payloads selectively in response to specific physiological conditions. This capability provides controlled and sustained drug release, which is pivotal for treating chronic CNS disorders [115]. Besides, the development of NCs that respond to disease progression markers could enable dynamic, adaptive treatments, particularly for neurodegenerative diseases [116, 117]. As personalised medicine advances, future NCs-based therapies will likely be customised to the molecular and genetic profiles of individual patients [118]. This approach will involve customising the size, composition, and surface properties of drug NCs to match the unique characteristics of BBB permeability and different CNS diseases. Advances in pharmacogenomics and biomarker identification will further support the development of individualised NCs-based treatments, particularly for complex CNS conditions with varying pathological presentations [118].

There is a growing demand for long-acting NCs formulations that provide sustained drug release over extended periods, reducing the requirement for frequent dosing, particularly for chronic conditions like epilepsy, multiple sclerosis, and chronic pain. Innovations in prolonged drug NCs release could involve combination with biodegradable polymers, ensuring steady drug release and safety in the brain [38, 115]. It has the potential to significantly improve patient adherence and ease the burden of long-term medication regimens. Additionally, NCs also play a role in theranostics, where they can be used for both therapeutic and diagnostic purposes [119]. Combining imaging agents with drug-loaded NCs will allow for real-time tracking of drug distribution, monitoring treatment progress, and adjusting dosing regimens accordingly [36, 120]. This dual-function capability could be especially valuable for the treatment of brain tumours or neurodegenerative diseases, as monitoring drug efficacy and disease progression is crucial. For NCs-based therapies translated from the lab to clinical applications, regulatory frameworks will need to adapt to the unique challenges posed by nanomedicines [36]. Future efforts will involve the establishment of clearer guidelines to ensure consistency, safety, and efficacy in large-scale NCs manufacturing. Advances in nanotechnology will make it more cost-effective and scalable, even accessible to the widespread populations [36].

4 Conclusion

Drug delivery to the CNS remains one of the most significant challenges in modern medicine, primarily due to the restrictive nature of the BBB. Although the BBB serves as a critical defence mechanism protecting the brain from harmful substances, it also impedes the effective delivery of therapeutics, particularly hydrophilic and macromolecular drugs. Advances in nanotechnology, specifically the development of drug NCs, have emerged as a transformative solution to overcome these barriers. By improving drug solubility, stability, and bioavailability, NCs offer a highly adaptable platform for CNS-targeted therapies.

NCs leverage sophisticated mechanisms such as receptor-mediated and adsorptive-mediated transcytosis to penetrate the BBB. Their ability to optimise surface properties facilitates targeted drug delivery to specific CNS regions, ensuring controlled release while minimising systemic toxicity. Moreover, NCs address conventional limitations of NP-based therapies, such as low drug loading and potential toxicity. Innovative delivery methods, including intranasal administration, oral formulations, and MN technology, further enhance the versatility and efficacy of NCs-based therapies. Surface modifications such as PEGylation, charge tuning, and ligand attachment improve the specificity and efficiency of drug delivery, making NCs particularly effective for treating complex CNS disorders like Alzheimer's disease, Parkinson's disease, epilepsy, and brain cancers.

Although challenges remain, such as achieving long-term stability, scalability, and navigating interindividual variability, NCs-based therapies represent a significant leap forward in CNS drug delivery. With advancements in precision targeting, personalised nanomedicine, and theranostic applications, NCs are poised to revolutionise the treatment landscape for CNS conditions. These innovations pave the way for more effective, safer, and tailored therapies, bringing us closer to overcoming the limitations of traditional CNS treatments and translating these promising technologies into widespread clinical use.

References

1. S. Patra, M. Singh, K. Wasnik, D. Pareek, P.S. Gupta, S. Mukherjee, P. Paik, Polymeric nanoparticle based diagnosis and Nanomedicine for treatment and development of vaccines for cerebral malaria: A review on recent advancement. ACS Appl. Bio Mater. **4**, 7342–7365 (2021). https://doi.org/10.1021/ACSABM.1C00635/ASSET/IMAGES/LARGE/MT1C00635_0010.JPEG
2. L. Thau, V. Reddy, P. Singh, Anatomy, central nervous system. BMJ **1**, 478–478 (2022). https://doi.org/10.1136/bmj.1.4293.478
3. A. Fisher, M. Memo, F. Stocchi, I. Hanin (Eds.), Advances in Alzheimer's and Parkinson's disease: insights, progress, and perspectives (2007)
4. S. Love, Demyelinating diseases. J. Clin. Pathol. **59**, 1151 (2006). https://doi.org/10.1136/JCP.2005.031195
5. D. Kuriakose, Z. Xiao, Pathophysiology and treatment of stroke: Present status and future perspectives. Int. J. Mol. Sci. **21**, 1–24 (2020). https://doi.org/10.3390/IJMS21207609
6. S. Li, I.P. Nguyen, K. Urbanczyk, Common infectious diseases of the central nervous system – clinical features and imaging characteristics. Quant. Imaging Med. Surg. **10**, 2227 (2020). https://doi.org/10.21037/QIMS-20-886
7. M.A. van der Marck, J.G. Kalf, I.H.W.M. Sturkenboom, M.J. Nijkrake, M. Munneke, B.R. Bloem, Multidisciplinary care for patients with Parkinson's disease. Parkinsonism Relat. Disord. **15**, S219–S223 (2009). https://doi.org/10.1016/S1353-8020(09)70819-3
8. M.G. Ewend, D.E. Morris, L.A. Carey, A.M. Ladha, S. Brem, Guidelines for the initial management of metastatic brain tumors: Role of surgery, radiosurgery, and radiation therapy. J. Natl. Compr. Cancer Netw. **6**, 505–514 (2008). https://doi.org/10.6004/JNCCN.2008.0038
9. T.K. Owonikoko, J. Arbiser, A. Zelnak, H.K.G. Shu, H. Shim, A.M. Robin, S.N. Kalkanis, T.G. Whitsett, B. Salhia, N.L. Tran, T. Ryken, M.K. Moore, K.M. Egan, J.J. Olson, Current approaches to the treatment of metastatic brain tumours. Nat. Rev. Clin. Oncol. **11**, 203–222 (2014). https://doi.org/10.1038/nrclinonc.2014.25
10. V.K. Gribkoff, L.K. Kaczmarek, The need for new approaches in CNS drug discovery: Why drugs have failed, and what can be done to improve outcomes. Neuropharmacology **120**, 11 (2016). https://doi.org/10.1016/J.NEUROPHARM.2016.03.021
11. M.B. McGuckin, J. Wang, R. Ghanma, N. Qin, S.D. Palma, R.F. Donnelly, A.J. Paredes, Nanocrystals as a master key to deliver hydrophobic drugs via multiple administration routes. J. Control. Release **345**, 334–353 (2022). https://doi.org/10.1016/J.JCONREL.2022.03.012
12. W.M. Pardridge, Drug transport across the blood–brain barrier. J. Cereb. Blood Flow Metab. **32**, 1959 (2012). https://doi.org/10.1038/JCBFM.2012.126
13. S. Gupta, S. Dhanda, R. Sandhir, Anatomy and physiology of blood-brain barrier, in *Brain Targeted Drug Delivery Systems: A Focus on Nanotechnology and Nanoparticulates*, (2019), pp. 7–31. https://doi.org/10.1016/B978-0-12-814001-7.00002-0
14. R. Daneman, A. Prat, The blood–brain barrier. Cold Spring Harb. Perspect. Biol. **7** (2015). https://doi.org/10.1101/CSHPERSPECT.A020412

15. P. Ballabh, A. Braun, M. Nedergaard, The blood–brain barrier: an overview: Structure, regulation, and clinical implications. Neurobiol. Dis. **16**, 1–13 (2004). https://doi.org/10.1016/J.NBD.2003.12.016
16. C.P. Profaci, R.N. Munji, R.S. Pulido, R. Daneman, The blood-brain barrier in health and disease: Important unanswered questions. J. Exp. Med. **217** (2020). https://doi.org/10.1084/JEM.20190062
17. F. Fernández-Klett, J. Priller, The fibrotic scar in neurological disorders. Brain Pathol. **24**, 404 (2014). https://doi.org/10.1111/BPA.12162
18. J. Gao, Z. Xia, S. Gunasekar, C. Jiang, J.M. Karp, N. Joshi, Precision drug delivery to the central nervous system using engineered nanoparticles. Nat. Rev. Mater. **9**, 567–588 (2024). https://doi.org/10.1038/s41578-024-00695-w
19. E. Henke, R. Nandigama, S. Ergün, Extracellular matrix in the tumor microenvironment and its impact on cancer therapy. Front. Mol. Biosci. **6**, 160 (2019). https://doi.org/10.3389/FMOLB.2019.00160
20. J. Liu, H. Cabral, P. Mi, Nanocarriers address intracellular barriers for efficient drug delivery, overcoming drug resistance, subcellular targeting and controlled release. Adv. Drug Deliv. Rev. **207**, 115239 (2024). https://doi.org/10.1016/J.ADDR.2024.115239
21. V. Agrahari, P.A. Burnouf, T. Burnouf, V. Agrahari, Nanoformulation properties, characterization, and behavior in complex biological matrices: Challenges and opportunities for brain-targeted drug delivery applications and enhanced translational potential. Adv. Drug Deliv. Rev. **148**, 146–180 (2019). https://doi.org/10.1016/J.ADDR.2019.02.008
22. K. Bossers, K.T.S. Wirz, G.F. Meerhoff, A.H.W. Essing, J.W. Van Dongen, P. Houba, C.G. Kruse, J. Verhaagen, D.F. Swaab, Concerted changes in transcripts in the prefrontal cortex precede neuropathology in Alzheimer's disease. Brain **133**, 3699–3723 (2010). https://doi.org/10.1093/BRAIN/AWQ258
23. E.D. Louis, P.L. Faust, J.P.G. Vonsattel, L.S. Honig, A. Rajput, A. Rajput, R. Pahwa, K.E. Lyons, W.G. Ross, R.J. Elble, C. Erickson-Davis, C.B. Moskowitz, A. Lawton, Torpedoes in Parkinson's disease, Alzheimer's disease, essential tremor, and control brains. Mov. Disord. **24**, 1600–1605 (2009). https://doi.org/10.1002/MDS.22567
24. M.N. Pangalos, L.E. Schechter, O. Hurko, Drug development for CNS disorders: strategies for balancing risk and reducing attrition. Nat. Rev. Drug Discov. **6**, 521–532 (2007). https://doi.org/10.1038/nrd2094
25. W. Löscher, H. Potschka, Drug resistance in brain diseases and the role of drug efflux transporters. Nat. Rev. Neurosci. **6**, 591–602 (2005). https://doi.org/10.1038/nrn1728
26. M.I. Alam, S. Beg, A. Samad, S. Baboota, K. Kohli, J. Ali, A. Ahuja, M. Akbar, Strategy for effective brain drug delivery. Eur. J. Pharm. Sci. **40**, 385–403 (2010). https://doi.org/10.1016/J.EJPS.2010.05.003
27. E. Nance, S.H. Pun, R. Saigal, D.L. Sellers, Drug delivery to the central nervous system. Nat. Rev. Mater. **7**, 314–331 (2021). https://doi.org/10.1038/s41578-021-00394-w
28. S. Krol, R. Macrez, F. Docagne, G. Defer, S. Laurent, M. Rahman, M.J. Hajipour, P.G. Kehoe, M. Mahmoudi, Therapeutic benefits from nanoparticles: the potential significance of nanoscience in diseases with compromise to the blood brain barrier. Chem. Rev. **113**, 1877–1903 (2013). https://doi.org/10.1021/CR200472G
29. N. Zhao, N.L. Francis, H.R. Calvelli, P.V. Moghe, Microglia-targeting nanotherapeutics for neurodegenerative diseases. APL Bioeng **4**, 030902 (2020). https://doi.org/10.1063/5.0013178
30. M. Battaglini, A. Marino, M. Montorsi, A. Carmignani, M.C. Ceccarelli, G. Ciofani, Nanomaterials as microglia modulators in the treatment of central nervous system disorders. Adv. Healthc. Mater. **13**, 2304180 (2024). https://doi.org/10.1002/ADHM.202304180
31. S. Fakhri, S. Abdian, S.N. Zarneshan, S.Z. Moradi, M.H. Farzaei, M. Abdollahi, Nanoparticles in combating neuronal dysregulated signaling pathways: Recent approaches to the Nanoformulations of phytochemicals and synthetic drugs against neurodegenerative diseases. Int. J. Nanomed. **17**, 299 (2022). https://doi.org/10.2147/IJN.S347187

32. L. Zhang, F.X. Gu, J.M. Chan, A.Z. Wang, R.S. Langer, O.C. Farokhzad, Nanoparticles in medicine: Therapeutic applications and developments. Clin. Pharmacol. Ther. **83**, 761–769 (2008). https://doi.org/10.1038/SJ.CLPT.6100400
33. A. Biswas, A.T. Nagaraja, M.J. McShane, Fabrication of nanocapsule carriers from multilayer-coated vaterite calcium carbonate nanoparticles. ACS Appl. Mater. Interfaces **6**, 21193–21201 (2014). https://doi.org/10.1021/AM5061195
34. M.R. Gigliobianco, C. Casadidio, R. Censi, P. di Martino, Nanocrystals of poorly soluble drugs: Drug bioavailability and physicochemical stability. Pharmaceutics **10** (2018). https://doi.org/10.3390/PHARMACEUTICS10030134
35. S. Mule, P. Khairnar, R. Shukla, Recent advances in nanocrystals heralding greater potential in brain delivery. Part. Part. Syst. Charact. **39**, 2200087 (2022). https://doi.org/10.1002/PPSC.202200087
36. J.U.A.H. Junghanns, R.H. Müller, Nanocrystal technology, drug delivery and clinical applications. Int. J. Nanomed. **3**, 295 (2008). https://doi.org/10.2147/IJN.S595
37. P. Kocbek, S. Baumgartner, J. Kristl, Preparation and evaluation of nanosuspensions for enhancing the dissolution of poorly soluble drugs. Int. J. Pharm. **312**, 179–186 (2006). https://doi.org/10.1016/J.IJPHARM.2006.01.008
38. E. Zingale, A. Bonaccorso, C. Carbone, T. Musumeci, R. Pignatello, Drug nanocrystals: Focus on brain delivery from therapeutic to diagnostic applications. Pharmaceutics **14**, 691 (2022). https://doi.org/10.3390/PHARMACEUTICS14040691
39. C. Wu, B. Li, Y. Zhang, T. Chen, C. Chen, W. Jiang, Q. Wang, T. Chen, Intranasal delivery of paeoniflorin nanocrystals for brain targeting. Asian J Pharm Sci **15**, 326–335 (2020). https://doi.org/10.1016/J.AJPS.2019.11.002
40. T. Chen, C. Li, Y. Li, X. Yi, S.M.Y. Lee, Y. Zheng, Oral delivery of a nanocrystal formulation of Schisantherin a with improved bioavailability and brain delivery for the treatment of Parkinson's disease. Mol. Pharm. **13**, 3864–3875 (2016). https://doi.org/10.1021/ACS.MOLPHARMACEUT.6B00644
41. S. Xiong, W. Liu, D. Li, X. Chen, F. Liu, D. Yuan, H. Pan, Q. Wang, S. Fang, T. Chen, Oral delivery of Puerarin nanocrystals to improve brain accumulation and anti-parkinsonian efficacy. Mol. Pharm. **16**(4), 1444–1455 (2019). https://doi.org/10.1021/acs.molpharmaceut.8b01012
42. A. Bonaccorso, M.R. Gigliobianco, R. Pellitteri, D. Santonocito, C. Carbone, P. Di Martino, G. Puglisi, T. Musumeci, Optimization of curcumin nanocrystals as promising strategy for nose-to-brain delivery application. Pharmaceutics **12**(5), 476 (2020)
43. J. Hao, J. Zhao, S. Zhang, T. Tong, Q. Zhuang, K. Jin, W. Chen, H. Tang, Fabrication of an ionic-sensitive in situ gel loaded with resveratrol nanosuspensions intended for direct nose-to-brain delivery. Colloids Surf. B Biointerfaces **147**, 376–386 (2016). https://doi.org/10.1016/J.COLSURFB.2016.08.011
44. J. Udabe, S. Martin-Saldaña, Y. Tao, M. Picchio, A. Beloqui, A.J. Paredes, M. Calderón, Unveiling the potential of surface polymerized drug nanocrystals in targeted delivery. ACS Appl. Mater. Interfaces **16**, 47124–47136 (2024). https://doi.org/10.1021/ACSAMI.4C07669
45. M.T.A. Abbate, I.K. Ramöller, A.H. Sabri, A.J. Paredes, A.J. Hutton, P.E. McKenna, K. Peng, J.A. Hollett, H.O. McCarthy, R.F. Donnelly, Formulation of antiretroviral nanocrystals and development into a microneedle delivery system for potential treatment of HIV-associated neurocognitive disorder (HAND). Int. J. Pharm. **640**, 123005 (2023). https://doi.org/10.1016/J.IJPHARM.2023.123005
46. S. Xiong, W. Liu, Y. Zhou, Y. Mo, Y. Liu, X. Chen, H. Pan, D. Yuan, Q. Wang, T. Chen, Enhancement of oral bioavailability and anti-parkinsonian efficacy of resveratrol through a nanocrystal formulation. Asian J. Pharm. Sci. **15**, 518 (2019). https://doi.org/10.1016/J.AJPS.2019.04.003
47. H.P. Patel, P.S. Chaudhari, P.A. Gandhi, B.V. Desai, D.T. Desai, P.P. Dedhiya, B.A. Vyas, F.A. Maulvi, Nose to brain delivery of tailored clozapine nanosuspension stabilized using (+)-alpha-tocopherol polyethylene glycol 1000 succinate: Optimization and in vivo

pharmacokinetic studies. Int. J. Pharm. **600**, 120474 (2021). https://doi.org/10.1016/J.IJPHARM.2021.120474
48. X. Dong, Current strategies for brain drug delivery. Theranostics **8**, 1481 (2018). https://doi.org/10.7150/THNO.21254
49. Z. Ma, H. Zhang, Y. Wang, X. Tang, Development and evaluation of intramuscularly administered nano/microcrystal suspension. Expert Opin. Drug Deliv. **16**, 347–361 (2019). https://doi.org/10.1080/17425247.2019.1588248
50. N. Das, P. Prabhu, Emerging avenues for the management of cerebral malaria. J. Pharm. Pharmacol. **74**, 800–811 (2022). https://doi.org/10.1093/JPP/RGAC003
51. M. Riemma Pierre, F. Rossetti, Microneedle-based drug delivery systems for transdermal route. Curr. Drug Targets **15**, 281–291 (2014). https://doi.org/10.2174/13894501113146660232
52. R.F. Donnelly, T.R.R. Singh, D.I.J. Morrow, A.D. Woolfson, Microneedle-mediated transdermal and intradermal drug delivery (2012). https://doi.org/10.1002/9781119959687
53. K. Ita, Transdermal delivery of drugs with microneedles – potential and challenges. Pharmaceutics **7**, 90 (2015). https://doi.org/10.3390/PHARMACEUTICS7030090
54. S.V. Dhuria, L.R. Hanson, W.H. Frey, Intranasal delivery to the central nervous system: mechanisms and experimental considerations. J. Pharm. Sci. **99**, 1654–1673 (2010). https://doi.org/10.1002/JPS.21924
55. K. Selvaraj, K. Gowthamarajan, V.V.S.R. Karri, Nose to brain transport pathways an overview: potential of nanostructured lipid carriers in nose to brain targeting. Artif. Cells Nanomed. Biotechnol. **46**, 2088–2095 (2018). https://doi.org/10.1080/21691401.2017.1420073
56. L. Xue, J. Ding, Y. Liu, Y. Ma, C. Yang, W. Wang, Y. Wang, Strategies and methods of nanocrystal technology for targeting drug delivery. J. Nanopart. Res. **26**, 1–19 (2024). https://doi.org/10.1007/S11051-024-06025-Z/FIGURES/8
57. K.R. Gajbhiye, A. Pawar, K.R. Mahadik, V. Gajbhiye, PEGylated nanocarriers: A promising tool for targeted delivery to the brain. Colloids Surf. B Biointerfaces **187**, 110770 (2020). https://doi.org/10.1016/J.COLSURFB.2019.110770
58. P.L. Turecek, M.J. Bossard, F. Schoetens, I.A. Ivens, PEGylation of biopharmaceuticals: A review of chemistry and nonclinical safety information of approved drugs. J. Pharm. Sci. **105**, 460–475 (2016). https://doi.org/10.1016/J.XPHS.2015.11.015
59. J.S. Suk, Q. Xu, N. Kim, J. Hanes, L.M. Ensign, PEGylation as a strategy for improving nanoparticle-based drug and gene delivery. Adv. Drug Deliv. Rev. **99**, 28–51 (2016). https://doi.org/10.1016/J.ADDR.2015.09.012
60. H. Otsuka, Y. Nagasaki, K. Kataoka, PEGylated nanoparticles for biological and pharmaceutical applications. Adv. Drug Deliv. Rev. **55**, 403–419 (2003). https://doi.org/10.1016/S0169-409X(02)00226-0
61. Y. Xiao, Q. Liu, A.J. Clulow, T. Li, M. Manohar, E.P. Gilbert, L. de Campo, A. Hawley, B.J. Boyd, PEGylation and surface functionalization of liposomes containing drug nanocrystals for cell-targeted delivery. Colloids Surf. B Biointerfaces **182**, 110362 (2019). https://doi.org/10.1016/J.COLSURFB.2019.110362
62. J.T. Huckaby, S.K. Lai, PEGylation for enhancing nanoparticle diffusion in mucus. Adv. Drug Deliv. Rev. **124**, 125–139 (2018). https://doi.org/10.1016/j.addr.2017.08.010
63. L. Lu, Q. Xu, J. Wang, S. Wu, Z. Luo, W. Lu, Drug nanocrystals for active tumor-targeted drug delivery. Pharmaceutics **14**, 797 (2022). https://doi.org/10.3390/PHARMACEUTICS14040797
64. M. Zhou, X. Zhang, C. Yu, X. Nan, X. Chen, X. Zhang, Shape regulated anticancer activities and systematic toxicities of drug nanocrystals in vivo. Nanomedicine **12**, 181–189 (2016). https://doi.org/10.1016/J.NANO.2015.09.006
65. M. Guo, M. Wei, W. Li, M. Guo, C. Guo, M. Ma, Y. Wang, Z. Yang, M. Li, Q. Fu, L. Yang, Z. He, Impacts of particle shapes on the oral delivery of drug nanocrystals: Mucus permeation,

transepithelial transport and bioavailability. J. Control. Release **307**, 64–75 (2019). https://doi.org/10.1016/J.JCONREL.2019.06.015
66. F. Guo, J. Shang, H. Zhao, K. Lai, Y. Li, Z. Fan, Z. Hou, G. Su, Cube-shaped theranostic paclitaxel prodrug nanocrystals with surface functionalization of SPC and MPEG-DSPE for imaging and chemotherapy. Colloids Surf. B Biointerfaces **160**, 649–660 (2017). https://doi.org/10.1016/J.COLSURFB.2017.10.013
67. M.A. Saleh, J.M.M. Mohamed, J.J. Ruby, S. Kanthiah, Y.F. Alanazi, K.A. Majrashi, S.M. Alshahrani, M.A. Eladl, F.S. Alaryani, M. El-Sherbiny, F. Menaa, Preparation of Memantine-loaded chitosan nanocrystals: In vitro and ex vivo toxicity analysis. Crystals **13**, 21 (2023). https://doi.org/10.3390/CRYST13010021
68. P. Mali, A.P. Sherje, Cellulose nanocrystals: Fundamentals and biomedical applications. Carbohydr. Polym. **275**, 118668 (2022). https://doi.org/10.1016/J.CARBPOL.2021.118668
69. M. Bai, M. Yang, J. Gong, H. Xu, Z. Wei, Progress and principle of drug nanocrystals for tumor targeted delivery. AAPS PharmSciTech **23**, 1–21 (2021). https://doi.org/10.1208/S12249-021-02200-W
70. F. Zhang, Y.A. Lin, S. Kannan, R.M. Kannan, Targeting specific cells in the brain with nanomedicines for CNS therapies. J. Control. Release **240**, 212–226 (2016). https://doi.org/10.1016/J.JCONREL.2015.12.013
71. J.D. Byrne, T. Betancourt, L. Brannon-Peppas, Active targeting schemes for nanoparticle systems in cancer therapeutics. Adv. Drug Deliv. Rev. **60**, 1615–1626 (2008). https://doi.org/10.1016/J.ADDR.2008.08.005
72. V. Patel, V. Chavda, J. Shah, Nanotherapeutics in neuropathologies: Obstacles, challenges and recent advancements in CNS targeted drug delivery systems. Curr. Neuropharmacol. **19**, 693–710 (2020). https://doi.org/10.2174/1570159X18666200807143526/CITE/REFWORKS
73. Y.H. Bae, K. Park, Targeted drug delivery to tumors: Myths, reality and possibility. J. Control. Release **153**, 198 (2011). https://doi.org/10.1016/J.JCONREL.2011.06.001
74. Z. Chai, D. Ran, L. Lu, C. Zhan, H. Ruan, X. Hu, C. Xie, K. Jiang, J. Li, J. Zhou, J. Wang, Y. Zhang, R.H. Fang, L. Zhang, W. Lu, Ligand-modified cell membrane enables the targeted delivery of drug nanocrystals to glioma. ACS Nano **13**, 5591–5601 (2019). https://doi.org/10.1021/ACSNANO.9B00661
75. J. Zhao, Y. Liu, L. Wang, Y. Zhou, J. Du, Y. Wang, Functional and modified nanocrystals technology for target drug delivery. J. Nanosci. Nanotechnol. **18**, 5207–5221 (2018). https://doi.org/10.1166/JNN.2018.15421
76. Y. Lu, J. Qi, X. Dong, W. Zhao, W. Wu, The in vivo fate of nanocrystals. Drug Discov. Today **22**, 744–750 (2017). https://doi.org/10.1016/J.DRUDIS.2017.01.003
77. P. Khairnar, M. Handa, R. Shukla, Nanocrystals: An approachable delivery system for anticancer therapeutics. Curr. Drug Metab. **23**, 603–615 (2022). https://doi.org/10.2174/1389200223666220610165850/CITE/REFWORKS
78. S. Mura, J. Nicolas, P. Couvreur, Stimuli-responsive nanocarriers for drug delivery. Nat. Mater. **12**, 991–1003 (2013). https://doi.org/10.1038/nmat3776
79. G. Liu, J.F. Lovell, L. Zhang, Y. Zhang, Stimulus-responsive Nanomedicines for disease diagnosis and treatment. Int. J. Mol. Sci. **21**, 6380 (2020). https://doi.org/10.3390/IJMS21176380
80. Y.W. Hu, Y.Z. Du, N. Liu, X. Liu, T.T. Meng, B.L. Cheng, J.B. He, J. You, H. Yuan, F.Q. Hu, Selective redox-responsive drug release in tumor cells mediated by chitosan based glycolipid-like nanocarrier. J. Control. Release **206**, 91–100 (2015). https://doi.org/10.1016/J.JCONREL.2015.03.018
81. V. Torchilin, Multifunctional and stimuli-sensitive pharmaceutical nanocarriers. Eur. J. Pharm. Biopharm. **71**, 431–444 (2009). https://doi.org/10.1016/J.EJPB.2008.09.026
82. A. Shafiu Kamba, M. Ismail, T.A. Tengku Ibrahim, Z.A.B. Zakaria, A pH-sensitive, biobased calcium carbonate aragonite nanocrystal as a novel anticancer delivery system. Biomed. Res. Int. **2013** (2013). https://doi.org/10.1155/2013/587451

83. I.S. Mohammad, W. He, L. Yin, A smart paclitaxel-disulfiram Nanococrystals for efficient MDR reversal and enhanced apoptosis. Pharm. Res. **35** (2018). https://doi.org/10.1007/S11095-018-2370-0
84. G. Huang, J. Xie, S. Shuai, S. Wei, Y. Chen, Z. Guan, Q. Zheng, P. Yue, C. Wang, Nose-to-brain delivery of drug nanocrystals by using Ca2+ responsive deacetylated gellan gum based in situ-nanogel. Int. J. Pharm. **594**, 120182 (2021). https://doi.org/10.1016/J.IJPHARM.2020.120182
85. C. Chen, L. Wang, F. Cao, X. Miao, T. Chen, Q. Chang, Y. Zheng, Formulation of 20(S)-protopanaxadiol nanocrystals to improve oral bioavailability and brain delivery. Int. J. Pharm. **497**, 239–247 (2016). https://doi.org/10.1016/J.IJPHARM.2015.12.014
86. N. Mittapelly, M. Thalla, G. Pandey, V.T. Banala, S. Sharma, A. Arya, S. Mishra, K. Mitra, S. Shukla, P.R. Mishra, Long acting Ionically paired Embonate based nanocrystals of donepezil for the treatment of Alzheimer's disease: a proof of concept study. Pharm. Res. **34**, 2322–2335 (2017). https://doi.org/10.1007/S11095-017-2240-1
87. Q. Wang, A.Y. Sun, J. Pardeike, R.H. Müller, A. Simonyi, G.Y. Sun, Neuroprotective effects of a nanocrystal formulation of sPLA2 inhibitor PX-18 in cerebral ischemia/reperfusion in gerbils. Brain Res. **1285**, 188–195 (2009). https://doi.org/10.1016/J.BRAINRES.2009.06.022
88. A. Mhaske, S. Shukla, K. Ahirwar, K.K. Singh, R. Shukla, Receptor-assisted nanotherapeutics for overcoming the blood–brain barrier. Mol. Neurobiol. **61**, 8702–8738 (2024). https://doi.org/10.1007/S12035-024-04015-9
89. H. Choudhury, M. Pandey, P.X. Chin, Y.L. Phang, J.Y. Cheah, S.C. Ooi, K.K. Mak, M.R. Pichika, P. Kesharwani, Z. Hussain, B. Gorain, Transferrin receptors-targeting nanocarriers for efficient targeted delivery and transcytosis of drugs into the brain tumors: a review of recent advancements and emerging trends. Drug Deliv. Transl. Res. **8**, 1545–1563 (2018). https://doi.org/10.1007/S13346-018-0552-2/TABLES/2
90. Y. Qi, T. Zhang, C. Jing, S. Liu, C. Zhang, P.J.J. Alvarez, W. Chen, Nanocrystal facet modulation to enhance transferrin binding and cellular delivery. Nat. Commun. **11**, 1–10 (2020). https://doi.org/10.1038/s41467-020-14972-z
91. R.P. Moura, C. Martins, S. Pinto, F. Sousa, B. Sarmento, Blood-brain barrier receptors and transporters: an insight on their function and how to exploit them through nanotechnology. Expert Opin. Drug Deliv. **16**, 271–285 (2019). https://doi.org/10.1080/17425247.2019.1583205
92. F. Fay, B.L. Sanchez-Gaytan, D.P. Cormode, T. Skajaa, E.A. Fisher, Z.A. Fayad, W.J.M. Mulder, Nanocrystal Core lipoprotein Biomimetics for imaging of lipoproteins and associated diseases. Curr. Cardiovasc. Imaging Rep. **6**, 45–54 (2013). https://doi.org/10.1007/S12410-012-9181-1/FIGURES/4
93. Z.X. Phuna, B.P. Panda, N.K.H. Shivashekaregowda, P. Madhavan, Recent development in nanocrystal based drug delivery for neurodegenerative diseases: Scope, challenges, current and future prospects. J Drug Deliv Sci Technol **68**, 102921 (2022). https://doi.org/10.1016/J.JDDST.2021.102921
94. Y. Liu, Y. Ma, J. Xu, Y. Chen, J. Xie, P. Yue, Q. Zheng, M. Yang, Apolipoproteins adsorption and brain-targeting evaluation of baicalin nanocrystals modified by combination of Tween80 and TPGS. Colloids Surf. B Biointerfaces **160**, 619–627 (2017). https://doi.org/10.1016/J.COLSURFB.2017.10.009
95. R.A. Razzak, G.J. Florence, F.J. Gunn-Moore, Approaches to CNS drug delivery with a focus on transporter-mediated Transcytosis. Int. J. Mol. Sci. **20**, E3108 (2019). https://doi.org/10.3390/IJMS20123108
96. R. Gauro, M. Nandave, V.K. Jain, K. Jain, Advances in dendrimer-mediated targeted drug delivery to the brain. J. Nanopart. Res. **23**, 1–20 (2021). https://doi.org/10.1007/S11051-021-05175-8/TABLES/2
97. F. Hervé, N. Ghinea, J.M. Scherrmann, CNS delivery via adsorptive transcytosis. AAPS J. **10**, 455–472 (2008). https://doi.org/10.1208/S12248-008-9055-2/TABLES/4

98. Z. Tian, Y. Mai, T. Meng, S. Ma, G. Gou, J. Yang, Nanocrystals for improving oral bioavailability of drugs: Intestinal transport mechanisms and influencing factors. AAPS PharmSciTech **22**, 1–17 (2021). https://doi.org/10.1208/S12249-021-02041-7/FIGURES/5
99. P.L. Tuma, A.L. Hubbard, Transcytosis: crossing cellular barriers. Physiol. Rev. **83**, 871–932 (2003). https://doi.org/10.1152/PHYSREV.00001.2003
100. D. Yenurkar, M. Nayak, S. Mukherjee, Recent advances of nanocrystals in cancer theranostics. Nanoscale Adv **5**, 4018–4040 (2023). https://doi.org/10.1039/D3NA00397C
101. Z. Hussain, H.E. Thu, I. Elsayed, M.A.S. Abourehab, S. Khan, M. Sohail, R.M. Sarfraz, M.A. Farooq, Nano-scaled materials may induce severe neurotoxicity upon chronic exposure to brain tissues: A critical appraisal and recent updates on predisposing factors, underlying mechanism, and future prospects. J. Control. Release **328**, 873–894 (2020). https://doi.org/10.1016/J.JCONREL.2020.10.053
102. Y. Li, D. Maciel, J. Rodrigues, X. Shi, H. Tomás, Biodegradable polymer Nanogels for drug/nucleic acid delivery. Chem. Rev. **115**, 8564–8608 (2015). https://doi.org/10.1021/CR500131F
103. M. Mahmoudi, M.P. Landry, A. Moore, R. Coreas, The protein corona from nanomedicine to environmental science. Nat. Rev. Mater. **8**, 422–438 (2023). https://doi.org/10.1038/s41578-023-00552-2
104. Q. Yang, S.K. Lai, Engineering well-characterized PEG-coated nanoparticles for elucidating biological barriers to drug delivery. Methods Mol. Biol. **1530**, 125–137 (2017). https://doi.org/10.1007/978-1-4939-6646-2_8
105. R. Ganesh Pillai, A.N. Bezbaruah, Perceptions and attitude effects on nanotechnology acceptance: an exploratory framework. J. Nanopart. Res. **19** (2017). https://doi.org/10.1007/S11051-016-3733-2
106. L. Gao, G. Liu, J. Ma, X. Wang, L. Zhou, X. Li, Drug nanocrystals: In vivo performances. J. Control. Release **160**, 418–430 (2012). https://doi.org/10.1016/J.JCONREL.2012.03.013
107. B. Nowack, T.D. Bucheli, Occurrence, behavior and effects of nanoparticles in the environment. Environ. Pollut. **150**, 5–22 (2007). https://doi.org/10.1016/J.ENVPOL.2007.06.006
108. M.L. Etheridge, S.A. Campbell, A.G. Erdman, C.L. Haynes, S.M. Wolf, J. McCullough, The big picture on nanomedicine: The state of investigational and approved nanomedicine products. Nanomedicine **9**, 1–14 (2013). https://doi.org/10.1016/J.NANO.2012.05.013
109. Considerations for Drug Products that Contain Nanomaterials | FDA., https://www.fda.gov/drugs/cder-small-business-industry-assistance-sbia/considerations-drug-products-contain-nanomaterials. Accessed 18 Sep 2024 (n.d.)
110. F. Sonvico, A. Clementino, F. Buttini, G. Colombo, S. Pescina, S.S. Guterres, A.R. Pohlmann, S. Nicoli, Surface-modified nanocarriers for nose-to-brain delivery: From bioadhesion to targeting. Pharmaceutics **10**, 34 (2018). https://doi.org/10.3390/PHARMACEUTICS10010034
111. E. Joseph, G. Singhvi, Multifunctional nanocrystals for cancer therapy: a potential nanocarrier, in *Nanomaterials for Drug Delivery and Therapy*, (2019), pp. 91–116. https://doi.org/10.1016/B978-0-12-816505-8.00007-2
112. S. Jain, S. Lenaghan, V. Dia, Q. Zhong, Co-delivery of curcumin and quercetin in shellac nanocapsules for the synergistic antioxidant properties and cytotoxicity against colon cancer cells. Food Chem. **428**, 136744 (2023). https://doi.org/10.1016/J.FOODCHEM.2023.136744
113. T.G. Agnihotri, G.S. Jadhav, B. Sahu, A. Jain, Recent trends of bioconjugated nanomedicines through nose-to-brain delivery for neurological disorders. Drug Deliv. Transl. Res. **12**, 3104–3120 (2022). https://doi.org/10.1007/S13346-022-01173-Y/FIGURES/4
114. J. Ahmad, N. Haider, M.A. Khan, S. Md, N.A. Alhakamy, M.M. Ghoneim, S. Alshehri, S. Sarim Imam, M.Z. Ahmad, A. Mishra, Novel therapeutic interventions for combating Parkinson's disease and prospects of nose-to-brain drug delivery. Biochem. Pharmacol. **195**, 114849 (2022). https://doi.org/10.1016/J.BCP.2021.114849

115. R.S. Kalhapure, S. Palekar, K. Patel, J. Monpara, Nanocrystals for controlled delivery: state of the art and approved drug products. Expert Opin. Drug Deliv. **19**, 1303–1316 (2022). https://doi.org/10.1080/17425247.2022.2110579
116. B. Conklin, B.M. Conley, Y. Hou, M. Chen, K.B. Lee, Advanced theragnostics for the central nervous system (CNS) and neurological disorders using functional inorganic nanomaterials. Adv. Drug Deliv. Rev. **192**, 114636 (2023). https://doi.org/10.1016/J.ADDR.2022.114636
117. Y.J. Kang, E.G. Cutler, H. Cho, Therapeutic nanoplatforms and delivery strategies for neurological disorders. Nano Convergence **5**, 1–15 (2018). https://doi.org/10.1186/S40580-018-0168-8
118. B.D. Kevadiya, B.M. Ottemann, M. Ben Thomas, I. Mukadam, S. Nigam, J.E. McMillan, S. Gorantla, T.K. Bronich, B. Edagwa, H.E. Gendelman, Neurotheranostics as personalized medicines. Adv. Drug Deliv. Rev. **148**, 252–289 (2019). https://doi.org/10.1016/J.ADDR.2018.10.011
119. J.H. Ryu, S. Lee, S. Son, S.H. Kim, J.F. Leary, K. Choi, I.C. Kwon, Theranostic nanoparticles for future personalized medicine. J. Control. Release **190**, 477–484 (2014). https://doi.org/10.1016/J.JCONREL.2014.04.027
120. Y. Cheng, R.A. Morshed, B. Auffinger, A.L. Tobias, M.S. Lesniak, Multifunctional nanoparticles for brain tumor imaging and therapy. Adv. Drug Deliv. Rev. **66**, 42–57 (2014). https://doi.org/10.1016/J.ADDR.2013.09.006

Overcoming the Challenges of Mucosal Tissues Using Drug Nanocrystals

Marta G. Fuster, Yushi Tao, Martina Sangalli, Lucia Lopez-Vidal, and Alejandro J. Paredes

Abstract Nanocrystals (NCs) have emerged as a promising strategy for enhancing drug delivery through mucosal routes, addressing the inherent challenges posed by mucosal barriers in various systems such as the gastrointestinal, respiratory, genito-urinary, and ocular tracts. These barriers, characterized by a protective mucus layer, enzymatic degradation, limited permeability, and rapid clearance mechanisms, often hinder the effective absorption and therapeutic efficacy of drugs. By leveraging the unique properties of nanotechnology, NCs can significantly improve drug stability and bioavailability, allowing for targeted and sustained release at the site of action. Furthermore, strategies for decorating nanocarriers include both mucoadhesive coatings, which facilitate adhesion to the mucus layer, and mucopenetrative coatings, such as PEGylation, which enable deeper penetration and prolonged retention in mucosal tissues. Overall, ongoing advancements in NCs technology hold great promise for overcoming the barriers of mucosal drug delivery, potentially leading to more effective therapeutic interventions for a wide range of diseases. This chapter primarily focuses on introducing the various mucosal barriers of the human body, as well as the challenges they present and the treatments that have been developed. Special attention is given to nanocrystals as a drug delivery system and their role in therapeutic applications.

Keywords Nanocrystals · Nanosuspensions · Solid drug nanoparticles · Mucosal tissue · Targeted nanocrystals · Mucus layer

M. G. Fuster · Y. Tao · M. Sangalli · L. Lopez-Vidal · A. J. Paredes (✉)
School of Pharmacy, Queens University Belfast, Belfast, UK
e-mail: a.paredes@qub.ac.uk

A. J. Paredes (ed.), *Nanocrystals*, https://doi.org/10.1007/978-3-032-10238-6_8

1 Introduction

1.1 The Mucosal Barrier

The mucosa is distributed throughout the body, with each section specifically adapted to protect different tissues (Fig. 1) [1]. These surfaces exhibit unique characteristics tailored to the needs of the tissues they safeguard. Covering these surfaces, mucus serves as a viscoelastic gel that protects tissues exposed to the external environment by acting as a barrier. It traps foreign particles, pathogens, and dust, preventing infections and damage to underlying tissues. Composed mainly of water, glycoproteins (such as mucins), lipids, and salts, mucus helps maintain tissue hydration and facilitates the movement of materials along mucosal surfaces. It primarily consists of crosslinked and entangled mucin fibres, which are secreted by goblet cells and submucosal glands [2]. Mucins are large molecules, generally ranging from 0.5 to 40 MDa in size, formed by linking multiple mucin monomers, each approximately 0.3–0.5 MDa. These mucins are coated with a highly complex and varied array of glycosaminoglycans [3].

– *Gastrointestinal Mucosa*

The gastrointestinal mucosa is the most extensive surface connecting the human body with the external environment. It acts as a selective barrier, allowing nutrients to enter the body while preventing harmful substances from passing through the intestine [4]. This barrier is composed of four main components: the mechanical, chemical, immune, and biological barriers (Fig. 2). The mechanical barrier includes intestinal epithelial cells and tight junctions that maintain structural integrity. The chemical barrier, consisting of digestive acids, enzymes, and other antimicrobial molecules, helps prevent bacterial invasion. The immune barrier, formed by gut-associated lymphoid tissue (GALT), secretory immunoglobulin A (SIgA), and immune cells like macrophages and lymphocytes, maintains immune balance by recognizing and responding to foreign antigens. Finally, the biological barrier is formed by the intestinal microbiota, with anaerobic bacteria playing a dominant role

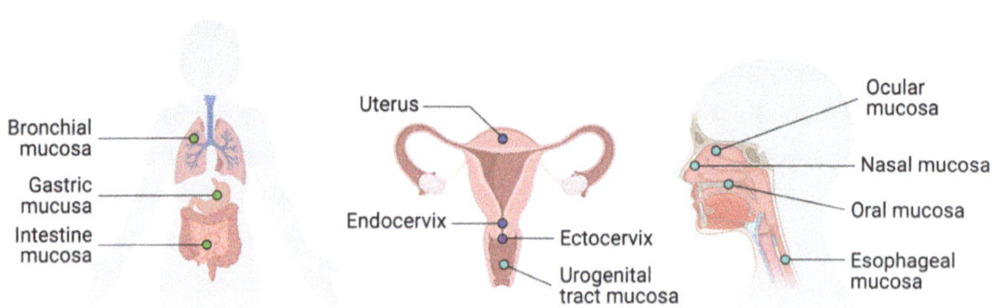

Fig. 1 Schematic representation of the distribution of various mucosal tissues in the human body. (Image was created using BioRender 2025)

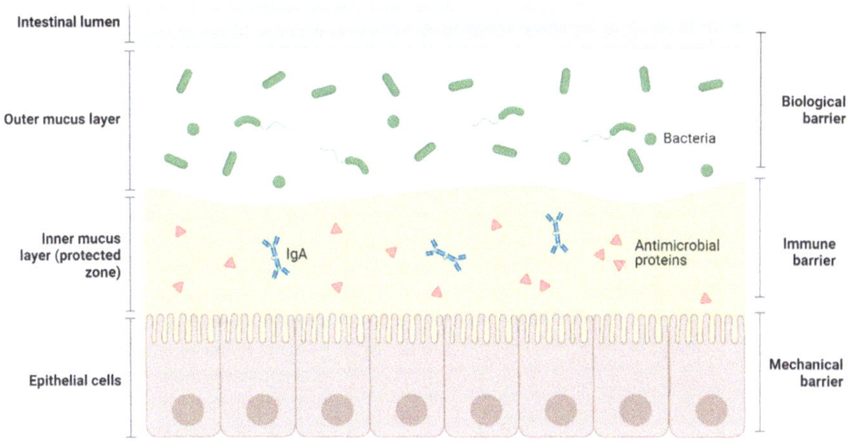

Fig. 2 Structure of the intestinal mucosal barrier: divided into the biological, immune, and mechanical barriers, with their different components schematically represented. (Image was created using BioRender 2025)

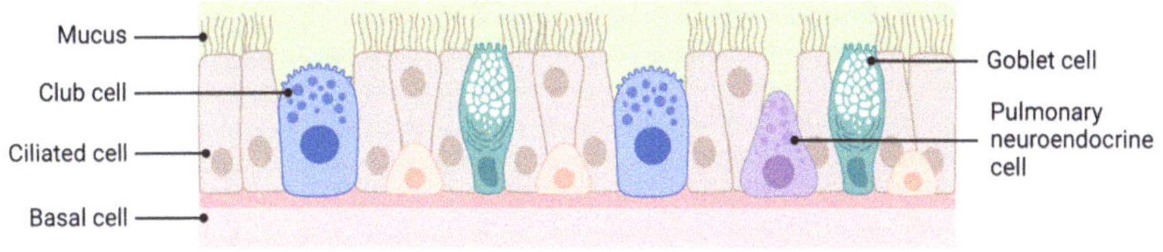

Fig. 3 Structure of the respiratory mucosa. (Adapted from BioRender.com)

in sustaining a stable microenvironment which serves to protect it from digestive enzymes and acidic pH. Together, these barriers maintain a balance between inflammatory responses and intestinal health [5].

– *Respiratory Mucosa*

The respiratory mucosa is lined with a protective mucosal layer that plays a crucial role in defending against inhaled particles, microorganisms, and other potentially harmful substances (Fig. 3). The mucus in this tract is continuously secreted, shed, and then either discarded or recycled, with a short lifespan that typically ranges from minutes to a few hours [3]. This rapid turnover, especially on surfaces with thinner mucus layers, likely evolved to enhance the protective functions of the respiratory mucosa. The mucosa covers the nasal, tracheal, and bronchial airways. In the nasal cavity, ciliary motion moves mucus from the anterior to the posterior regions, directing trapped particles and toxins towards the nasopharynx for removal. Generally, mucus in the respiratory tract flows at a rate of around 5 mm per minute, with full renewal occurring roughly every 20 min. Similarly, the respiratory tract's luminal gel layer, which is more solid-like, tends to refresh every 10–20 min, aiding in the efficient clearing of inhaled particles. This rapid secretion and clearance system helps the respiratory mucosa maintain its protective functions, contributing to pulmonary health and reducing exposure to harmful substances [6].

– *Oral Mucosa*

The oral mucosa is a moist, protective membrane that lines the oral cavity, and it serves multiple functions, including protection, secretion, and sensory perception. It is composed of three main types: lining mucosa, masticatory mucosa, and specialized mucosa, each having distinct characteristics. The oral mucosa is structured with three layers: a stratified squamous epithelium (either keratinized or non-keratinized), a connective tissue layer called the lamina propria, and, in some regions, a submucosa (Fig. 4) [7]. This mucosal barrier plays a critical role in defending against

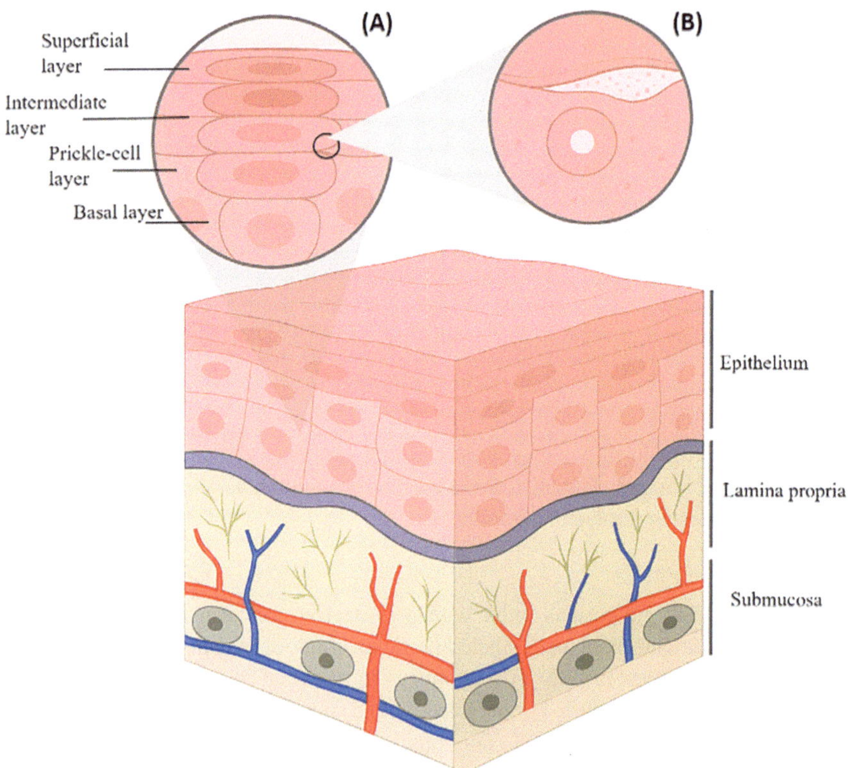

Fig. 4 Different layers of the non-keratinized epithelium during the differentiation process (**a**) and membrane-coating granules involved in the extrusion of amorphous material into the intercellular space (**b**). Illustration of the anatomical structures composing the lining oral mucosa. (Adapted from "*Nanostructured systems for transbuccal drug delivery*")

mechanical, chemical, and biological stressors, as well as enabling the perception of temperature, touch, and taste. Its regenerative capacity is high, with cells being replaced regularly. What sets the oral mucosa apart from other mucosal membranes is its frequent exposure to food, mechanical abrasion, and harmful substances. For instance, the masticatory mucosa is keratinized to resist the stresses of chewing, while the lining mucosa is more vulnerable to injury.

– *Genitourinary Mucosa*

The genitourinary mucosa, which lines both the reproductive and urinary tracts, acts as a barrier against pathogens and other external agents. This mucosal layer is composed primarily of epithelial cells, which are reinforced by tight junctions to maintain barrier integrity. It includes various cell types, such as squamous and columnar epithelial cells, depending on the specific region (urinary or reproductive) [8]. The main component of this tract is mucin, which provides structural support and helps trap pathogens. The mucus layer also contains antimicrobial peptides, SIgA, and enzymes like lysozyme, which contribute to its immune function [9, 10]. Additionally, it has microorganisms, particularly in the vaginal area, where *Lactobacillus* species play a significant role in maintaining a low pH, which is inhospitable to many pathogens, regulating its thickness and mucus production, which varies throughout different stages of the menstrual cycle or life stages [11, 12]. This mucosal immunity is crucial in protecting against urinary and reproductive tract infections and is essential for overall genitourinary health.

– *Ocular Mucosa*

Finally, the ocular mucosa lines the surface of the eye and inner eyelids, creating a barrier that protects the eye from pathogens, debris, and environmental irritants. This mucosa is composed primarily of epithelial cells, including goblet cells that secrete mucins. The conjunctival epithelium also contains tight junctions, which help prevent the entry of pathogens while maintaining ocular hydration [13]. Key components of the ocular mucus include MUC1, MUC4, and MUC5AC mucins, secreted by conjunctival goblet cells, which form a gel-like layer essential for tear stability and surface lubrication [14]. The tear film, which is composed of water, lipids, proteins, and mucins provides continuous clearance of small particles and microbes, preventing infections [15]. Additionally, antimicrobial peptides, enzymes such as lysozyme, and SIgA enhance the ocular mucosa's immune defence. The turnover rate of the tear film is quite high, typically clearing nearly all molecules and particulates from the eye within minutes, which is essential for eye health and clarity of vision [16].

1.2 Challenges to Drug Delivery

Drug delivery via mucosal routes faces several significant challenges that limit the effectiveness of therapies targeting mucosal tissues. At first, the sticky and viscous nature of mucus helps trap foreign particles, including pathogens and inhaled debris, and also effectively captures and removes drug particles, particularly larger or charged molecules [17]. This is especially challenging for drugs intended for nasal, respiratory, or gastrointestinal delivery [18]. So, new muco-inert drug formulations need to reduce binding interaction with mucus. Many mucosal surfaces, such as those in the gastrointestinal and respiratory tracts, contain proteolytic enzymes that degrade proteins, peptides, and other biomolecules. For instance, the gastrointestinal tract secretes pepsin, trypsin, and other enzymes that break down protein-based drugs before they can be absorbed [19]. This enzymatic activity poses a major obstacle for biological drugs like peptide-based therapeutics and vaccines, as they may be broken down before reaching their target sites. Strategies to minimized enzymatic degradation include protective drug coatings, enzyme inhibitors, and encapsulation of drugs within nanoparticles. The epithelial layer of mucosal tissues serves as a barrier to prevent the passage of harmful substances into the bloodstream [20]. However, this layer also limits the permeability of many drugs, especially large or hydrophilic molecules [21]. Tight junctions between epithelial cells restrict passive diffusion, so molecules often require specific transport mechanisms to cross. Permeation enhancers, which temporarily open tight junctions, and nanocarrier systems that facilitate cellular uptake are being investigated to increase the permeability of drugs across mucosal barriers. Finally, mucosal tissues are designed for continuous clearance to maintain homeostasis and prevent infection. In the respiratory tract, for instance, mucociliary clearance propels mucus (and any trapped particles) upward towards the throat, where it can be swallowed or expelled [22]. Similarly, gastrointestinal motility moves contents along the digestive tract, continuously clearing substances and limiting the contact time for drug absorption. This rapid clearance reduces drug bioavailability, especially for formulations requiring sustained release. Strategies to overcome rapid clearance include bioadhesive drug formulations, which adhere to mucosal surfaces longer, and controlled-release systems designed to withstand clearance mechanisms for extended therapeutic action. Developing effective mucosal drug delivery systems requires addressing these challenges through innovative drug formulations, protective encapsulation techniques, and permeability enhancers that can improve both drug stability and bioavailability across mucosal barriers.

2 Nanoparticles in Mucosal Drug Delivery and the Role of Nanocrystals

Nanotechnology has emerged as a powerful approach for overcoming the inherent barriers in mucosal drug delivery, offering new strategies to improve therapeutic outcomes across a range of diseases. Mucosal surfaces, including those of the respiratory, gastrointestinal, and genitourinary tracts, pose unique challenges for drug delivery due to the presence of the mucus layer, enzymatic degradation, rapid clearance mechanisms, and limited permeability of the epithelial layer, as mentioned earlier. To address these issues, the use of nanoparticles (NPs) has gained significant interest, as they provide unique advantages, including enhanced stability, improved bioavailability, and the potential for targeted delivery. Recent advancements have shown that different NPs can be designed to penetrate or bypass the mucus barrier, protect therapeutic agents from enzymatic degradation, and improve their retention time, thereby significantly increasing the efficacy of mucosal drug delivery [3, 23]. As previously discussed, one of the primary challenges in mucosal drug delivery is overcoming the mucus barrier. This barrier effectively traps and removes foreign particles, including many therapeutic agents, which limits their residence time on the mucosal surface. To tackle this issue, research has focused on developing muco-inert or mucoadhesive NPs that can either avoid or anchor within the mucus layer, thus allowing prolonged interaction with the underlying epithelial cells [24]. Andreas Bernkop-Schnürch et al. [18] extensively studied mucoadhesive drug delivery systems, showing that thiolated polymers, or thiomers, exhibit increased mucoadhesive properties, thereby enhancing the residence time of NPs on the mucosal surface. Thiomers can form covalent bonds with mucins, which stabilize the carrier within the mucus and allow for a more controlled release of the drug payload. Various types of NPs have also been investigated for their potential in mucosal drug delivery, offering specific benefits depending on the therapeutic application. Polymeric NPs, for example, are highly customizable and can be modified to achieve mucoadhesion or mucus penetration. Also, studies have demonstrated that chitosan-based NPs exhibit strong mucoadhesive properties, allowing for extended contact with the mucosal surface [25]. Other promising NPs include lipid-based systems, such as solid lipid NPs (SLNPs) and nanostructured lipid carriers (NLCs), which have shown improved drug stability and biocompatibility [21]. In addition, inorganic NPs, such as gold and silica, have been explored for their ability to facilitate mucosal drug delivery. These particles can be engineered to possess specific surface charges or coatings that enhance their interaction with mucosal surfaces [26]. For example, coating silica NPs with polyethylene glycol (PEG) has been shown to improve mucus penetration by reducing interactions with mucin fibres, thus enhancing drug delivery to the underlying tissues [27]. The strategies for enhancing mucosal penetration of different NPs continue to evolve, with a strong focus on optimizing surface properties to either evade or interact favourably with the mucus layer. "Muco-penetrating particles" (MPPs) represent a key area of research, where NPs are designed with hydrophilic or muco-inert coatings, such as PEG, that minimize binding to mucins and allow rapid transit through the mucus layer.

Within the NPs, NCs are gaining significant attention as an advanced platform for delivering therapeutic agents across mucosal barriers [28], offering new possibilities in treating a wide range of diseases, including neurological disorders, cancers, and inflammatory conditions. Due to their unique physical and chemical characteristics, such as high drug-loading capacity, increased dissolution rates, and the ability to bypass conventional drug delivery limitations, NCs represent an innovative and versatile approach in mucosal drug delivery. Their primary advantage is the enhancement of the dissolution rate and bioavailability of poorly water-soluble drugs. This improvement is due to the high surface area of the nanocrystals, which leads to increased solubility and, consequently, more efficient cellular uptake [29]. Compared to other NPs, NCs exhibit higher drug loading, as they are typically composed solely of the drug and a stabilizing agent. This makes them a suitable choice for formulations intended to deliver high concentrations of active pharmaceutical ingredients, especially in challenging biological environments like mucosal surfaces. As discussed earlier, one of the critical challenges in mucosal drug delivery is the effective transport of drugs through mucosal barriers. These barriers act as protective layers that naturally prevent the entry of foreign substances, including therapeutic agents. NCs, however, offer a promising solution. Various studies have shown that the unique properties of NCs can be tailored for targeted delivery, allowing drugs to reach specific sites while minimizing side effects and optimizing therapeutic outcomes.

Parkinson's disease (PD) has been a focal point for NC-based therapies due to the need for efficient brain-targeting delivery systems. For example, Liu et al. [30] investigated the use of Ginkgolide B NCs (GB-NCs) as a therapeutic approach for PD. These NCs were developed through an antisolvent precipitation method, yielding a formulation with high dissolution rates, enhanced cellular uptake, and substantial permeability across physiological barriers. The GB-NCs were able to cross mucosal barriers effectively, showing promise in animal models, where oral administration improved dopamine levels and reduced PD symptoms. The neuroprotective effects observed highlight the potential of NCs in providing symptomatic relief in PD while mitigating further neurodegeneration. Similarly, Wu et al. [31] developed paeoniflorin (PA) NCs designed for intranasal delivery, targeting the challenges of brain bioavailability in PD treatment. The PA-NCs were stabilized with D-α-tocopherol polyethylene glycol succinate (TPGS), a stabilizer that enhances permeability and cellular uptake. The rod-shaped PA-NCs demonstrated significant nasal mucosa permeation and increased brain concentration in pharmacokinetic studies. By overcoming the nasal epithelial barrier, these NCs achieved enhanced brain targeting, suggesting that intranasal administration could serve as a viable alternative to invasive methods like injections, offering patients a more accessible treatment option. Adding to the range of brain-targeted delivery systems, Huang et al. [32] explored harmine NCs incorporated into a deacetylated gellan gum (DGG) in situ gel for intranasal administration. Using homogenization and spray-drying techniques, the harmine NCs showed enhanced bioadhesion and sustained release, with a 25-fold increase in brain bioavailability compared to oral administration. This formulation improved cognitive function in animal models, demonstrating the advantages of DGG as a mucoadhesive material that can prolong the residence time of NCs on the nasal mucosa and enhance drug delivery efficiency to the brain. The potential for NCs in delivering poorly soluble drugs to treat neurological disorders is further exemplified in a study by Zhu et al. [33], which investigated an NC-based armodafinil hydrogel (NCsG) designed for intranasal administration. The armodafinil NCs demonstrated significant cognitive and wakefulness enhancement, especially in sleep-deprived subjects. The study highlighted the potential of NCs to effectively penetrate nasal mucosa, delivering therapeutic agents that increase brain-derived neurotrophic factor in the hippocampus, a protein linked to cognitive function.

Another study by Duan et al. [34] developed an advanced delivery system for cervical cancer by integrating paclitaxel (PTX) NCs into a polydopamine (PDA) and PEG matrix, using an electrospun-nanofiber membrane for vaginal administration. This approach, which aimed to enhance drug retention and minimize mucosal irritation, demonstrated deeper penetration and improved tumor inhibition in cervical cancer models. PEGylation and PDA modifications improved the NCs' mucoadhesive properties and decreased mucus entrapment, showing the promise of localized NC-based therapy for treating cancers at mucosal sites. In other cancer-related applications, NCs have been formulated for the treatment of metastatic melanoma. Lv et al. [35] developed docetaxel NCs modified with TAT peptides to target cervical cancer via vaginal administration. To reduce mucus entrapment, the NCs were coated with polydopamine (PDA) and PEGylated. The formulation showed enhanced cellular uptake and cytotoxicity in cervical cancer cells, and when dispersed in a thermosensitive gel, it demonstrated prolonged vaginal retention and deeper mucosal penetration compared to unmodified nanocrystals. NCs also show potential for treating inflammatory and infectious diseases by enhancing drug penetration across gastrointestinal barriers. Zhou et al. [36] introduced mPEG-modified dendritic mesoporous silica nanoparticles (DMSNPs) loaded with andrographolide NCs for oral administration. This formulation was tailored to improve mucosal permeability and intestinal absorption of andrographolide, an anti-inflammatory agent. The mPEG-DMSNPs effectively bypassed mucus barriers, leading to higher bioavailability and notable anti-inflammatory effects in vivo, as seen through reduced levels of inflammatory cytokines such as IL-1β, IL-6, and TNF-α. This system underscores the potential of NCs for managing chronic inflammatory diseases, where enhanced oral bioavailability and targeted intestinal release are desirable. In the context of nasal inflammatory conditions, Masuda et al. [37] developed mometasone furoate (MF) NCs as a nasal spray alternative to improve local drug absorption in conditions like rhinitis. Produced through bead milling, these NCs ranged from 80 to 200 nm in size, showing enhanced local absorption compared to conventional MF nasal sprays. Additionally, the study experimented with an in situ gelling system to control formulation viscosity, which further increased retention time on the nasal mucosa. While this gel system improved gel formation and residence time, it reduced drug diffusion, indicating a trade-off

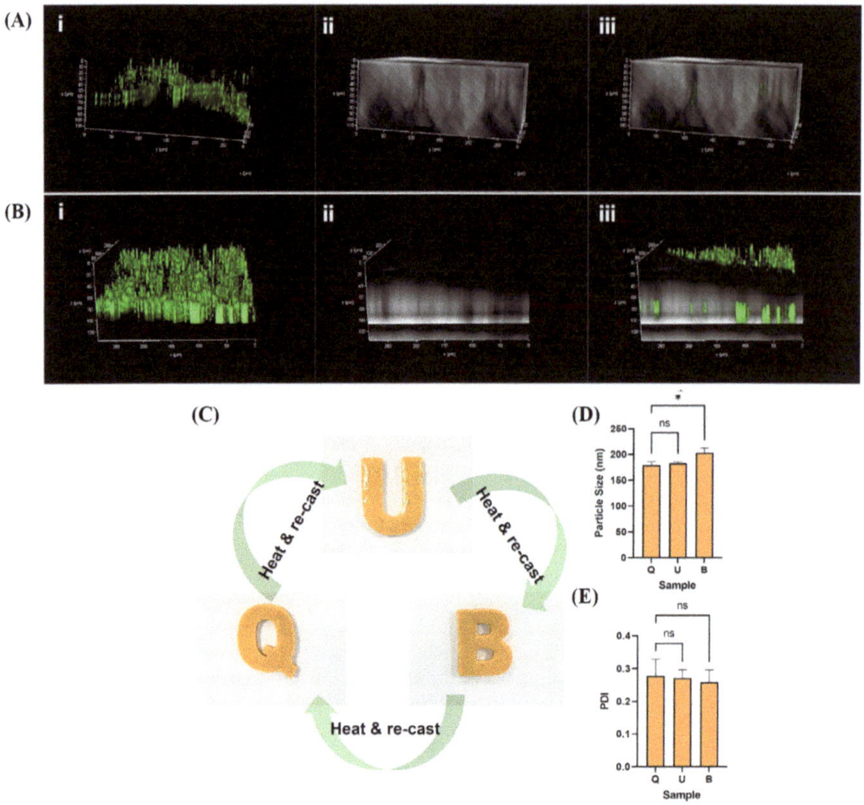

Fig. 5 Drug delivery to porcine mucosa using eutectogels loaded with curcumin physical mixtures (PM) or nanocrystals (NCs). (**a**) Multiphoton microscopy of neonatal porcine mucosa treated with Egel-20-PM, showing (i) green fluorescent channel, (ii) transmitted light channel, and (iii) merged channels. (**b**) Multiphoton microscopy of neonatal porcine mucosa treated with Egel-20-NCs, showing (i) green fluorescent channel, (ii) transmitted light channel, and (iii) merged channels. (**c**) Multiple casting of Egel-20-NCs into Q-U-B-shaped moulds and subsequent NC redispersion. (**d**) Mean particle size of CUR-NCs after redispersion of Egel-20-NCs cast three times. (**e**) PDI of CUR-NCs after redispersion of Egel-20-NCs cast three times. Data are expressed as mean ± SD ($n = 3$). (Illustration adapted from *Bioadhesive eutectogels supporting drug nanocrystals for long-acting delivery to mucosal tissues*)

between retention and bioavailability that could be optimized for specific therapeutic needs. Wu et al. [31] tested an in situ gelling system that adjusted the NC formulation's viscosity for better retention on the nasal mucosa, though this approach required optimization to balance drug diffusion rates with mucosal adhesion. A similar approach was taken with NC-based eutectogels containing curcumin for treating sinusitis and bacterial infections [38]. The most visually striking evidence of the efficacy of the nanocrystal-based eutectogel is provided by multiphoton microscopy images (Fig. 5a and b), which clearly demonstrate the enhanced mucosal penetration of curcumin. Specifically, Fig. 5a shows that the drug nanocomposite eutectogel (Egel-20-NCs) enables drug penetration up to ~100 μm into neonatal mucosa, indicating deeper tissue integration. In contrast, Fig. 5b illustrates the formulation containing the physical mixture of the components (Egel-PM), which reaches only ~60 μm of tissue depth. This visual comparison confirms that reducing particle size markedly improves drug deposition and tissue penetration. Moreover, Fig. 5c–e depict repeated casting of Egel-20-NCs into molds, demonstrating the material's stability and reproducibility across multiple manufacturing cycles, thereby supporting its potential for scalable production and broader drug delivery applications. Shen et al. [39] developed NCs with an environment-responsive probe to track their fate using a water-initiated aggregation-caused quenching mechanism. The in vivo behaviour of quercetin hybrid NCs after oral administration presented 12–16 h of gastrointestinal retention instead of dissolving immediately. Particle size was found to significantly affect absorption, with 550 nm being the best balance of retention and absorption. Cellular studies also revealed both cell-mediated absorption and trans-epithelial pathways, with the mucus layer acting as a barrier.

For ocular treatments, Romero et al. [40] developed a novel cationic NCs formulation containing dexamethasone acetate (0.05%) and polymyxin B (0.10%) using a self-developed wet bead milling technique. The NCs, ranging from 200 to 250 nm, showed enhanced solubility, increased bioavailability, and prolonged retention in the eye. The formulation used cetylpyridinum chloride (CPC) and benzalkonium chloride (BAC) to create a positive charge, which improved mucoadhesion (Table 1).

3 Conclusion

NCs have emerged as a multifaceted platform capable of addressing complex drug delivery challenges, particularly for drugs requiring effective mucosal penetration. Their high drug-loading capacity, customizable surfaces, and adaptability to various therapeutic requirements make NCs suitable for targeting specific tissues and providing localized or systemic therapeutic effects. From neurology to oncology, as well as in the treatment of inflammation and infectious diseases, NCs offer broad-spectrum applications across mucosal sites. With continued research on material modifications and integration with mucoadhesive systems, NCs have the potential to further expand the landscape of targeted, minimally invasive, and effective therapies. The inclusion of specialized modifications, such as PEGylation, RBC coating, and peptide functionalization, furthers the possibilities of NCs to meet the exacting needs of different therapeutic contexts, ultimately paving the way for personalized and site-specific treatments in mucosal drug delivery.

Table 1 Nanocrystal-based drug delivery systems for mucosal administration: Influence of particle type, surface modification, and administration route on tissue penetration and therapeutic efficacy

Carrier type	Particle size	Zeta potential	Stabiliser	Administration route	Tissue	Strategy to pass the mucosa	Application	Key characteristics	References
NCs	139.6 nm	−23.2 mV	TPGS	Nasal	Nasal epithelium	NCs features: small particle size, negative zeta potential and the use of TPGS enhanced solubility, cellular uptake, and permeability through the mucosa	Parkinson's disease, cognitive enhancement	Enhanced solubility: (87.14% cumulative drug release)	[31]
	181.26 nm	−41.4 mV	TPGS	Nasal	Nasal epithelium	Bioadhesive gel: the DGG interacts with calcium ions of the nasal cavity, forming a bioadhesive gel that enhances the mucosal adhesion, the retention time, and the absorption of the NCs through the mucosal barrier into the brain	Parkinson's disease, cognitive enhancement	Superior mucosal penetration: with enhanced brain bioavailability (25-fold increase)	[32]
NCs in multilayer core shell NPs	380 nm; (internal layer) 421 nm (core shell system)	+34 mV (internal layer) Core shell system (pH dependent) at pH 1.2 and 4.5: −47 ± 3 mV (due to the presence of the CAP outer	Chitosan	Oral	Intestinal mucosa	Polymeric coating: The polymeric shell prevents NCs from early dissolution in the upper GIT The pH-triggered surface charge reversal makes the nanoparticles "smart," enabling them to	Ulcerative colitis	Colon targeting: pH-sensitive release, targeted colonic delivery, high colonic drug accumulation (~80% of the administered dose localized in the colon), and superior therapeutic efficacy	[41]

		layer) At pH 7.2: +34 ± 4 mV (due to charge reversal after dissolution of the CAP layer)				concentrate and release the drug in a targeted manner within the inflamed tissues of the colon		with minimal upper GIT leakage (<10% detected in the stomach and small intestine)	
PEGylated NCs	178.2 nm (NCs) 385.6 nm (whole system)	−1.74 mV (NCs only) −14.7 (whole system)	Pluronic F127	Vaginal	Vaginal epithelium	NCs PEGylation: minimizes interaction with mucus, enabling deeper penetration	Cervical cancer therapy	Quick release of the NCs: achieving a cumulative drug release of over 80% within 72 h Enhanced penetration with a fluorescence intensity area under the curve (AUC) of approximately 2.5-fold greater than uncoated NCs Prolonged vaginal retention of up to 7 h	[34]
PEGylated NCs	Below 500 nm	+ 9.13 mV	TPGS [35]	Vaginal	Vaginal epithelium	Nanofiber integration: embedding NCs in electrospun nanofibres ensures prolonged retention at the mucosal surface.	Cervical cancer therapy	Enhanced penetration: The modification significantly improved mucosal penetration, with a 2.5-fold increase in permeation through the cervical mucus compared to non-modified NCs	[35]

(continued)

Table 1 (continued)

Carrier type	Particle size	Zeta potential	Stabiliser	Administration route	Tissue	Strategy to pass the mucosa	Application	Key characteristics	References
Cationic NCs	230–260 nm	+40–50 mV	Cetylpyridinium chloride	Ocular	Ocular mucosa	Nanocrystals: the small size improves adhesion and bioavailability while the z-potential allows electrostatic interactions with the mucin layer and significantly reduce the clearance by the lacrimal fluid	Ocular inflammation	Increased Mucoadhesion: confirmed by a decreasing of the NCs Z-potential (from −30 mV to −15 mV upon contact with mucin solutions Sustained release and retention: Approximately 80% of the drug remains available for release over time Cytotoxicity testing showed a reactivity grade of 0, indicating no cytotoxic effects on fibroblast cells	[40]
Eutectogels with NCs	94.8 nm	Greater than −10 mV	P188	Mucosal (nasal, vaginal, Oral)	General mucosal tissues	Synergistic combination of drug NCs and a eutectic mixture supported into a gelatin matrix to obtain stretchable and adhesive eutectogels for long-acting mucosal delivery	Antimicrobial, anti-inflammatory	Sustained drug release: 80% of the encapsulated drug being released over a 48-h period High mucosal deposition: 2.5-fold increase in drug concentration at the mucosal site	[38]

References

1. M.M. France, J.R. Turner, The mucosal barrier at a glance. J. Cell Sci. **130**, 307–314 (2017). https://doi.org/10.1242/jcs.193482
2. D.J. Thornton, J.K. Sheehan, From mucins to mucus: toward a more coherent understanding of this essential barrier. Proc. Am. Thorac. Soc. **1**, 54–61 (2004). https://doi.org/10.1513/pats.2306016
3. S.K. Lai, Y.Y. Wang, J. Hanes, Mucus-penetrating nanoparticles for drug and gene delivery to mucosal tissues. Adv. Drug Deliv. Rev. **61**, 158–171 (2009). https://doi.org/10.1016/j.addr.2008.11.002
4. S. Chen, H. Zhu, Y. Luo, The gut-mediated function of polyphenols: Opinions on functional food development for nonalcoholic fatty liver disease. Curr. Opin. Food Sci. **49** (2023). https://doi.org/10.1016/j.cofs.2022.100972
5. K. Wang, L.Y. Wu, C.Z. Dou, X. Guan, H.G. Wu, H.R. Liu, Research advance in intestinal mucosal barrier and pathogenesis of Crohn's disease. Gastroenterol. Res. Pract. **2016** (2016). https://doi.org/10.1155/2016/9686238
6. M.R. Knowles, R.C. Boucher, Mucus clearance as a primary innate defense mechanism for mammalian airways. J. Clin. Invest. **109**, 571–577 (2002). https://doi.org/10.1172/JCI0215217
7. S. Groeger, J. Meyle, Oral mucosal epithelial cells. Front. Immunol. **10**, 1–22 (2019). https://doi.org/10.3389/fimmu.2019.00208
8. M. Abolbashari, A. Atala, J.J. Yoo, Chapter 36 – genitourinary system, in *Translational Regenerative Medicine*, ed. by A. Atala, J.G. Allickson, (Academic Press, Boston, 2015), pp. 495–505. https://doi.org/10.1016/B978-0-12-410396-2.00036-0
9. S. Srinivasan, Regulation of body fat in C. elegans. Annu. Rev. Physiol., 161–178 (2015)
10. M. Imai, T. Watanabe, M. Kiso, N. Nakajima, S. Yamayoshi, K. Iwatsuki-horimoto, M. Hatta, S. Yamada, M. Ito, Y. Sakai, M. Shirakura, E. Takashita, S. Fujisaki, R. Mcbride, J. Thompson, K. Takahashi, T. Maemura, H. Mitake, S. Chiba, G. Zhong, S. Fan, K. Oishi, A. Yasuhara, K. Takada, T. Nakao, S. Fukuyama, M. Yamashita, T.J.S. Lopes, T. Odagiri, S. Watanabe, Y. Shu, J.C. Paulson, Y. Kawaoka, L. Jolla, A highly pathogenic avian H7N9 influenza virus isolated from a human is lethal in some ferrets infected via respiratory droplets Masaki. Cell Host & Microbe **22**, 615–626 (2018). https://doi.org/10.1016/j.chom.2017.09.008.A
11. S.S. Witkin, I.M. Linhares, Why do lactobacilli dominate the human vaginal microbiota? Int. J. Obstetr. Gynaecol. **124**, 606–611 (2017). https://doi.org/10.1111/1471-0528.14390
12. J. Ravel, P. Gajer, Z. Abdo, G.M. Schneider, S.S.K. Koenig, S.L. McCulle, S. Karlebach, R. Gorle, J. Russell, C.O. Tacket, R.M. Brotman, C.C. Davis, K. Ault, L. Peralta, L.J. Forney, Vaginal microbiome of reproductive-age women. Proc. Natl. Acad. Sci. USA **108**, 4680–4687 (2011). https://doi.org/10.1073/pnas.1002611107
13. M. Saßmannshausen, J.S. Steinberg, R. Fimmers, M. Pfau, S. Thiele, M. Fleckenstein, F.G. Holz, S. Schmitz-Valckenberg, Structure-function analysis in patients with intermediate age-related macular degeneration. Invest. Ophthalmol. Vis. Sci. **59**, 1599–1608 (2018). https://doi.org/10.1167/iovs.17-22712
14. R.D. Yuanwei Fan, M. Graham, Membrane-associated mucins of the human ocular surface in health and disease. Physiol. Behav. **176**, 139–148 (2016). https://doi.org/10.1016/j.jtos.2021.03.003.Membrane-Associated
15. Y. Hori, Secreted mucins on the ocular surface. Invest. Ophthalmol. Vis. Sci. **59**, DES151–DES156 (2018). https://doi.org/10.1167/iovs.17-23623
16. M.D.P. Willcox, P. Argüeso, G.A. Georgiev, J.M. Holopainen, G.W. Laurie, T.J. Millar, E.B. Papas, J.P. Rolland, T.A. Schmidt, U. Stahl, T. Suarez, L.N. Subbaraman, O. Uçakhan, L. Jones, TFOS DEWS II tear film report. Ocul. Surf. **15**, 366–403 (2017). https://doi.org/10.1016/j.jtos.2017.03.006
17. A. Fortuna, K. Schindowski, F. Sonvico, Editorial: Intranasal drug delivery: Challenges and opportunities. Front. Pharmacol. **13**, 10–12 (2022). https://doi.org/10.3389/fphar.2022.868986

18. A. Bernkop-Schnürch, S. Dünnhaupt, Chitosan-based drug delivery systems. Eur. J. Pharm. Biopharm. **81**, 463–469 (2012). https://doi.org/10.1016/j.ejpb.2012.04.007
19. S. Golshani, A. Vatanara, M. Amin, Recent advances in Oral Mucoadhesive drug delivery. J. Pharm. Pharm. Sci. **25**, 201–217 (2022). https://doi.org/10.18433/jpps32705
20. M. Du, An overview on transmucosal permeability and formulation. J. Dev. Drugs **13**, 1–2 (2024)
21. E. Mazzinelli, I. Favuzzi, A. Arcovito, R. Castagnola, G. Fratocchi, A. Mordente, G. Nocca, Oral mucosa models to evaluate drug permeability. Pharmaceutics. **15**, 1–16 (2023). https://doi.org/10.3390/pharmaceutics15051559
22. D. Wu, Y. Xiang, Role of mucociliary clearance system in respiratory diseases. J. Central South Univer. Med. Sci. **48**, 275–284 (2023). https://doi.org/10.11817/j.issn.1672-7347.2023.220372
23. H. Villetti, An overview on mucosal drug delivery for enhanced therapeutic outcomes research & reviews. Drug Deliv. **8**, 21–22 (2024). https://doi.org/10.4172/resrevdrugdeliv.8.1.006
24. F. Laffleur, A. Bernkop-Schnürch, Strategies for improving mucosal drug delivery. Nanomedicine **8**, 2061–2075 (2013). https://doi.org/10.2217/nnm.13.178
25. K. Zhao, Y. Xie, X. Lin, W. Xu, The Mucoadhesive nanoparticle-based delivery system in the development of mucosal vaccines. Int. J. Nanomedicine **17**, 4579–4598 (2022). https://doi.org/10.2147/IJN.S359118
26. M. Ouellette, F. Masse, M. Lefebvre-Demers, Q. Maestracci, P. Grenier, R. Millar, N. Bertrand, M. Prieto, É. Boisselier, Insights into gold nanoparticles as a mucoadhesive system. Sci. Rep. **8**, 1–15 (2018). https://doi.org/10.1038/s41598-018-32699-2
27. G.S. Irmukhametova, G.A. Mun, V.V. Khutoryanskiy, Thiolated mucoadhesive and PEGylated nonmucoadhesive organosilica nanoparticles from 3-mercaptopropyltrimethoxysilane. Langmuir **27**, 9551–9556 (2011). https://doi.org/10.1021/la201385h
28. M.B. McGuckin, J. Wang, R. Ghanma, N. Qin, S.D. Palma, R.F. Donnelly, A.J. Paredes, Nanocrystals as a master key to deliver hydrophobic drugs via multiple administration routes. J. Control. Release **345**, 334–353 (2022). https://doi.org/10.1016/j.jconrel.2022.03.012
29. Z. Xin, B. Prasad, N. Kumar, H. Shivashekaregowda, Recent development in nanocrystal based drug delivery for neurodegenerative diseases: Scope, challenges, current and future prospects. J. Drug Deliv. Sci. Technol. **68**, 102921 (2022). https://doi.org/10.1016/j.jddst.2021.102921
30. J. Liu, L. Tu, M. Cheng, J. Feng, Y. Jin, Mechanisms for oral absorption enhancement of drugs by nanocrystals. J. Drug Deliv. Sci. Technol. **56**, 101607 (2020). https://doi.org/10.1016/j.jddst.2020.101607
31. C. Wu, B. Li, Y. Zhang, T. Chen, C. Chen, W. Jiang, Q. Wang, T. Chen, Intranasal delivery of paeoniflorin nanocrystals for brain targeting. Asian J. Pharm. Sci. **15**, 326–335 (2020). https://doi.org/10.1016/j.ajps.2019.11.002
32. G. Huang, J. Xie, S. Shuai, S. Wei, Y. Chen, Z. Guan, Q. Zheng, P. Yue, C. Wang, Nose-to-brain delivery of drug nanocrystals by using Ca 2 + responsive deacetylated gellan gum based in situ-nanogel. Int. J. Pharm. **594**, 120182 (2021). https://doi.org/10.1016/j.ijpharm.2020.120182
33. S. Zhu, S. Zhang, L. Pang, G. Ou, L. Zhu, J. Ma, R. Li, Y. Liu, L. Wang, L. Wang, L. Du, Y. Jin, Effects of armodafinil nanocrystal nasal hydrogel on recovery of cognitive function in sleep-deprived rats. Int. J. Pharm. **597** (2021). https://doi.org/10.1016/j.ijpharm.2021.120343
34. H. Duan, H. Chen, C. Qi, F. Lv, J. Wang, Y. Liu, Z. Liu, Y. Liu, A novel electrospun nanofiber system with PEGylated paclitaxel nanocrystals enhancing the transmucus permeability and in situ retention for an efficient cervicovaginal cancer therapy. Int. J. Pharm. **650**, 123660 (2024). https://doi.org/10.1016/j.ijpharm.2023.123660
35. F. Lv, J. Wang, H. Chen, L. Sui, L. Feng, Z. Liu, Y. Liu, G. Wei, W. Lu, Enhanced mucosal penetration and efficient inhibition efficacy against cervical cancer of PEGylated docetaxel nanocrystals by TAT modification. J. Control. Release **336**, 572–582 (2021). https://doi.org/10.1016/j.jconrel.2021.07.008

36. W. Zhou, B. Li, R. Min, Z. Zhang, G. Huang, Y. Chen, B. Shen, Q. Zheng, P. Yue, Mucus-penetrating dendritic mesoporous silica nanoparticle loading drug nanocrystal clusters to enhance permeation and intestinal absorption. Biomater. Sci. **11**, 1013–1030 (2022). https://doi.org/10.1039/d2bm01404a
37. S. Masuda, S. Deguchi, F. Ogata, J. Yoshitomi, H. Otake, K. Kanai, N. Kawasaki, N. Nagai, Nasal absorption enhancement of Mometasone Furoate nanocrystal dispersions. Int. J. Nanomed. **18**, 5685–5699 (2023). https://doi.org/10.2147/IJN.S430952
38. M.B. Bianchi, C. Zhang, E. Catlin, G. Sandri, M. Calderón, E. Larrañeta, R.F. Donnelly, M.L. Picchio, A.J. Paredes, Bioadhesive eutectogels supporting drug nanocrystals for long-acting delivery to mucosal tissues. Materials Today Bio. **17** (2022). https://doi.org/10.1016/j.mtbio.2022.100471
39. C. Shen, Y. Yang, B. Shen, Y. Xie, J. Qi, X. Dong, W. Zhao, W. Zhu, W. Wu, H. Yuan, Y. Lu, Self-discriminating fluorescent hybrid nanocrystals: Efficient and accurate tracking of translocation via oral delivery. Nanoscale **10**, 436–450 (2018). https://doi.org/10.1039/c7nr06052a
40. G.B. Romero, C.M. Keck, R.H. Müller, N.A. Bou-Chacra, Development of cationic nanocrystals for ocular delivery. Eur. J. Pharm. Biopharm. **107**, 215–222 (2016). https://doi.org/10.1016/j.ejpb.2016.07.005
41. M.A. Oshi, J. Lee, M. Naeem, N. Hasan, J. Kim, H.J. Kim, E.H. Lee, Y. Jung, J.W. Yoo, Curcumin nanocrystal/pH-responsive polyelectrolyte multilayer Core-Shell nanoparticles for inflammation-targeted alleviation of ulcerative colitis. Biomacromolecules **21**, 3571–3581 (2020). https://doi.org/10.1021/acs.biomac.0c00589
42. Z. Chai, D. Ran, L. Lu, C. Zhan, H. Ruan, X. Hu, et al., Ligand-modified cell membrane enables the targeted delivery of drug nanocrystals to glioma. ACS Nano **13**(5), 5591–5601 (2019)
43. W. Geldenhuys, D. Wehrung, A. Groshev, A. Hirani, V. Sutariya, Brain-targeted delivery of doxorubicin using glutathione-coated nanoparticles for brain cancers. Pharm. Dev. Technol. **20**, 497–506 (2015). https://doi.org/10.3109/10837450.2014.892130
44. A. Sarnes, M. Kovalainen, M.R. Häkkinen, T. Laaksonen, J. Laru, J. Kiesvaara, J. Ilkka, O. Oksala, S. Rönkkö, K. Järvinen, J. Hirvonen, L. Peltonen, Nanocrystal-based per-oral itraconazole delivery: Superior in vitro dissolution enhancement versus Sporanox® is not realized in in vivo drug absorption. J. Control. Release **180**, 109–116 (2014). https://doi.org/10.1016/j.jconrel.2014.02.016

Management of Ocular Diseases Using Nanocrystal-based Formulations

Juan Matias Pernochi Scerbo, Julieta Ribotta, Santiago D. Palma, and María Lina Formica

Abstract Pharmacological ocular therapy is one of the great challenges in pharmaceutical technology, particularly for drugs with poor aqueous solubility, as they must overcome the anatomical and physiological barriers of the eye. Drug nanocrystals (NCs) are pure drug particles in the nanometer range that present a large surface area, providing high apparent saturation solubility, enhanced dissolution rates, and improved adhesive properties. These advantages lead to improved drug corneal penetration, prolonged retention in ocular tissues, and higher bioavailability with reduced toxicity and patient discomfort compared to conventional eye drops. In turn, NCs are versatile systems, as they can be incorporated into secondary drug delivery systems, such as in situ gelling systems or microneedles. Whether for treating inflammation, infections, or glaucoma, NCs have shown higher therapeutic effects than conventional treatments. Furthermore, NCs have also proven to be safe, well tolerated, and less toxic than available formulations. This technology has been demonstrated to be simple, cost-effective, and scalable. Several formulations are available as NCs or nanosuspensions (NSs), including those for ocular applications. In this chapter, we discuss the specific challenges of pharmacological ocular therapy, highlighting the advantages of NCs for ocular drug delivery. Furthermore, we review the latest developments in nanocrystal (NC) formulations for ocular applications, including both laboratory-scale preclinical studies and approved ocular formulations.

Keywords Ocular · Cornea · Nanocrystal · Nanosuspension · Drug delivery

J. M. Pernochi Scerbo · J. Ribotta · S. D. Palma · M. L. Formica (✉)
Unidad de Investigación y Desarrollo en Tecnología Farmacéutica (UNITEFA), CONICET and Departamento de Ciencias Farmacéuticas, Facultad de Ciencias Químicas, Universidad Nacional de Córdoba, Ciudad Universitaria, Córdoba, Argentina
e-mail: lina.formica@unc.edu.ar

A. J. Paredes (ed.), *Nanocrystals*, https://doi.org/10.1007/978-3-032-10238-6_9

Abbreviations

NCs	nanocrystals
NSs	nanosuspensions
DDSs	drug delivery systems
NDDSs	nanoparticle drug delivery systems
ZP	zeta potential
HPH	high-pressure homogenization
MNs	microneedles
IOP	intraocular pressure

1 Introduction

The development of ocular drug delivery systems (DDSs) is one of the major challenging tasks for pharmaceutical scientists due to the eye's anatomical and physiological barriers limiting their access to the desired pharmacological target site. Anatomical barriers like the cornea, conjunctiva, and sclera are highly effective at restricting the access of foreign substances to the eye, while physiological barriers, such as tear film turnover and blinking, can wash out topically administered formulations before they dissolve and permeate the cornea. Moreover, the available biological fluid on the ocular surface is reduced, hindering drug dissolution. In addition, many drugs used to treat eye diseases have low aqueous solubility and low ocular permeability. The primary objective of ocular drug delivery is to enhance drug bioavailability and prolong its retention in ocular tissues, ensuring therapeutic efficacy while minimizing the need for frequent administrations. Although this can be partially addressed by administering the drug directly into the tissue of concern via invasive administration routes like intraocular injections [1], it may result in further complications such as intraocular bleeding, pain, and discomfort, which can lead to poor patient compliance [2]. Therefore, developing efficient DDSs for ocular disease management is imperative.

For ocular therapies to be effective, it is essential to develop innovative strategies that overcome the anatomical and physiological barriers of the eye. In this regard, nanomedicine has emerged as a key strategy in designing more efficient DDSs by improving the bioavailability of therapeutic molecules in ocular tissues. Nanomedicine is the use of nanotechnology in the field of medicine, utilizing nanoscale systems for diagnosis, monitoring, treatment, and disease prevention. It involves the design and evaluation of nanoparticle drug delivery systems (NDDSs), which are structures at the nanometric scale, where both the active pharmaceutical ingredient and the carrier system contribute to their specialized functionality. As nanoscale tools ranging from 1 to 1000 nm, NDDSs can facilitate early disease detection, targeted drug delivery, and controlled release, optimizing treatment outcomes [3, 4]. NDDSs exhibit unique physical, chemical, and biological properties due to their nanometer size and surface characteristics [5, 6], which contribute to improved therapeutic efficacy. These characteristics determine particle interactions with ocular tissues, directly affecting their retention time and drug bioavailability. In particular, particle size plays a crucial role in their clearance from the ocular surface, as dynamic ocular barriers influence the elimination of topically administered drugs [7].

Several NDDSs have been developed for ophthalmic treatments, demonstrating great potential to enhance the penetration of therapeutic agents across ocular tissues. These systems can sustain drug release and prolong residence time in eye tissues, thereby increasing bioavailability and reducing the frequency of administration, and, in some cases, eliminating the need for invasive procedures [8, 9]. Furthermore, these systems can protect drugs from degradation and metabolism, improve their stability, extend their shelf life, and reduce toxicity and adverse effects.

In this context, drug nanocrystals (NCs) offer advantages that can improve the ocular bioavailability of poorly soluble drugs. They are crystalline nanoparticles (<1 μm) formed only by the pure drug and stabilized by a layer of either polymers or surfactants [10]. NCs are often obtained as suspensions (nanosuspensions, NSs), with further solvent removal leading to redispersible nanocrystal (NC) powder [11]. Compared to other NDDSs, NCs have a high drug loading, since they are composed primarily of the active pharmaceutical ingredient, nearly 100% with excipients for stabilization. Their reduced particle size and larger surface area can increase the drug dissolution rate, saturation solubility, and adhesive properties, which can prolong their ocular residence time and enhance corneal permeation, thereby increasing bioavailability. Moreover, this technology can improve the safety and tolerability of current ocular treatments [12]. These advantages are particularly significant in ophthalmic therapy, where the acceptable instillation volume is typically less than 50 μL, making NCs an attractive option for enhancing drug bioavailability, especially for poorly soluble drugs [13].

This chapter explores the advantages of NC technology in ocular DDSs, highlighting its effectiveness in overcoming ocular barriers, as a promising candidate for future clinical studies and large-scale production. Additionally, it reviews several studies regarding their efficacy in treating conditions such as inflammation, infection, and glaucoma, along with a description of both approved formulations and those currently undergoing clinical trials.

2 Ocular Barriers

Overcoming the eye's natural anatomical and physiological barriers remains one of the greatest challenges in ocular therapy, as these structures are meant to prevent the penetration of foreign substances to preserve ocular integrity (Fig. 1). The human eye consists of two major compartments: the anterior and posterior segments. The anterior segment comprises the front third of the eyeball and includes the cornea, iris, conjunctiva, ciliary body, lens, and a chamber filled with aqueous humor [14]. In

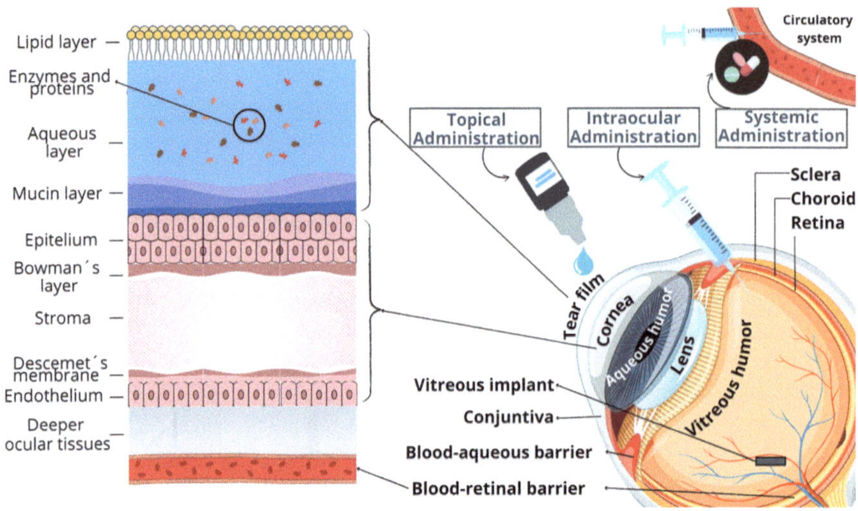

Fig. 1 Ocular drug delivery. Conventional administration routes and barriers

contrast, the posterior segment constitutes the remaining two-thirds and contains a fluid-filled cavity with vitreous humor, surrounded by three layers: the sclera, choroid, and retina [15]. Numerous ocular disorders can lead to severe vision loss, affecting either one or both segments. The most common conditions include age-related macular degeneration, diabetic macular edema, cataract, uveitis, and glaucoma [16]. The conventional routes for ocular drug delivery include topical, systemic, and intraocular administration [17].

The tear film is the first barrier for topical ocular drug administration. This film is a protective layer produced by the lacrimal Meibomian gland. It is composed of lipid, water, and mucin layers, with proteins and enzymes dissolved in the aqueous part. These are capable of binding to and degrading drugs, limiting their effectiveness [18]. Following instillation, excess liquid is rapidly drained by the nasolacrimal duct. Moreover, the continuous turnover of lacrimal fluid (around 1 μL/min) and potential systemic absorption via the capillaries in the conjunctival sac further reduces the drug concentration on the eye surface [19]. Beyond the tear film, the cornea represents another critical barrier. This transparent multilayered structure consists of five layers: epithelium (multiple cell layers), Bowman's membrane, stroma, Descemet's membrane, and endothelium (single cell layer). The corneal epithelium acts as a dense physical barrier. It is continuously renewed through mitosis, restricting the intracellular passage of substances, particularly hydrophilic molecules [15, 20]. The conjunctiva, a multilayered epithelium and stroma with fewer intercellular gaps than the corneal epithelium, also limits the penetration of hydrophilic drugs [21]. Beneath the sclera, the choroid further restricts drug access to the retina [22], which is a highly vascularized layer with rapid blood flow. Meanwhile, the sclera, composed primarily of connective tissue, provides structural support and protection to the inner ocular components [23]. The blood-aqueous barrier is formed by the nonpigmented epithelium of the ciliary body and includes the iris vessel endothelium with tight junctions. It regulates active and paracellular transport, restricting drug movement from systemic circulation into the anterior chamber [24]. Furthermore, physiological mechanisms, including tear exchange, nasolacrimal drainage, and blinking continuously, clear particles from the eye surface, significantly limiting the penetration of therapeutic substances [25].

3 Ocular Drug Administration Routes

Given the challenges posed by the anatomical and physiological organization of the eye, topically instilled drugs face significant difficulties in penetrating ocular barriers and reaching their action site. As a result, it is estimated that less than 5% of the instilled dose is absorbed into the eye [21]. Therefore, frequent drug administration is often required to maintain therapeutic levels, which can increase the risk of potential side effects and poor patient compliance with the dosage schedule [26]. To overcome these drawbacks, two main approaches have been employed in topical therapies. On one side, several strategies to increase drug residence time in the eye were explored, using different excipients to increase viscosity. On the other side, the use of penetration enhancers to improve corneal permeability was studied, such as surfactants, bile acids, salts, among others. Although these strategies have improved the ocular performance of several drugs, the low solubility in tear fluid remains a challenge for many others [23]. Despite these limitations, topical installation is the most preferred route for ocular drug delivery due to its simplicity, convenience, and self-administration [18, 27]. Consequently, topically administered drugs account for over 90% of marketed formulations in this field [28].

When targeting the posterior segment of the eye, the challenge becomes even more complex, as drugs must not only penetrate the ocular surface but also reach the deeper layers of the eye to achieve therapeutic concentrations. Most topically applied drugs fail to achieve therapeutic levels in these deeper tissues [29]. Systemically administered drugs also struggle to reach therapeutic concentrations within ocular tissues due to the blood–retinal barrier (BRB), which restricts the diffusion of drugs from the systemic circulation to the retina [30]. This barrier is formed by the tight junctions of cells from the vascular endothelium of the retinal vessels and the retinal pigment epithelium [31]. Thus, high drug doses are required to achieve effective levels in the eye via systemic administration, which can lead to severe systemic side effects [32]. As a result, oral and intravenous ocular therapies are not often preferred in ophthalmology. However, researchers are also studying the transitory BRB-breakdown to deliver drugs to the eye and access intraocular tissues effectively [33].

Ocular injections are commonly employed for drug delivery in the posterior segment, such as peribulbar, sub-tenon, subconjunctival, intracameral, and intravitreal injections. The first three involve the injection in close proximity to the eyeball (periocular administration), making drugs more likely to permeate through the sclera [34]. Intravitreal injection is considered the most effective for achieving therapeutic concentrations in the retina and choroid [35]. However, this administration route is highly invasive and can lead to complications such as vitreous hemorrhage, increased ocular pressure, endophthalmitis, and even retinal detachment [36, 37]. These potential complications contribute to patient discomfort and reduced compliance with the treatment. In addition to ocular injections, several intravitreal inserts have been approved by the United States Food and Drug Administration (FDA), such as Retisert® (Bausch and Lomb, NY, USA) and Ozurdex® (Allergan, Inc., Dublin, Republic of Ireland). They have been developed as effective intravitreal DDSs capable of achieving a prolonged sustained release [38].

4 Ophthalmic Formulations

Due to the complex physio-anatomy of the eye, pharmaceutical formulations must meet several characteristics for ocular drug delivery. Ideally, ophthalmic formulations should be isotonic and present a neutral pH. The pH of ocular formulations typically ranges from 3.5 to 8.5, as more acidic or alkaline formulations could cause damage in the eyes. The osmolality of open eyes ranges from 231 to 446 mOsm/kg; thus, formulations should present values within this range to avoid irritation. Refractive index, surface tension, and viscosity should also be considered. A high refractive index may cause blurred vision, while low surface tension can lead to irritation or pain. Although higher viscosity can prolong drug retention times, it may also produce patient discomfort [39]. Moreover, all ocular formulations must be sterile.

According to the biopharmaceutical classification system (BCS), aqueous solubility and membrane permeability are the most critical factors influencing drug bioavailability. When a drug presents low aqueous solubility, its absorption is limited by its dissolution rate [40]. Drugs are classified into four groups based on solubility and intestinal permeability: Group I (high solubility and permeability), Group III (high solubility and low permeability), Group II (low solubility and high permeability), and Group IV (low solubility and low permeability). Among these groups, Groups II and IV pose the greater challenges to pharmaceutical scientists. Approximately 70% of new drug candidates and 40% of currently marketed immediate-release oral drugs are classified as practically insoluble [41]. Different approaches have been taken to enhance solubility and, therefore, bioavailability of poorly soluble drugs (BCS II and IV). These include prodrug formation, cocrystals, cyclodextrins, eutectic mixtures, solid dispersions, and NC [42]. Although BCS only addresses intestinal drug permeability, it provides insight into what might occur in other biological membranes such as those of the eye.

Ocular drug therapies are predominantly administered in liquid form. Ideally, the active ingredient must exhibit adequate solubility in both lipid and aqueous media and possess good permeability across ocular tissues. It must also be potent enough to reach therapeutic levels after the administration of small volumes (30–50 μL for eye drops and 50–100 μL for intravitreal injections) [43]. Currently, topical ocular therapies for poorly aqueous soluble drugs (BSC II and IV) primarily consist of suspensions (particle sizes above 1 μm), which have limited ability to overcome ocular barriers [44]. This results in low drug concentrations at the target site, leading to frequent administrations and/or choosing invasive routes of administration such as intraocular injections. Traditional approaches to enhance drug solubility and bioavailability include the use of cosolvents [45], cyclodextrins [46], or micronization [47]. Additionally, several DDSs have been developed, including in situ gels [48], ointments, and inserts [49]. However, their application remains limited due to their poor ocular bioavailability and the eye barriers.

5 Nanoparticle Drug Delivery Systems for Ocular Delivery

In this context, NDDSs have gained significant attention over the past decades as systems capable of effectively delivering drugs into ocular tissues, particularly in the posterior segment. NDDSs are platforms in the nanometer range (10^{-9} m) that can offer several advantages over conventional DDSs, such as suspensions, ointments, and implants. These carriers provide controlled and sustained release by increasing the nanoparticle residence time in the eye, enabling drug-targeted delivery to specific sites, enhancing their membrane permeability, and accessibility to challenging locations within the body, like the posterior segment of the eye. Therefore, an improved efficacy and reduced adverse effects can be observed [29].

Several NDDSs were developed over the past years, including vesicular carriers (liposomes, niosomes), dendrimers, microemulsions, polymeric nanoparticles, lipid nanocapsules, solid lipid nanoparticles, nanostructured lipid carriers, microcrystals, and NCs, among others [16, 50]. Particular attention has been given to NCs as platforms capable of effectively increasing the bioavailability of poorly aqueous soluble drugs, which are associated with a higher dissolution rate and saturation solubility [20].

For NDDSs, particle size (with narrow size distribution) is a crucial parameter for ocular delivery. Particles exceeding a certain size range may cause an uncomfortable, scratchy sensation due to their large particle size, besides presenting less bioavailability [51]. Smaller particle sizes (50–200 nm) are associated with higher uptakes and improved patient compliance due to better tolerability [52], though particles around 20 nm have been reported to undergo rapid clearance from the periocular tissues [53]. A zeta potential (ZP) of approximately 20–40 mV (absolute value) is desired for formulation stability. Furthermore, cationic formulations may enhance corneal retention times as they will have a stronger interaction with the negatively charged corneal mucin [54].

Several NDDSs are already on the market in ophthalmic formulations, such as Visudyne® (Verteporfin, Liposome), Cequa® (Cyclosporine A, Micelles), Inveltys® (Loteprednol etabonate, drug NCs), and Durezol® (Difluprednate, Nanoemulsion), among others. Alongside NDDSs, the development of new excipients has played a crucial role in enhancing these systems. Synthetic and natural viscosity-enhancing polymers (polyethylene oxides, polyvinyl alcohols, polyesters, polymethacrylates, polyolefins, collagen, dextran, and gelatin), along with mucoadhesive polymers (cellulose derivatives, hyaluronic acid, chitosan, and guar gum), have been extensively studied to increase the residence time of ophthalmic drugs [55].

6 Nanocrystals: Features and Preparation Methods

NCs are particles in the submicron range (<1 μm) of pure drug and stabilized by a surfactant layer. They are typically produced in a liquid suspension to form drug NSs, whereas solvent removal by freeze- or spray-drying allows obtaining solid redispersible NCs. Drug NSs are colloidal dispersions of submicron drug particles, generally defined as very finely dispersed biphasic colloids containing solid drug particles smaller than 1 μm. While some variation exists in the literature regarding the definitions of drug NS and drug NCs [56], these terms are often used accordingly if NCs were obtained as nanosized drug particles dispersed in a liquid, or they were dried as a powder form that can then be redispersed in a liquid. Despite their name, in some cases, NCs may not be in a purely crystalline state, but rather in a partially crystalline or even amorphous state. They are composed almost entirely of pure drug, with a minimum amount of excipients such as surfactants or polymers used for stabilization [12].

The physicochemical properties of particles in the nanometer range differ from those of microparticles, due to their reduced particle size, which provides significant advantages for ocular drug delivery (Fig. 2). Thus, drug NCs exhibit a greater

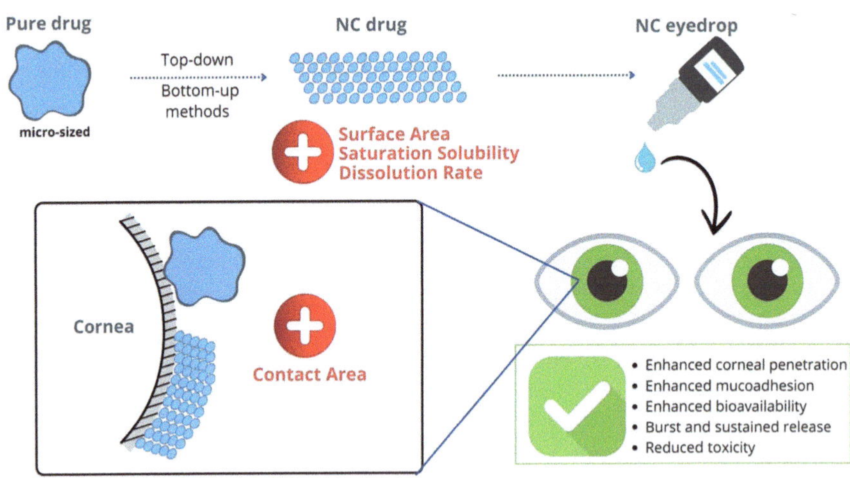

Fig. 2 Diagram illustrating the process of converting a micro-sized pure drug into a nanocrystal (NC) drug using top-down and bottom-up methods. The NC drug is then formulated into an eye drop. The diagram highlights increased surface area, saturation solubility, and dissolution rate. An inset shows the NC drug's enhanced contact area with the cornea. Benefits listed include enhanced corneal penetration, mucoadhesion, bioavailability, burst and sustained release, and reduced toxicity

surface area, leading to an increase in drug saturation solubility, as described by the Ostwald–Freundlich equation (Eq. 1).

$$\log \frac{Cs}{C} = \frac{2\sigma V}{2.303RT\rho r} \tag{1}$$

where C_s is the saturation solubility, C the solubility of the drug formed by large particles, σ the interfacial tension of the substance, V the molar volume, ρ the drug's density, r the radius of the particle, R the gas constant (8.314 J/mol.K), and T the absolute temperature. While this effect is not significant for macroparticles, it becomes highly significant at the nanoscale. A decreased size, increased surface area, and saturation solubility lead to a greater dissolution rate, as explained by the Noyes–Whitney equation (Eq. 2).

$$\frac{dx}{dt} = \frac{DA}{h}(Cs - Ct) \tag{2}$$

where dx/dt is the dissolution rate, D is the diffusion coefficient, h is the diffusional distance, A the surface area, C_s is the drug saturation solubility in the diffusion layer, and C_t is the drug concentration in the medium at the time [68].

NC fabrication methods can be broadly classified into three categories: top-down, bottom-up, or a combination of both. Top-down techniques, also known as nanonization, involve the mechanical breakdown of larger drug particles in the microscale by the application of high-energy inputs until they reach the nanoscale [58]. These methods, widely used in industrial settings, rely on mechanical forces generated by specialized equipment to reduce particle size. Top-down technology is suitable for drugs that are insoluble in both aqueous and organic phases, offering a rapid and scalable process commonly used for marketed drug NC production [59].

One of the most widely used top-down approaches is media milling, where coarse drug particles are mixed with excipients and micrometric beads of hard and dense materials, and subjected to high-energy milling to achieve size reduction. The milling process typically employs small pearls, with their diameter and material chosen based on the desired particle size and product specifications. Common media milling beads include zirconia, stainless steel, glass, or resin glass, with ball sizes ranging from 0.1 to 0.6 mm [60]. NCs produced via media milling exhibit a well-defined shape and narrow size distribution. In turn, media milling is an easily scalable process and potentially eco-friendly. However, this technique generates significant heat, which can impact crystal formation and morphology, often leading to amorphous forms prone to Ostwald ripening, where uncontrolled precipitation or crystallization results in particle growth or agglomeration [58]. Additionally, a significant drawback of this technique is the potential material erosion, which can lead to product contamination [61].

High-pressure homogenization (HPH) is another widely used top-down technique that operates based on microfluidization or piston-gap homogenization principles. HPH reduces particle size through shear and collision forces generated during repeated homogenization cycles, effectively breaking down drug particles to the nanoscale [62]. This technique enables the mass production of NCs with uniform particle size. However, prolonged homogenization times can cause sample contamination, and the high-energy input required is a significant drawback [63].

Melt emulsification, primarily used for producing solid lipid nanoparticles, is also employed for NC preparation. In this method, coarse drug particles are mixed with stabilizers in an aqueous solution and heated until melted. The resulting mixture is then homogenized to form an emulsion, which is subsequently cooled down to obtain NCs [64]. However, its applicability is limited to heat-sensitive drugs. Another NC preparation technique is laser fragmentation, where a laser beam is focused on a drug suspension to directly break down particles to the nanoscale [65].

Bottom-up techniques involve drug dissolution in a solvent (typically organic), followed by controlled precipitation and solvent evaporation. Compared to top-down methods, precipitation-based techniques are considered simpler and more cost-effective [66] and can be performed at low temperatures, making them suitable for producing NCs of heat-sensitive drugs. However, scalability remains a challenge due to issues with product reproducibility.

The antisolvent precipitation procedure is the most widely used bottom-up method. In this process, the drug is fully dissolved in a solvent before being introduced into a non-solvent that contains stabilizers, causing NC precipitation. This is followed by solvent removal via evaporation or lyophilization, resulting in a dry nanocrystalline drug product [67]. To achieve an optimal particle structure and prevent excessive growth into the micrometer range, it is essential to regulate processing conditions and incorporate stabilizers such as surfactants [68]. Ultrasound can also facilitate a controlled drug crystallization, yielding more homogeneous and finer NCs [69].

A major drawback is the use of organic solvents, which must be removed after NCs precipitation, increasing the production costs [68]. To address this issue, alternative bottom-up methods were developed. For instance, the use of supercritical fluids either as solvents or antisolvents, depending on the drug's solubility, eliminates the need for organic solvents. This approach results in high-purity products. However, this technique requires a large volume of supercritical fluids and is only suitable for drugs that are soluble under these conditions [70]. The properties of NCs obtained via bottom-up methods are influenced by various process parameters, including the type of solvent and antisolvent, the rate and ratio of solvent addition to the antisolvent, drug concentration, stabilizer type and quantity, and mixing conditions, among others [23]. In order to further improve the stability of NSs, drying methods like freeze- or spray-drying may be applied using a cryoprotectant, such as trehalose, mannitol, and sucrose, among others [71].

Combined approaches are also employed, integrating a bottom-up method as a pretreatment step, followed by a top-down high-energy process, typically HPH, to further reduce particle size. The combined process allows for obtaining NCs with fewer homogenization cycles and, thus, lower energy costs. In the second step, HPH is used to strengthen the nanoparticles generated in the first step, enhancing their thermodynamic stability [72].

When formulating NCs, the chosen approach depends on the formulation's specific objectives, including scalability, costs, and process time. Regarding NCs for oral drug delivery, most commercially approved formulations are developed via top-down methods [73]. Despite the majority of NCs in scientific research being formulated by top-down methods for ocular drug delivery, many bottom-up formulations were also studied.

The effect of bottom-up and top-down formulation methods on the production of NSs for ocular delivery was investigated [74]. Thus, hydrocortisone NSs were obtained by microfluidic nanoprecipitation (bottom-up) and wet milling (top-down). Both milled and precipitated drug NS showed increased intensity and absorption compared to the drug solution. Interestingly, despite having a similar ZP around −2 mV for both nano-formulations, stability was superior in milled NS, with no changes in particle size for 2 months at room temperature. In contrast, precipitated drug NS showed a nearly 100% increase in size. The author attributed this difference to the amorphous nature of precipitated particles compared with NS obtained by wet milling, which maintained the crystalline structure of hydrocortisone.

7 Nanocrystals in Ocular Drug Delivery

NCs were first developed as a strategy to enhance the bioavailability of poorly water-soluble oral drugs. NSs can be administered via different routes (parenteral, oral, ocular, pulmonary, and topical, among others) either in liquid or solid forms obtained through processes such as freeze-drying, spray-drying, or spray–freezing, enabling diverse dosage forms including powders, pellets, tablets, capsules, films, or gels [71]. Moreover, NCs have shown that they can overcome many challenges associated with ophthalmic drug delivery. Examples of different ocular NC formulations, their formulation methods, and their advantages are presented in Table 1. Compared with conventional suspensions, the reduced size and increased surface area of NCs enhance their saturation solubility and dissolution rate following topical application. This also improves corneal penetration, thereby boosting bioavailability with minimal to no irritation, unlike larger particles in the micrometer range. Additionally, NCs help prevent the high osmolarity often associated with conventional solutions [97]. Compared to other drug-loaded nanocarriers, NCs present high drug loading (almost 100%), which gives them the capacity to efficiently deliver drugs to the desired site, as fewer particles are required to obtain therapeutic effects [57]. Even though many NDDSs have been shown to greatly permeate the cornea, they face the challenge of low drug loading capacities in comparison to the almost 100% drug loading of NCs. The NC preparation does not require high concentrations of surfactants or other lipid/polymeric substances for drug solubilization, which can be a significant source of toxicity [23]. NCs can improve particle interactions with corneal tissues since they have a higher surface area than drugs in the micrometer range, which leads to more contact points. This improved interaction could lead to

Table 1 Nanocrystal formulations in development for ocular drug delivery

Nanocrystal drug	Method of formulation	In vivo studies	Objectives	Highlights	References
Hydrocortisone, prednisolone, dexamethasone	**Top-down:** high-pressure homogenization	Ocular bioavailability/male healthy albino rabbits	Comparison of micro- vs nanocrystals	High absorption, intensity and duration of nanocrystals	[75]
Hydrolyzable dye	**Bottom-up:** precipitation	Ocular penetration/male Sprague–Dawley rats	Comparison of micro- vs nanocrystals	High ocular penetration	[76]
Fluorometholone	**Bottom-up:** precipitation	Ocular penetration/New Zealand female white rabbit	Comparison of micro- vs nanocrystals	Higher ocular penetration and longer release of the drug into the aqueous humor	[77]
Hydrocortisone	**Top-down:** wet milling **Bottom- up:** microfluidic precipitation	Ocular bioavailability/male healthy albino rabbits	Comparison of microfluidic nanoprecipitation (bottom-up) and wet milling (top-down) formulation methods	High stability for milled nanosuspension	[74]
Fluconazole	**Bottom-up:** antisolvent evaporation method (precipitation)	–	Characterisation of nanocrystals	Increased solubility of nano-crystal vs untreated drug	[78]
Indomethacin	**Top-down:** bead milling	Image analysis of corneal wound healing/male Wistar rats Transcorneal penetration/adult Japanse albino rabbits	Corneal toxicity	Improved penetration and better tolerance in vitro and in vivo	[79]
Ciclosporin A	**Top-down:** media milling method	–	Corneal toxicity	Low toxicity of nanocrystal in comparison to Restasis® in a Draize test	[80]
Disulfiram	**Top-down:** bead mill method	Efficacy studies/intraocular pressure enhanced male Japanese rabbits	Hypotensive effect	Increased corneal residence time and penetration	[98]
Acetazolamide	**Bottom-up:** antisolvent precipitation followed by sonication	Hypotensive efficacy in an steroid glaucoma model/adult male New Zealand albino rabbits	Hypotensive effect	Higher reduction of intraocular pressure and longer duration of action compared to the same drug solution	[81]

Acetazolamide	**Top-down:** wet bead milling and spray drying	Male albino New Zealand rabbits	Hypotensive effect	Higher reduction of intraocular pressure and longer duration of action compared to the same drug suspension and Azopt®	[82]
Triamcinolone acetonide	**Top-down:** bead milling	Endotoxin-induced uveitis model/albino male white rabbits	Anti-inflammatory efficacy	Higher efficacy in an endotoxin-induced uveitis model compared to Fortcinolona 40®	[83]
Brinzolamide	**Top-down:** rapid wet milling	Hypertension model/seven-month-old male Wistar rats	Hypotensive effect	Higher efficacy in an ocular hypertensive model compared to Azopt®	[10]
Moxifloxacin	**Top-down:** wet bead milling	Ocular infection model and pharmacokinetics/male Sprague Dawley rats	Antimicrobial efficacy	Higher efficacy in vivo compared to Vigamox®	[84]
Dexamethasone acetate	**Top-down:** wet bead milling	–	Enhancing mucoadhesion (cationic nanocrystal)	Significant electrostatic interaction in vitro	[54]
Tedizolid phosphate	**Bottom-up:** antisolvent precipitation followed by homogenization and sonication	Ocular pharmacokinetics/healthy New Zealand white rabbits	Enhancing mucoadhesion (cationic nanocrystal)	Increased retention times and bioavailability of nanocrystal compared to aqueous solution ex vivo and in vivo	[85, 86]
Olopatadine	**Bottom-up:** spray-drying	Retention time study/healthy female and male Anatolian merino sheep	Enhanced mucoadhesion (cationic nanocrystal)	Longer retention times compared to olopatadine solutions in vivo	[87]
Econazole	**Bottom-up:** spray drying	Bioavailability/albino rabbits	Enhance mucoadhesion (chitosan solution)	Enhanced retention times compared to nanocrystals suspended in an isotonic buffer in vivo	[88]

(continued)

Table 1 (continued)

Nanocrystal drug	Method of formulation	In vivo studies	Objectives	Highlights	References
Triamcinolone acetonide	**Top-down:** wet bead milling	–	Enhanced mucoadhesion (soy protein films)	Excellent mucoadhesive strength and sustained release for 5 days in vitro	[89]
Triamcinolone acetonide	**Bottom-up:** solvent-antisolvent precipitation, sonication, homogenization	Ocular pharmacokinetics/ male New Zealand rabbits	Enhancing mucoadhesion (in situ gelling system)	Increased drug concentration and sustained release compared to traditional triamcinolone nanocrystals in vivo	[90]
Posaconazole	**Top-down:** microfluidization	–	Enhancing mucoadhesion (in situ gelling system)	Higher retention times ex vivo and greater antifungal activity in vitro against *Candida Albicans* compared to itraconazole aqueous solution	[91]
Itraconazole	**Top-down:** media milling method with ultra-small-scale device	–	Enhancing mucoadhesion (in situ gelling system)	Higher concentration and retention times ex vivo and greater antifungal activity ex vivo against *Candida Albicans* compared to itraconazole aqueous solution	[92]
Triamcinolone Acetonide	**Top-down:** wet milling technique	–	Enhanced transscleral permeation (nanocrystal loaded microneedles)	Faster dissolution of nanocrystals loaded microneedles compared to triamcinolone loaded microneedles	[93]
Deferasirox	**Top-down:** wet media milling	–	Enhanced transscleral permeation (nanocrystal-loaded microneedles)	Faster dissolution of nanocrystals loaded microneedles compared to triamcinolone loaded microneedles	[94]

Forskolin	**Top-down:** high-pressure homogenization with Ultraturrax®	Efficacy studies/adult glaucomatous New Zealand rabbits	Enhanced anti-glaucoma efficacy (in situ gelling system)	Higher intraocular pressure suppressing effects compared to a forskolin suspension in an in vivo dexamethasone-induced ocular hypertensive model	[95]
Tranilast	**Top-down:** bead mill method	–	Transcorneal penetration and toxicity	Enhanced corneal penetration and less toxicity compared to Rizaben® in vitro	[96]

prolonged ocular retention time, sustained drug release effect, enhanced corneal drug permeation, and increased bioavailability. NCs present a dual release profile, involving an immediate or burst release effect followed by a sustained release effect. The burst release effect is associated with an increased drug dissolution rate and saturation solubility, leading to a higher concentration gradient and diffusive flux into the eye. The sustained release profile involves the NCs that are not dissolved immediately in the available fluid and remain on the ocular surface, acting as a drug reservoir [54].

Lastly, NCs can be produced using simple and cost-effective procedures in comparison with other nanoparticle production techniques. Since organic solvents can be avoided for top-down methods, greater chances of process scalability can be achieved and, eventually, commercialization of NC-based formulations can be achieved. This is supported by the many NC-based formulations that have been approved for use in humans over the last decades, including the ocular field. To our knowledge, there are currently four formulations in the market with NC-based technology for ocular applications, developed using top-down methods. These include Ilevro® (Nepafenac), Inveltys® (Loteprednol etabonate), Eysuvis® (Loteprednol etabonate), and Eyenovia® (Clobetasol propionate). Another NC formulation, NCT03926026 (Fluticasone propionate), is under clinical trial [98]. These will be further analyzed ahead. In addition, NCs showed great advantages in ocular drug delivery, including their reduced size, less toxicity, and enhanced adhesive properties, as described in Table 1.

7.1 *Nanocrystals Versus Microcrystalline Suspensions*

It seems important to mention that drug powder micronization was first explored by achieving particles in the micrometer range (>1 μm). However, this approach proved unable to surpass the bioavailability challenges of many poorly soluble drugs, as the resulting surface area was not large enough to significantly enhance their dissolution rate. As a consequence, this kind of technology moved from micronization to nanonization of drugs [11, 62] and studies comparing micro- and NC systems were carried out. In vivo studies in healthy male albino rabbits showed that three NSs of hydrocortisone, prednisolone, and dexamethasone exhibited statistically significantly higher drug absorption, intensity, and duration of action than microsuspensions and solutions of the same drugs [75]. In another study, the superiority of a hydrolyzable-dye NS against micron-sized particles was explored, where the first ones showed about a 50-fold higher ocular penetration in an in vivo study with male Sprague–Dawley rats [76]. A few years later, in in vivo studies in New Zealand, white female rabbits evidenced that fluorometholone NS attained higher drug ocular penetration and longer drug release into the aqueous humor compared to the same drug microsuspension. Fluorometholone NS presented a 1.6 times faster drug dissolution rate than microsuspensions [77].

7.2 Ocular Tolerability and Toxicity

Since the eye is a specialized and sensitive organ, toxicity is a major issue to address when researching in the field of ocular drug delivery. Despite nanomedicine advancing at great steps every year, with over 3000 nanomaterials being used by pharmaceutical companies, the data about their toxicology is scarce and their impact on biological systems is far from being completely understood [99]. NDDSs should not cause any toxicity; thus, their development should always be accompanied by toxicological studies. Many variables can influence the toxicity of NDDSs in ocular tissues, including size, shape, charge, composition, and dose, among others [100]. In comparison to larger particles, NCs interact with biological tissues differently. NCs penetrate deep into body tissues and have longer retention times in them due to their capacity to avoid the reticuloendothelial system. Smaller particles (<100 nm) will present even longer retention times. The bio-persistency and toxicity that may arise from this technology should be carefully assessed. Nevertheless, evidence showed that NC formulations are safe and well tolerated, with less toxicity in comparison to conventional drug products [101]. Toxicity in the eye principally occurs through cellular oxidative stress, mitochondrial or DNA damage, lipid peroxidation, and chromosome aberration [102]. For NSs, the formulation together with their excipients (stabilizing agents) must be safe, stable, and nontoxic. To evaluate toxicity in the eye, in silico, in vitro (MTT assay, Hen's Egg Test-Chorioallantoic Membrane test), ex vivo, or in vivo studies can be conducted. The in vivo Draize test performed in rabbits is recognized as the international standard for evaluating acute ocular toxicity. New Zealand white rabbits are the most commonly used animals for this test due to their large eyes, well-defined anatomy and physiology, cost, and availability [103].

Cyclosporin A is a potent immunomodulator that acts selectively and locally when administered to the ocular surface. Given that Restasis®, an ophthalmic emulsion that contains a fine dispersion of Cyclosporin A 0.05%, reported ocular burning as a severe side effect in 17% of patients, a NS of Cyclosporin A was developed via a wet milling method and evaluated for stability and corneal toxicity in male New Zealand rabbits' eyes [80]. Draize test and Schirmer tear test were conducted, revealing that the drug NS exhibited lower toxicity than Restasis®, which can be attributed to the absence of solubilized drug and oil in the NS, as well as lower surfactant concentration.

In another study, the ocular tolerance of Indomethacin NCs was studied and compared with commercially available eye drops of this nonsteroidal anti-inflammatory drug, which reports limited clinical use due to its adverse effects such as burning sensation and irritation. Indomethacin NC presented enhanced corneal penetration compared with commercially available indomethacin eye drops, both in in vitro and in vivo studies performed in adult Japanese albino rabbit corneas. Moreover, they demonstrated better tolerance in both rat and human corneal epithelial cells compared to commercially available indomethacin eye drops, as well as pranoprofen, diclofenac, bromfenac, and nepafenac eye drops. This improved tolerance may be primarily due to the lower corneal accumulation of indomethacin-NCs upon instillation, as their corneal penetration reduces drug retention compared to conventional solutions [79]. Similarly, in an in vitro transcorneal penetration study, Tranilast NS showed enhanced corneal penetration compared to the commercially available tranilast eye drops Rizaben®. Tranilast NS was better tolerated by human corneal epithelial cells than Rizaben® eye drops in a corneal toxicity study by the same authors [96].

7.3 Mucoadhesive Properties

Mucoadhesion is defined as the ability of materials to adhere to and interact with the body's mucosal membranes. After an ocular application, a higher material interaction with ocular tissues can prolong the formulation residence time in the eye [104]. Conventional topical forms are rapidly washed out from the cornea, due to many eye mechanisms such as blinking, lachrymal fluid, and its turnover. While conventional enhanced viscous systems like gels or ointments can improve retention, they often cause discomfort and blurred vision [105]. Additionally, patients usually face problems related to their administration, as the application of an exact amount and product contamination, since they must use their fingers to correctly place the formulation in the eye [43]. Therefore, an improved mucoadhesion was explored as an alternative strategy to enhance corneal retention time without compromising patient comfort by increasing the viscosity of the system through the use of mucoadhesive polymers, the formation of in situ gelling systems, or the incorporation of cationic charges to the systems [106, 107]. One of the most common in vitro methods to evaluate the adhesion between NDDSs and mucin is through ZP measurements. Corneal mucin, with a relatively high negative charge related to the prevalence of anionic sialic acid groups, can interact with cationic nanoparticles, causing changes in their ZP value [108]. Such changes help to indicate the presence of mucoadhesive interactions, but not their strength, between a wide range of cationic nanoparticles and mucin based on reduction or reversal of the native cationic nanoparticle charge upon mucin binding [104]. Other methods include tensile and shear strength measurements, chip-based systems, rheological measurements, and atomic force microscopy, among others [91].

Conventional NCs alone present higher retention time in the eye due to their high surface-area-to-volume ratio. Therefore, they provide a longer retention time after penetrating into the mucus layers of the tear film and a higher penetration property in the cornea [109]. Nevertheless, efforts to improve this residence time and enhance the ocular bioavailability of drugs were addressed. Arguing that a positively charged system could present better interaction with the negatively charged eye surface, leading to longer retention times in the cornea and therefore increasing the drug permeation and bioavailability, cationic NCs made up of dexamethasone acetate (hydrophobic) and polymyxin B (hydrophilic) were developed [54]. The positive charge was generated using two FDA-approved cationic preservatives such as cetylpyridinium chloride and benzalkonium chloride. Although further studies are needed, an in vitro mucin-adhesion study determined that there was a significant electrostatic interaction between the mucin layer and these NCs, which may lead to increased retention times in vivo. Furthermore, no in vitro toxicity was observed in a study in fibroblast cell culture.

Another approach to enhancing mucoadhesion is the hydration of certain polymers, such as chitosan, which become adhesive upon contact with mucosal membranes [110]. Econazole NCs were developed [88] via a nano-spray-drying technique and suspended either in an isotonic buffer pH 7.4 or a 0.1% chitosan HCl solution. The inclusion of chitosan aimed to enhance retention time following topical administration by increasing formulation viscosity and providing a positive charge, which promotes electrostatic interactions with the negatively charged ocular surface. Mucoadhesion studies showed a decrease in the positive ZP of econazole NCs from 42 to 31 mV, indicating interaction with mucin. In vivo retention studies performed in albino rabbits showed increased retention time for econazole NC suspended in chitosan. None of the tested formulations presented irritation in the Draize test evaluation after 4 days of continued instillation in rabbit eyes. Similarly, cationic olopatadine NCs for ocular allergic disease were developed via a spray-drying method. In vivo retention studies in healthy female and male Anatolian Merino sheep using Schirmer tear strips demonstrated that olopatadine NCs exhibited longer retention times compared to olopatadine solutions [87].

In situ gelling systems are reversible phase transition systems (solution-gel), triggered by factors such as pH or temperature. They usually present a pseudoplastic behavior, ideal to minimize interference with blinking, and are an excellent strategy to enhance viscosity without the previously mentioned drawbacks. Additionally, NDDSs can be incorporated into in situ gelling systems [111]. For instance, in situ gelling formulations of the previously mentioned indomethacin-NCs were developed to prolong drug retention time after ocular instillation [112]. When these NCs come in contact with body temperature (approximately 37 °C) after their instillation on the ocular surface as a solution, they transform into a gel due to a property given by a thermoresponsive polymer (Pluronic®F-127). In this way, the formulation demonstrated higher corneal permeation in adult Japanese albino rabbits compared to standard indomethacin-NCs, likely due to its increased viscosity, which extends the precorneal residence time. Moreover, the system showed no corneal toxicity after being repetitively administered to Japanese rabbit eyes for 1 month.

7.4 *Nanocrystals for Ocular Pathologies*

The abovementioned advantages characteristic of NCs—increased bioavailability, enhanced corneal retention time, diminished toxicity—demonstrate the superiority of drug NCs over conventional drug forms such as suspensions that are currently available on the market and/or microcrystals. The efficacy of NCs was also studied in several ocular therapeutic areas, such as anti-inflammatory, antimicrobial, antifungal, and as an alternative to decrease the intraocular pressure (IOP) for treating glaucoma.

Ocular inflammation can affect different eye tissues and can emerge from infectious and noninfectious processes. Corticosteroids are the first-line treatment in many ocular pathologies with inflammatory processes. All corticosteroids contain a common steroid nucleus of 21 carbon atoms and can be modified to make them more hydrophilic or lipophilic [113]. While topical application is the widely accepted treatment and is often sufficient for managing anterior segment inflammation, corticosteroids are primarily administered via injection for posterior segment inflammation. This is because topical, subconjunctival, and systemic administration often fail to achieve therapeutic concentrations in the vitreous and retina or are associated with significant adverse effects, particularly in long-term treatments [114]. Uveitis is responsible for 15% of the causes of total blindness in Western countries. It describes a group of conditions characterized by the inflammation of the middle layer of the eye (uvea), which includes the iris, ciliary body, choroid, and surrounding tissues such as the retina, optic disc, and vitreous [115]. In noninfectious uveitis, corticosteroids are the first line of treatment [116]. Triamcinolone acetonide is the corticosteroid of preference for ocular injections, as its low solubility in crystalline form enables it to maintain therapeutic concentrations for an extended period [117].

Triamcinolone acetonide-NCs were developed to improve ocular corticosteroid therapy [83]. These presented a dissolution rate at least 40 times greater than the micronized drug. Compared to Fortcinolona 40®, triamcinolone acetonide-NCs achieved the most effective subconjunctival treatment in an in vivo endotoxin-induced uveitis model, alleviating the inflammatory response in the anterior chamber and iris of the eyes. Notably, a better spread of particles in the subconjunctival space was observed for triamcinolone acetonide-NCs in comparison to the commercially available formulation, which formed a defined deposit. It may be associated with an adequate particle dispersion of NCs and their reduced particle size. As mentioned before, invasive routes are often required for drug delivery into the posterior segment of the eye due to their poor solubility and/or permeability properties. Another double system comprising NCs was developed by incorporating these triamcinolone acetonide-NCs into soy protein films as a potential alternative to invasive administration methods [118]. This dual system presented excellent mucoadhesive strength using an Instron Universal Testing Instrument and a 5-day sustained release in an in vitro study on simulated tear fluid using the dialysis method. In another study, in situ gelling systems loaded with other triamcinolone acetonide-NCs were also developed for topical ocular delivery [119]. A pharmacokinetic study in male New Zealand rabbit eyes evidenced that NCs incorporated in in situ gelling systems showed 1.3 times higher concentrations and sustained drug concentration for longer times than conventional triamcinolone acetonide-NCs, and 3.2 times higher concentrations than a suspension of triamcinolone acetonide containing hydroxypropyl-β-cyclodextrins, when these formulations were applied [120].

Microneedles (MNs) are innovative, noninvasive DDSs designed to achieve therapeutic effects comparable to intraocular injections while being painless, convenient, and patient-friendly administration devices. Ranging in height from 25 to 2000 μm, MNs can be manufactured in diverse shapes from a wide range of materials, allowing adaptability to specific therapeutic needs [121]. Triamcinolone acetonide-NCs-loaded MNs were developed for ocular therapy [122]. This drug was selected as a model hydrophobic molecule in order to test this co-system for transscleral delivery of hydrophobic drugs. Triamcinolone acetonide-NCs-loaded MNs were successfully developed and could effectively deliver the drug to the deeper scleral tissue in an ex vivo study. Triamcinolone acetonide-NCs-loaded MNs dissolution was around 3 min after being pierced in the excised porcine sclera, with around 4.5-fold higher concentrations of triamcinolone acetonide in the scleral tissue after 5 min of application in comparison to drug-loaded MNs, which took more than 8 min to dissolve. Toxicity studies were carried out in vitro, with human retinal pigment epithelial cells and with the Hen's Egg Test-Chorioallantoic Membrane test, showing that this system was biocompatible with ocular tissues. Very similarly, deferasirox-NCs-loaded MNs were developed as a minimally invasive system for intraocular delivery, designed to treat retinal disorders associated with excess iron, which can induce oxidative stress in the eye [94]. Through a wet media milling technique and utilizing poly(vinyl alcohol) as a stabilizer, deferasirox-NCs were successfully loaded into MNs. Deferasirox exhibited a fourfold increase in solubility in deferasirox-NS compared to the pure drug. Deferasirox-NCs-loaded MNs exhibited 5 times greater deposition of deferasirox in porcine sclera after 5 min, in contrast to deferasirox-loaded MNs. This system also showed good biocompatibility after testing in human retinal pigment epithelial cells.

Ocular infection involves tissue inflammation caused by any pathogen that invades the eye, including bacteria, fungi, and viruses, among others. In this context, first-line treatment typically involves the topical administration of antibiotics and antifungal drugs, often complemented with adjuvant therapy such as corticosteroids to control inflammatory processes [123]. Ocular fungal infections are a significant cause of morbidity and blindness, with keratitis (corneal inflammation) being the most frequent one [124]. One of the main long-term challenges related to the management of ocular fungal infections is designing DDSs that provide adequate drug distribution without affecting intraocular structures and/or producing systemic side effects. Additionally, systemic drug exposure must also be addressed to ensure sustained disease management [97]. The majority of antifungal drugs used in the ocular field belong to the azole class, but polyene and allylamine antifungals are also utilized [125].

Regarding antifungal ocular therapies with NCs, different fluconazole NSs stabilized with different agents were developed and characterized [78]. These systems exhibited a solubility increase up to 5.7-fold compared with the untreated drug. In an ex vivo transcorneal study using goat corneas, Kollicoat® IR stabilized NSs showed the highest extent of drug release, the greatest flux, and the highest permeability coefficient. In another study, an in situ gelling NS of posaconazole was formulated using microfluidization (top-down) with gellan gum as the gelling agent. The developed posaconazole NS was found to be nonirritant in a HET-CAM test. Ex vivo corneal retention studies showed posaconazole retention for up to 24 h in the cornea, while in an in vitro study against *Candida albicans* using the agar diffusion method, posaconazole NS showed a greater zone of inhibition compared to an itraconazole marketed formulation [91]. Moreover, thermosensitive and mucoadhesive in situ gel itraconazole-NCs were developed for the treatment of fungal keratitis [126]. The production technique utilized was media milling with an ultrasmall-scale device. Pluronic®F-127 was used as a stabilizer, and its combination with Pluronic®F-68 and hydroxypropyl methylcellulose was optimized using a central composite design to formulate the in situ gelling system. In an ex vivo ocular kinetic study using a porcine eye model infected with *Candida albicans*, itraconazole-NCs dispersed in the thermosensitive gel exhibited approximately 4 times higher concentrations and retention times than a conventional itraconazole eye drop. Antifungal efficacy was tested using the same infected eye model, revealing that the itraconazole-NCs in situ gelling system led to a 93% reduction in the *Candida albicans* population after 48 h, compared to a 59% reduction achieved by the conventional itraconazole eye drop.

Despite antibiotic eye drops being the most common treatment for bacterial ophthalmic infections, currently available formulations are cleared from the eye, leading to short corneal residence time. Hence, frequent administration is required. Therapeutic regimens with antibiotic eye drops can last for up to 7 days, which may reduce patient compliance and, consequently, treatment efficacy [127]. Therefore, ensuring patient adherence to treatment is essential in antibiotic ocular therapy.

NCs composed of a moxifloxacin (hydrophilic) and pamoic acid complex (hydrophobic), together with Pluronic®F-127 as a stabilizer, were developed [84]. This novel formulation proved to be more effective than the commercially available formulation Vigamox®, when a single drop was administered at the time of *Staphylococcus aureus* inoculation in rats. When used after 24 h of bacterial inoculation, a single drop of moxifloxacin NS was as effective as Vigamox® administered 3 times a day. In this case, the increased efficacy and sustained release of the NC formulation enabled a more efficient dosing schedule, improving patient compliance. Also, two separate studies were conducted to develop tedizolid phosphate NCs for topical ocular application [120]. The tedizolid phosphate NCs demonstrated significantly enhanced ($p < 0.05$) antimicrobial activity against all the tested Gram-positive bacteria—*Bacillus subtilis, Streptococcus pneumonia, Staphylococcus aureus,* and *MRSA (SA-6538)*—using the agar diffusion method [86]. Moreover, in vivo pharmacokinetics studies in healthy New Zealand white rabbit eyes and ex vivo transcorneal permeation studies showed increased retention times and bioavailability of tedizolid phosphate NCs in comparison to a tedizolid phosphate aqueous suspension. This can be attributed to the high superficial area of NCs and the cationic nature of tedizolid phosphate NCs.

NCs technology was also explored for glaucoma treatment. Glaucoma comprises a group of progressive optic neuropathies characterized by the degeneration of retinal ganglion cells and the retinal nerve fiber layer, leading to changes in the optic nerve [128]. The primary treatment goal is to slow down optic nerve degeneration. Since elevated IOP is the most significant risk factor that accelerates disease progression, lowering IOP remains the only proven therapeutic strategy in clinical practice [129]. Disulfiram-NCs were developed to enhance the drug's hypotensive efficacy [130]. Corneal penetration and in vivo retention studies in male Japanese rabbits demonstrated increased corneal residence time and penetration of disulfiram-NCs. Compared with a commercially available timolol eye drop (Timoptol®), disulfiram-NCs exhibited significantly higher viability in human corneal epithelial cells. After increasing the IOP of rabbits by placing them in darkness for several hours, both disulfiram-NCs and disulfiram aqueous suspension significantly decreased IOP compared to a saline solution, with disulfiram-NCs exhibiting a superior effect. In another study, a stable acetazolamide NSs was developed to enhance the ocular delivery of this poorly soluble hypotensive drug [81]. Acetazolamide NS were prepared using hyaluronic acid salt and poly-δ-glutamic acid as stabilizing polymers by an antisolvent precipitation method coupled with sonication, followed by a spray-drying process to obtain dried acetazolamide-NCs. In vivo studies, using a steroid-induced glaucoma model in adult male New Zealand albino rabbits, demonstrated that redispersed acetazolamide NS achieved a greater and longer-lasting IOP reduction compared to an acetazolamide solution. Notably, only hyaluronic acid salt was able to stabilize spray-dried NSs and maintain their properties for 6 months. Similarly, three ophthalmic NS formulations of brinzolamide were developed via the wet milling technique and compared to Azopt® in a modern ocular hypertensive model utilizing 7-month-old male Wistar rats [131]. In this regard, formulation I (pH = 7.5) contained benzalkonium chloride, the same as Azopt®, formulation II (pH = 7.5) contained polysorbate 80 as a penetration enhancer, and formulation III (pH = 4.5) contained benzalkonium chloride. All three brinzolamide NS formulations achieved at least equal IOP reduction compared to the marketed product, but only formulation III demonstrated a statistically significant IOP reduction 60 min after topical administration. In another study, acetazolamide-NCs with a particle size around 300 nm, prepared by the wet bead milling technique followed by spray-drying, demonstrated increased drug saturation concentration and higher adhesive properties than normal-sized acetazolamide powder. Acetazolamide-NCs showed a higher hypotensive effect than the same drug suspension and Azopt® in normotensive male albino New Zealand rabbits after topical ocular application. Azopt® showed a maximum IOP decrease after 1 h of instillation, while acetazolamide-NCs achieved it at 4 h. Particularly, acetazolamide-NCs presented an equal effect to Azopt® after 1 h of instillation and a significantly lower IOP reduction after 8 and 24 h [82]. Finally, an in situ gel-forming NS of forskolin was developed to enhance the hypotensive efficacy of labdane diterpenoid derived from the Indian *Coleus* plant [95]. This in situ *gelling* system containing forskolin-NCs showed a greater and longer-lasting IOP suppressing effect than a forskolin suspension (12 h vs 6 h) in an in vivo study with dexamethasone-induced ocular hypertensive New Zealand rabbits.

7.5 Commercially Available Ocular Nanocrystals

Up to this point, we have focused on formulations that are still in development and undergoing testing at the laboratory level, meaning they are in the preclinical stage. This phase encompasses the full characterization of pharmaceutical systems, as well as studies conducted in vitro, ex vivo, in silico, and in vivo. Before advancing to human trials, all these studies must be submitted to regulatory agencies for evaluation. The final steps before commercialization involve clinical trials (Phases I, II, and III) to establish the drug's safety and efficacy. Lastly, Phase IV or post-marketing trials are carried out to identify less frequent adverse reactions [132]. The transition of NDDSs from the laboratory and experimental stages to approved and commercially available medications faces many challenges, primarily related to scalability and quality control [133]. To the best of our knowledge, four NC-based ocular formulations have been approved by the FDA, while only one is undergoing clinical trials.

Approved in 2012, Ilevro® (Novartis, Nepafenac) demonstrated in Phase III clinical studies [134] that a single-drop formulation of nepafenac NC formulation (0.3%) was equivalent in efficacy and safety to a conventional formulation (0.1%) administered 3 times daily for the treatment of pain and inflammation following cataract surgery. Similarly, the efficacy and safety of Inveltys® (Kala Pharmaceutics, KP-121 1%, loteprednol etabonate) for the same pathology were confirmed against a placebo in Phase III studies [135]. Twice a day administration of KP-121 1% was found to be effective, safe, and well tolerated, offering patients a potent anti-inflammatory product with less frequent dosing. Pharmacokinetic studies in male New Zealand albino rabbits showed threefold higher concentrations of KP-121 1% in ocular tissues and aqueous humor than the commercially available Lotemax® formulation [136]. Eysuvis® (Kala Pharmaceutics, KP1–121 0.25%, loteprednol etabonate) was approved for human use for the short-term treatment (up to 2 weeks) of dry eye disease. In Phase III clinical trials, this formulation rapidly and significantly reduced signs and symptoms of dry eye disease, with minimal adverse effects [137]. In addition to these corticosteroid-approved formulations, KAL821, another NC formulation composed of a receptor kinase inhibitor, was developed as a potential treatment for age-related macular degeneration using the wet milling method. Its pharmacokinetics were studied in Göttingen mini-pigs, where it demonstrated enhanced drug penetration into the posterior segment of the eye. Both KAL821 and the approved KP-121 formulation were developed using a mucus-penetrating particle technology, in which the drug is milled in the presence of a mucus penetrating particle until the particles reach the nanometer range [138]. APP13007 is a Clobetasol Propionate (0.05%) NC formulation developed by Formosa Pharmaceuticals using their proprietary nanomilling technology, and was later approved for commercialization as Eyenovia® in 2024. In Phase III clinical trials [139] completed in 2022, the efficacy and safety of this formulation were evaluated for treating pain and inflammation following cataract surgery, after the administration of one drop twice daily for 14 days. This product demonstrated rapid inflammation and pain relief, as well as safety and comfort for patients compared to a placebo. Once again, NC technology enabled the development of a more convenient dosing regimen for this condition.

In 2019, a Phase II clinical trial evaluating the safety and tolerability of fluticasone propionate NC ophthalmic suspension (0.1%) for the treatment of acute exacerbations of blepharitis was completed [140]. Previously, its safety and tolerability were studied in an in vivo study with male and female Beagle dogs [141].

8 Final Remarks

There is no greater validation for DDSs than its approval for human use after successful clinical trials. Since the introduction of NCs in the 1990s, several formulations utilizing this technology have been approved for commercialization. Remarkably, there was only a short gap of approximately 10 years between the discovery of NCs and the first commercially available product in the 2000s. NC technology has proven highly successful in the pharmaceutical field, with numerous formulations currently available on the market.

Initially, NCs addressed challenges related to oral drug administration, but their application has since expanded to other routes of administration. However, this technology is far from being fully elucidated or exploited. The field of ocular drug delivery using NCs is rapidly advancing and is expected to continue growing, as evidenced by the numerous recent studies and the success of NCs in oral drug formulations. In ocular drug delivery, NCs have demonstrated superiority not only over microcrystal formulations but also over currently available treatments. As seen in the previously mentioned studies, the drug's saturation solubility and dissolution rate are augmented due to small particle size and high superficial area. Moreover, several studies across various pathologies, including inflammatory, infectious, and glaucomatous diseases, have positioned NCs as promising therapeutic ocular DDSs due to their advantages associated with drug bioavailability, mucoadhesion, sustained release, toxicity, and efficacy.

Four NC formulations have been approved for ocular use so far, three of which were developed using top-down milling methods. Although information on the fourth one is unavailable, it was likely produced using similar techniques. As discussed in this chapter, numerous preclinical laboratory studies are underway, further highlighting the advantages of NC technology for ocular drug delivery. Thus, NC-based formulations are promising therapeutic alternatives for ocular drug delivery due to their present attributes to move toward a pharmaceutical product. In this regard, their simple composition, smaller particle size than the limited size for ophthalmic formulations, and the possibility of being sterilized and included in other pharmaceutical dosage forms are key issues in the ophthalmological industry, as well as the use of scalable preparation processes. Overall, NC technology stands out as a promising and innovative approach for overcoming the challenges associated with ocular drug delivery, offering significant advantages in both drug administration and therapeutic efficacy.

References

1. R. Zhang, J. Qian, X. Li, Y. Yuan, Treatment of experimental autoimmune uveoretinitis with intravitreal injection of infliximab encapsulated in liposomes. Br. J. Ophthalmol. **101**(12), 1731–1738 (2017)
2. E.Y. Chew, A.R. Glassman, R.W. Beck, N.M. Bressler, G.E. Fish, F.L. Ferris, J.L. Kinyoun, Diabetic retinopathy clinical research network, ocular side effects associated with peribulbar injections of triamcinolone acetonide for diabetic macular edema. Retina Phila. Pa **31**, 284–289 (2011). https://doi.org/10.1097/IAE.0b013e3181f049a8
3. A.Z. Mirza, F.A. Siddiqui, Nanomedicine and drug delivery: A mini review. Int. Nano Lett. **4**, 94 (2014). https://doi.org/10.1007/s40089-014-0094-7
4. C. Fornaguera, M.J. García-Celma, Personalized nanomedicine: A revolution at the nanoscale. J. Pers. Med. **7**, 12 (2017). https://doi.org/10.3390/jpm7040012
5. Y. Inokuchi, K. Hironaka, T. Fujisawa, Y. Tozuka, K. Tsuruma, M. Shimazawa, H. Takeuchi, H. Hara, Physicochemical properties affecting retinal drug/Coumarin-6 delivery from nanocarrier systems via Eyedrop administration. Invest. Ophthalmol. Vis. Sci. **51**, 3162–3170 (2010). https://doi.org/10.1167/iovs.09-4697
6. B. Mahaling, D.S. Katti, Physicochemical properties of core–shell type nanoparticles govern their spatiotemporal biodistribution in the eye. Nanomed Nanotechnol. Biol. Med. **12**, 2149–2160 (2016). https://doi.org/10.1016/j.nano.2016.05.017
7. D.A. Srinivasarao, G. Lohiya, D.S. Katti, Fundamentals, challenges, and nanomedicine-based solutions for ocular diseases. WIREs Nanomed. Nanobiotechnol. **11**, e1548 (2019). https://doi.org/10.1002/wnan.1548
8. D.A. Srinivasarao, S.S. Reddy, G.B. Reddy, D.S. Katti, Spatio-temporal control on the delivery of triamcinolone acetonide using polymeric nanoparticles reduces steroid induced cataract. Int. J. Pharm. **568**, 118474 (2019). https://doi.org/10.1016/j.ijpharm.2019.118474
9. T. Iwase, J. Fu, T. Yoshida, D. Muramatsu, A. Miki, N. Hashida, L. Lu, B. Oveson, et al., Sustained delivery of a HIF-1 antagonist for ocular neovascularization. J. Control. Release **172**, 625–633 (2013). https://doi.org/10.1016/j.jconrel.2013.10.008
10. A. Tuomela, J. Hirvonen, L. Peltonen, Stabilizing agents for drug nanocrystals: Effect on bioavailability. Pharmaceutics **8** (2016). https://doi.org/10.3390/pharmaceutics8020016
11. J.-U.A.H. Junghanns, R.H. Müller, Nanocrystal technology, drug delivery and clinical applications. Int. J. Nanomedicine **3**, 295–310 (2008). https://doi.org/10.2147/ijn.s595
12. M. Malamatari, K.M.G.G. Taylor, S. Malamataris, D. Douroumis, K. Kachrimanis, Pharmaceutical nanocrystals: production by wet milling and applications. Drug Discov. Today **23**, 534–547 (2018). https://doi.org/10.1016/j.drudis.2018.01.016
13. H. Chen, Recent developments in ocular drug delivery. J. Drug Target. **23**, 597–604 (2015). https://doi.org/10.3109/1061186X.2015.1052073
14. A. Lens, S.C. Nemeth, J.K. Ledford, *Ocular Anatomy and Physiology* (SLACK Incorporated, 2008)
15. L.E. Downie, S. Bandlitz, J.P.G. Bergmanson, J.P. Craig, D. Dutta, C. Maldonado-Codina, W. Ngo, J.S. Siddireddy, J.S. Wolffsohn, BCLA CLEAR – anatomy and physiology of the anterior eye. Contact Lens Anterior Eye **44**, 132–156 (2021). https://doi.org/10.1016/j.clae.2021.02.009
16. S. Wadhwa, R. Paliwal, S.R. Paliwal, S.P. Vyas, Hyaluronic acid modified chitosan nanoparticles for effective management of glaucoma: Development, characterization, and evaluation. J. Drug Target. **18**, 292–302 (2010). https://doi.org/10.3109/10611860903450023

17. V. Gote, S. Sikder, J. Sicotte, D. Pal, Ocular drug delivery: Present innovations and future challenges. J. Pharmacol. Exp. Ther. **370**, 602–624 (2019). https://doi.org/10.1124/jpet.119.256933
18. D.R. Janagam, L. Wu, T.L. Lowe, Nanoparticles for drug delivery to the anterior segment of the eye. Adv. Drug Deliv. Rev. **122**, 31–64 (2017). https://doi.org/10.1016/j.addr.2017.04.001
19. V. Singh, R. Ahmad, T. Heming, The challenges of ophthalmic drug delivery: A review. Int. J. Drug Discov. **3**(1), 56–62 (2011)
20. M.C.C. Peters, E. dos Santos Neto, L.M. Monteiro, M.N. Yukuyama, M.G.M. Machado, I.F. de Oliveira, M.H.A. Zanin, R. Löbenberg, N. Bou-Chacra, Advances in ophthalmic preparation: The role of drug nanocrystals and lipid-based nanosystems. J. Drug Target. **28**, 259–270 (2020). https://doi.org/10.1080/1061186X.2019.1663858
21. S. Gorantla, V.K. Rapalli, T. Waghule, P.P. Singh, S.K. Dubey, R.N. Saha, G. Singhvi, Nanocarriers for ocular drug delivery: Current status and translational opportunity. RSC Adv. **10**, 27835–27855 (2020). https://doi.org/10.1039/D0RA04971A
22. P.M. Hughes, J. Shen, Ocular surface anatomy and physiology: Impact on product development, in *Ophthalmic Product Development: From Bench to Bedside*, ed. by S. Neervannan, U.B. Kompella, (Springer International Publishing, Cham, 2021), pp. 15–37. https://doi.org/10.1007/978-3-030-76367-1_2
23. O.P. Sharma, V. Patel, T. Mehta, Nanocrystal for ocular drug delivery: Hope or hype. Drug Deliv. Transl. Res. **6**, 399–413 (2016). https://doi.org/10.1007/s13346-016-0292-0
24. Y. Weng, J. Liu, S. Jin, W. Guo, X. Liang, Z. Hu, Nanotechnology-based strategies for treatment of ocular disease. Acta Pharm. Sin. B **7**, 281–291 (2017). https://doi.org/10.1016/j.apsb.2016.09.001
25. J.K. Patel, V. Sutariya, J.R. Kanwar, Y.V. Pathak (eds.), *Drug Delivery for the Retina and Posterior Segment Disease* (Springer International Publishing, Cham, 2018). https://doi.org/10.1007/978-3-319-95807-1
26. R.R. Joseph, S.S. Venkatraman, Drug delivery to the eye: What benefits do Nanocarriers offer? Nano **12**, 683–702 (2017). https://doi.org/10.2217/nnm-2016-0379
27. C.A. Cutolo, S. Barabino, C. Bonzano, C.E. Traverso, The use of topical corticosteroids for treatment of dry eye syndrome. Ocul. Immunol. Inflamm. **27**, 266–275 (2019). https://doi.org/10.1080/09273948.2017.1341988
28. A. Patel, Ocular drug delivery systems: An overview. World J. Pharmacol. **2**, 47 (2013). https://doi.org/10.5497/wjp.v2.i2.47
29. Y. Diebold, M. Calonge, Applications of nanoparticles in ophthalmology. Prog. Retin. Eye Res. **29**, 596–609 (2010). https://doi.org/10.1016/j.preteyeres.2010.08.002
30. E. Sánchez-López, M. Espina, S. Doktorovova, E.B. Souto, M.L. García, Lipid nanoparticles (SLN, NLC): Overcoming the anatomical and physiological barriers of the eye – Part II – ocular drug-loaded lipid nanoparticles. Eur. J. Pharm. Biopharm. **110**, 58–69 (2017). https://doi.org/10.1016/j.ejpb.2016.10.013
31. Y. Chen, F. Wang, H.A.E. Benson, Effect of formulation factors on incorporation of the hydrophilic peptide dalargin into PLGA and mPEG-PLGA nanoparticles. Pept. Sci. **90**, 644–650 (2008). https://doi.org/10.1002/bip.21013
32. Y.C. Kim, B. Chiang, X. Wu, M.R. Prausnitz, Ocular delivery of macromolecules. J. Control. Release **190**, 172–181 (2014). https://doi.org/10.1016/j.jconrel.2014.06.043
33. M.L. Occhiutto, F.R. Freitas, R.C. Maranhao, V.P. Costa, Breakdown of the blood-ocular barrier as a strategy for the systemic use of nanosystems. Pharmaceutics **4**, 252–275 (2012). https://doi.org/10.3390/pharmaceutics4020252
34. Y.-Y. Cong, B. Fan, Z.-Y. Zhang, G.-Y. Li, Implantable sustained-release drug delivery systems: A revolution for ocular therapeutics. Int. Ophthalmol. **43**, 2575–2588 (2023). https://doi.org/10.1007/s10792-023-02637-x
35. Y.E. Choonara, V. Pillay, M.P. Danckwerts, T.R. Carmichael, L.C. Du Toit, A review of implantable intravitreal drug delivery technologies for the treatment of posterior segment eye diseases. J. Pharm. Sci. **99**, 2219–2239 (2010). https://doi.org/10.1002/jps.21987

36. T.R. Thrimawithana, S. Young, C.R. Bunt, C. Green, R.G. Alany, Drug delivery to the posterior segment of the eye. Drug Discov. Today **16**, 270–277 (2011). https://doi.org/10.1016/j.drudis.2010.12.004
37. R. Varela-Fernández, V. Díaz-Tomé, A. Luaces-Rodríguez, A. Conde-Penedo, X. García-Otero, A. Luzardo-Álvarez, A. Fernández-Ferreiro, F. Otero-Espinar, Drug delivery to the posterior segment of the eye: Biopharmaceutic and pharmacokinetic considerations. Pharmaceutics **12**, 269 (2020). https://doi.org/10.3390/pharmaceutics12030269
38. H. Massa, P. Georgoudis, G.D. Panos, Dexamethasone Intravitreal implant (OZURDEX®) for macular edema secondary to noninfectious uveitis: A review of the literature. Ther. Deliv. **10**, 343–351 (2019). https://doi.org/10.4155/tde-2019-0024
39. S. Ahmed, M.M. Amin, S. Sayed, Ocular drug delivery: A comprehensive review. AAPS PharmSciTech **24**, 66 (2023). https://doi.org/10.1208/s12249-023-02516-9
40. J. Van Der Merwe, J. Steenekamp, D. Steyn, J. Hamman, The role of functional excipients in solid Oral dosage forms to overcome poor drug dissolution and bioavailability. Pharmaceutics **12**, 393 (2020). https://doi.org/10.3390/pharmaceutics12050393
41. Y. Kawabata, K. Wada, M. Nakatani, S. Yamada, S. Onoue, Formulation design for poorly water-soluble drugs based on biopharmaceutics classification system: Basic approaches and practical applications. Int. J. Pharm. **420**, 1–10 (2011). https://doi.org/10.1016/j.ijpharm.2011.08.032
42. L.D.O. Macedo, J.F. Masiero, N.A. Bou-Chacra, Drug nanocrystals in oral absorption: Factors that influence pharmacokinetics. Pharmaceutics **16**, 1141 (2024). https://doi.org/10.3390/pharmaceutics16091141
43. G.D. Novack, A.L. Robin, Ocular pharmacology. J. Clin. Pharmacol. **64**, 1068–1082 (2024). https://doi.org/10.1002/jcph.2451
44. V.K. Pawar, Y. Singh, J.G. Meher, S. Gupta, M.K. Chourasia, Engineered nanocrystal technology: In-vivo fate, targeting and applications in drug delivery. J. Control. Release **183**, 51–66 (2014). https://doi.org/10.1016/j.jconrel.2014.03.030
45. M.-K. Yeh, L.-C. Chang, A.H.-J. Chiou, Improving Tenoxicam solubility and bioavailability by Cosolvent system. AAPS PharmSciTech **10**, 166–171 (2009). https://doi.org/10.1208/s12249-009-9189-2
46. R. Salzillo, C. Schiraldi, L. Corsuto, A. D'Agostino, R. Filosa, M. De Rosa, A. La Gatta, Optimization of hyaluronan-based eye drop formulations. Carbohydr. Polym. **153**, 275–283 (2016). https://doi.org/10.1016/j.carbpol.2016.07.106
47. N. Rasenack, B.W. Müller, Micron-size drug particles: Common and novel Micronization techniques. Pharm. Dev. Technol. **9**, 1–13 (2004). https://doi.org/10.1081/PDT-120027417
48. R. Gonzalez-Pizarro, P. Carvajal-Vidal, L.H. Bellowa, A.C. Calpena, M. Espina, M.L. García, In-situ forming gels containing fluorometholone-loaded polymeric nanoparticles for ocular inflammatory conditions. Colloids Surf. B: Biointerfaces **175**, 365–374 (2019)
49. A. Lee, H.A. Blair, Dexamethasone Intracanalicular insert: A review in treating post-surgical ocular pain and inflammation. Drugs **80**, 1101–1108 (2020). https://doi.org/10.1007/s40265-020-01344-6
50. M.L. Formica, J.M. Pernochi Scerbo, H.G. Awde Alfonso, P.T. Palmieri, J. Ribotta, S.D. Palma, Nanotechnological approaches to improve corticosteroids ocular therapy. Methods **234**, 152–177 (2025). https://doi.org/10.1016/j.ymeth.2024.12.008
51. R.C. Nagarwal, S. Kant, P.N. Singh, P. Maiti, J.K. Pandit, Polymeric nanoparticulate system: A potential approach for ocular drug delivery. J. Control. Release **136**, 2–13 (2009). https://doi.org/10.1016/j.jconrel.2008.12.018
52. F. Dilnawaz, S.K. Sahoo, Nanotechnology-based ophthalmic drug delivery system, in *Focal Controlled Drug Delivery*, ed. by A.J. Domb, W. Khan, (Springer US, Boston, 2014), pp. 225–241. https://doi.org/10.1007/978-1-4614-9434-8_10
53. A.C. Amrite, U.B. Kompella, Size-dependent disposition of nanoparticles and microparticles following subconjunctival administration. J. Pharm. Pharmacol. **57**, 1555–1563 (2010). https://doi.org/10.1211/jpp.57.12.0005

54. G.B. Romero, C.M. Keck, R.H. Müller, N.A. Bou-Chacra, Development of cationic nanocrystals for ocular delivery. Eur. J. Pharm. Biopharm. **107**, 215–222 (2016). https://doi.org/10.1016/j.ejpb.2016.07.005
55. S. Ashique, N. Mishra, S. Mohanto, B.H.J. Gowda, S. Kumar, A.S. Raikar, P. Masand, A. Garg, P. Goswami, I. Kahwa, Overview of processed excipients in ocular drug delivery: Opportunities so far and bottlenecks. Heliyon **10**, e23810 (2024). https://doi.org/10.1016/j.heliyon.2023.e23810
56. J. Zhang, Z. Xie, N. Zhang, J. Zhong, Chapter 13 – nanosuspension drug delivery system: preparation, characterization, postproduction processing, dosage form, and application, in *Nanostructures for Drug Delivery*, ed. by E. Andronescu, A.M. Grumezescu, (Elsevier, 2017), pp. 413–443. https://doi.org/10.1016/B978-0-323-46143-6.00013-0
57. P.S. Chary, S. Shaikh, V. Bhavana, N. Rajana, R. Vasave, N.K. Mehra, Emerging role of nanocrystals in pharmaceutical applications: A review of regulatory aspects and drug development process. Appl. Mater. Today **40**, 102334 (2024). https://doi.org/10.1016/j.apmt.2024.102334
58. I.S. Mohammad, H. Hu, L. Yin, W. He, Drug nanocrystals: Fabrication methods and promising therapeutic applications. Int. J. Pharm. **562**, 187–202 (2019). https://doi.org/10.1016/j.ijpharm.2019.02.045
59. Q. Ran, M. Wang, W. Kuang, J. Ouyang, D. Han, Z. Gao, J. Gong, Advances of combinative nanocrystal preparation Technology for Improving the insoluble drug solubility and bioavailability. Crystals **12**, 1200 (2022). https://doi.org/10.3390/cryst12091200
60. B. Sudhakar, K. NagaJyothi, K.V. Ramana Murthy, Nanosuspensions as a versatile carrier based drug delivery system – an overview. Curr. Drug Deliv. **11**, 299–305 (2014)
61. F. Fontana, P. Figueiredo, P. Zhang, J.T. Hirvonen, D. Liu, H.A. Santos, Production of pure drug nanocrystals and nano co-crystals by confinement methods. Adv. Drug Deliv. Rev. **131**, 3–21 (2018). https://doi.org/10.1016/j.addr.2018.05.002
62. R. Shegokar, R.H. Müller, Nanocrystals: Industrially feasible multifunctional formulation technology for poorly soluble actives. Int. J. Pharm. **399**, 129–139 (2010). https://doi.org/10.1016/j.ijpharm.2010.07.044
63. L. Peltonen, J. Hirvonen, Pharmaceutical nanocrystals by nanomilling: Critical process parameters, particle fracturing and stabilization methods. J. Pharm. Pharmacol. **62**, 1569–1579 (2010). https://doi.org/10.1111/j.2042-7158.2010.01022.x
64. A.R. Fernandes, J. Dias-Ferreira, C. Cabral, M.L. Garcia, E.B. Souto, Release kinetics and cell viability of ibuprofen nanocrystals produced by melt-emulsification. Colloids Surf. B Biointerfaces **166**, 24–28 (2018). https://doi.org/10.1016/j.colsurfb.2018.03.005
65. J.-P. Sylvestre, M.-C. Tang, A. Furtos, G. Leclair, M. Meunier, J.-C. Leroux, Nanonization of megestrol acetate by laser fragmentation in aqueous milieu. J. Control. Release **149**, 273–280 (2011). https://doi.org/10.1016/j.jconrel.2010.10.034
66. T.G. İnal, R.N. Gürsoy, L. Oner, Nanocrystal technology for oral delivery of poorly water-soluble drugs. Fabad J. Pharm. Sci. **34** (2009)
67. H. De Waard, H.W. Frijlink, W.L.J. Hinrichs, Bottom-up preparation techniques for nanocrystals of lipophilic drugs. Pharm. Res. **28**, 1220–1223 (2011). https://doi.org/10.1007/s11095-010-0323-3
68. V.B. Junyaprasert, B. Morakul, Nanocrystals for enhancement of oral bioavailability of poorly water-soluble drugs. Asian J. Pharm. Sci. **10**, 13–23 (2015). https://doi.org/10.1016/j.ajps.2014.08.005
69. H.-K. Chan, P.C.L. Kwok, Production methods for nanodrug particles using the bottom-up approach. Adv. Drug Deliv. Rev. **63**, 406–416 (2011). https://doi.org/10.1016/j.addr.2011.03.011
70. P. Yue, Y. Liu, J. Xie, Y. Chen, M. Yang, Review and prospect on preparation technology of drug nanocrystals in the past thirty years. Yao Xue Xue Bao, 529–537 (2018)

71. S.G. Pınar, A.N. Oktay, A.E. Karaküçük, N. Çelebi, Formulation strategies of nanosuspensions for various administration routes. Pharmaceutics **15**, 1520 (2023). https://doi.org/10.3390/pharmaceutics15051520
72. B. Sun, Y. Yeo, Nanocrystals for the parenteral delivery of poorly water-soluble drugs. Curr. Opin. Solid State Mater. Sci. **16**, 295–301 (2012). https://doi.org/10.1016/j.cossms.2012.10.004
73. Y. Wang, X. Tan, X. Fan, L. Zhao, S. Wang, H. He, T. Yin, Y. Zhang, X. Tang, L. Jian, J. Jin, J. Gou, Current strategies for oral delivery of BCS IV drug nanocrystals: Challenges, solutions and future trends. Expert Opin. Drug Deliv. **18**, 1211–1228 (2021). https://doi.org/10.1080/17425247.2021.1903428
74. M.M. Ali, A. Sabnis, D. DeWitt, G. Troiano, J. Wright, M. Figueiredo, M. Figa, T.V.G. Hoven, Y.-H. Song, J. Auer, T.C. Campbell, Therapeutic polymeric nanoparticle compositions with high glass transition temperature or high molecular weight copolymers, US20110294717A1, 2011. https://patents.google.com/patent/US20110294717A1/en?oq=US+2011%2f0294717+A1+. Accessed 25 Nov 2024
75. M. Kassem, A. Abdelrahman, M. Ghorab, M. Ahmed, R. Khalil, Nanosuspension as an ophthalmic delivery system for certain glucocorticoid drugs. Int. J. Pharm. **340**, 126–133 (2007). https://doi.org/10.1016/j.ijpharm.2007.03.011
76. K. Baba, Y. Tanaka, A. Kubota, H. Kasai, S. Yokokura, H. Nakanishi, K. Nishida, A method for enhancing the ocular penetration of eye drops using nanoparticles of hydrolyzable dye. J. Control. Release **153**, 278–287 (2011). https://doi.org/10.1016/j.jconrel.2011.04.019
77. K. Baba, N. Hashida, M. Tujikawa, A.J. Quantock, K. Nishida, The generation of fluorometholone nanocrystal eye drops, their metabolization to dihydrofluorometholone and penetration into rabbit eyes. Int. J. Pharm. **592**, 120067 (2021). https://doi.org/10.1016/j.ijpharm.2020.120067
78. B.N. Aldosari, M.A. Ibrahim, Y. Alqahtani, A.E.S.F. Abou El Ela, Formulation and evaluation of fluconazole nanosuspensions: In vitro characterization and transcorneal permeability studies. Saudi Pharm. J. **32**, 102104 (2024). https://doi.org/10.1016/j.jsps.2024.102104
79. N. Nagai, Y. Ito, N. Okamoto, Y. Shimomura, A nanoparticle formulation reduces the corneal toxicity of indomethacin eye drops and enhances its corneal permeability. Toxicology **319**, 53–62 (2014). https://doi.org/10.1016/j.tox.2014.02.012
80. J.H. Kim, S.W. Jang, S.D. Han, H.D. Hwang, H.-G. Choi, Development of a novel ophthalmic ciclosporin A-loaded nanosuspension using top-down media milling methods. Pharm. Int J. Pharm. Sci. **66**, 491–495 (2011). https://doi.org/10.1691/ph.2011.0358
81. M. Donia, R. Osman, G.A.S. Awad, N. Mortada, Polypeptide and glycosaminoglycan polysaccharide as stabilizing polymers in nanocrystals for a safe ocular hypotensive effect. Int. J. Biol. Macromol. **162**, 1699–1710 (2020). https://doi.org/10.1016/j.ijbiomac.2020.07.306
82. H.G. Awde Alfonso, L.I. Tártara, A.J. Paredes, S.D. Palma, M.L. Formica, Enhanced in vivo performance of topical ocular acetazolamide nanocrystals: A novel approach for glaucoma treatment. Int. J. Pharm. **674**, 125440 (2025). https://doi.org/10.1016/j.ijpharm.2025.125440
83. M.L. Formica, H.G. Awde Alfonso, A.J. Paredes, M.E. Melian, N.M. Camacho, R. Faccio, L.I. Tártara, S.D. Palma, Development of triamcinolone Acetonide nanocrystals for ocular administration. Pharmaceutics **15**, 683 (2023). https://doi.org/10.3390/pharmaceutics15020683
84. A. Josyula, R. Omiadze, K. Parikh, P. Kanvinde, M.B. Appell, P. Patel, H. Saeed, Y. Sutar, N. Anders, P. He, P.J. McDonnell, J. Hanes, A.A. Date, L.M. Ensign, An ion-paired moxifloxacin nanosuspension eye drop provides improved prevention and treatment of ocular infection. Bioeng. Transl. Med. **6**, e10238 (2021). https://doi.org/10.1002/btm2.10238
85. M.A. Kalam, The potential application of hyaluronic acid coated chitosan nanoparticles in ocular delivery of dexamethasone. Int. J. Biol. Macromol. **89**, 559–568 (2016). https://doi.org/10.1016/j.ijbiomac.2016.05.016

86. A. Alshememry, M. Alkholief, M. Abul Kalam, M. Raish, R. Ali, S.S. Alhudaithi, M. Iqbal, A. Alshamsan, Perspectives of positively charged nanocrystals of Tedizolid phosphate as a topical ocular application in rabbits. Molecules **27**, 4619 (2022). https://doi.org/10.3390/molecules27144619
87. U.M. Güven, E. Yenilmez, Olopatadine hydrochloride loaded Kollidon® SR nanoparticles for ocular delivery: Nanosuspension formulation and in vitro–in vivo evaluation. J. Drug Deliv. Sci. Technol. **51**, 506–512 (2019). https://doi.org/10.1016/j.jddst.2019.03.016
88. A. Maged, A.A. Mahmoud, M.M. Ghorab, Nano spray drying technique as a novel approach to formulate stable Econazole nitrate Nanosuspension formulations for ocular use. Mol. Pharm. **13**, 2951–2965 (2016). https://doi.org/10.1021/acs.molpharmaceut.6b00167
89. P.A. Mercadal, H.G. Awde Alfonso, M.L. Formica, S.D. Palma, A. González, Crosslinked soy protein films containing nanosized triamcinolone acetonide for drug delivery applications. J. Drug Deliv. Sci. Technol. **105**, 106609 (2025). https://doi.org/10.1016/j.jddst.2025.106609
90. H. Khan, A. Mehtab, J. Ahmed, S.E. Lofland, K.V. Ramanujachary, T. Ahmad, Harvesting Green hydrogen by self propelling built-in electric field photo/electro-catalytic performance of DyCrO3 nanoparticles developed by reverse microemulsion route. ChemNanoMat (2023). https://doi.org/10.1002/cnma.202300091
91. P. Khare, M.M. Chogale, P. Kakade, V.B. Patravale, Gellan gum–based in situ gelling ophthalmic nanosuspension of Posaconazole. Drug Deliv. Transl. Res. **12**, 2920–2935 (2022). https://doi.org/10.1007/s13346-022-01155-0
92. A.D. Permana, A.J. Paredes, F. Volpe-Zanutto, Q.K. Anjani, E. Utomo, R.F. Donnelly, Dissolving microneedle-mediated dermal delivery of itraconazole nanocrystals for improved treatment of cutaneous candidiasis. Eur. J. Pharm. Biopharm. **154**, 50–61 (2020). https://doi.org/10.1016/j.ejpb.2020.06.025
93. L. Wu, W. Huang, K. Peng, Y. Wang, Q. Chen, B. Lu, Enhancing the stability, BBB permeability and neuroprotective activity of verbascoside in vitro using lipid nanocapsules in combination with menthol. Food Chem. **414**, 135682 (2023). https://doi.org/10.1016/j.foodchem.2023.135682
94. H.S. Faizi, M.I. Nasiri, Y. Wu, D. Mishra, R.F. Donnelly, M.U. Minhas, L.K. Vora, R.R. Singh Thakur, Deferasirox nanosuspension loaded dissolving microneedles for ocular drug delivery. Int. J. Pharm. **664**, 124614 (2024). https://doi.org/10.1016/j.ijpharm.2024.124614
95. S. Gupta, M.K. Samanta, A.M. Raichur, Dual-drug delivery system based on in situ gel-forming nanosuspension of Forskolin to enhance Antiglaucoma efficacy. AAPS PharmSciTech **11**, 322–335 (2010). https://doi.org/10.1208/s12249-010-9388-x
96. N. Nagai, H. Ono, M. Hashino, Y. Ito, N. Okamoto, Y. Shimomura, Improved corneal toxicity and permeability of Tranilast by the preparation of ophthalmic formulations containing its nanoparticles. J. Oleo Sci. **63**, 177–186 (2014). https://doi.org/10.5650/jos.ess13082
97. S. Jacob, A.B. Nair, J. Shah, Emerging role of nanosuspensions in drug delivery systems. Biomater. Res. **24**, 3 (2020). https://doi.org/10.1186/s40824-020-0184-8
98. F. Geng, X. Fan, Y. Liu, W. Lu, G. Wei, Recent advances in nanocrystal-based technologies applied for ocular drug delivery. Expert Opin. Drug Deliv. **21**, 211–227 (2024). https://doi.org/10.1080/17425247.2024.2311119
99. N.K. Mehra, D. Cai, L. Kuo, T. Hein, S. Palakurthi, Safety and toxicity of nanomaterials for ocular drug delivery applications. Nanotoxicology **10**, 836–860 (2016). https://doi.org/10.3109/17435390.2016.1153165
100. M.A. Kamaleddin, Nano-ophthalmology: Applications and considerations. Nanomed. Nanotechnol. Biol. Med. **13**, 1459–1472 (2017). https://doi.org/10.1016/j.nano.2017.02.007
101. M.A. Jahangir, S.S. Imam, A. Muheem, A. Chettupalli, F.A. Al-Abbasi, M.S. Nadeem, I. Kazmi, M. Afzal, S. Alshehri, Nanocrystals: Characterization overview, applications in drug delivery, and their toxicity concerns. J. Pharm. Innov. **17**, 237–248 (2022). https://doi.org/10.1007/s12247-020-09499-1

102. C. Yang, J. Yang, A. Lu, J. Gong, Y. Yang, X. Lin, M. Li, H. Xu, Nanoparticles in ocular applications and their potential toxicity. Front. Mol. Biosci. **9**, 931759 (2022). https://doi.org/10.3389/fmolb.2022.931759
103. R.J. Wilson, Y. Li, G. Yang, C.-X. Zhao, Nanoemulsions for drug delivery. Particuology **64**, 85–97 (2022). https://doi.org/10.1016/j.partic.2021.05.009
104. R.S. Dave, T.C. Goostrey, M. Ziolkowska, S. Czerny-Holownia, T. Hoare, H. Sheardown, Ocular drug delivery to the anterior segment using nanocarriers: A mucoadhesive/mucopenetrative perspective. J. Control. Release **336**, 71–88 (2021). https://doi.org/10.1016/j.jconrel.2021.06.011
105. R. Shaikh, T. Raj Singh, M. Garland, A. Woolfson, R. Donnelly, Mucoadhesive drug delivery systems. J. Pharm. Bioallied Sci. **3**, 89 (2011). https://doi.org/10.4103/0975-7406.76478
106. C.-H. Tsai, P.-Y. Wang, I.-C. Lin, H. Huang, G.-S. Liu, C.-L. Tseng, Ocular drug delivery: Role of degradable polymeric Nanocarriers for ophthalmic application. Int. J. Mol. Sci. **19**, 2830 (2018). https://doi.org/10.3390/ijms19092830
107. L. Chen, L. Mei, D. Feng, D. Huang, X. Tong, X. Pan, C. Zhu, C. Wu, Anhydrous reverse micelle lecithin nanoparticles/PLGA composite microspheres for long-term protein delivery with reduced initial burst. Colloids Surf. B Biointerfaces **163**, 146–154 (2018). https://doi.org/10.1016/j.colsurfb.2017.12.040
108. W. Chaiyasan, S.P. Srinivas, W. Tiyaboonchai, Crosslinked chitosan-dextran sulfate nanoparticle for improved topical ocular drug delivery. Mol. Vis. **21**, 1224 (2015)
109. M.B. McGuckin, J. Wang, R. Ghanma, N. Qin, S.D. Palma, R.F. Donnelly, A.J. Paredes, Nanocrystals as a master key to deliver hydrophobic drugs via multiple administration routes. J. Control. Release **345**, 334–353 (2022). https://doi.org/10.1016/j.jconrel.2022.03.012
110. I.A. Sogias, A.C. Williams, V.V. Khutoryanskiy, Why is chitosan Mucoadhesive? Biomacromolecules **9**, 1837–1842 (2008). https://doi.org/10.1021/bm800276d
111. H. Almeida, M.H. Amaral, P. Lobão, J.M.S. Lobo, *In situ* gelling systems: A strategy to improve the bioavailability of ophthalmic pharmaceutical formulations. Drug Discov. Today **19**, 400–412 (2014). https://doi.org/10.1016/j.drudis.2013.10.001
112. N. Nagai, T. Isaka, S. Deguchi, M. Minami, M. Yamaguchi, H. Otake, N. Okamoto, Y. Nakazawa, In situ gelling systems using Pluronic F127 enhance corneal permeability of indomethacin nanocrystals. Int. J. Mol. Sci. **21**, 7083 (2020). https://doi.org/10.3390/ijms21197083
113. B. Abadia, P. Calvo, A. Ferreras, F. Bartol, G. Verdes, L. Pablo, Clinical applications of dexamethasone for aged eyes. Drugs Aging **33**, 639–646 (2016). https://doi.org/10.1007/s40266-016-0392-z
114. C.M. Jermak, J.T. Dellacroce, J. Heffez, G.A. Peyman, Triamcinolone Acetonide in ocular therapeutics. Surv. Ophthalmol. **52**, 503–522 (2007). https://doi.org/10.1016/j.survophthal.2007.06.004
115. A.K. Denniston, G.N. Holland, A. Kidess, R.B. Nussenblatt, A.A. Okada, J.T. Rosenbaum, A.D. Dick, Heterogeneity of primary outcome measures used in clinical trials of treatments for intermediate, posterior, and panuveitis. Orphanet J. Rare Dis. **10**, 97 (2015). https://doi.org/10.1186/s13023-015-0318-6
116. R. José-Vieira, A. Ferreira, P. Menéres, B. Sousa-Pinto, L. Figueira, Efficacy and safety of intravitreal and periocular injection of corticosteroids in noninfectious uveitis: a systematic review. Surv. Ophthalmol. **67**, 991–1013 (2022). https://doi.org/10.1016/j.survophthal.2021.12.002
117. Y. Tao, J.B. Jonas, Intravitreal Triamcinolone. Ophthalmologica **225**, 1–20 (2011). https://doi.org/10.1159/000317909
118. M.A. Kalam, M. Iqbal, A. Alshememry, M. Alkholief, A. Alshamsan, Fabrication and characterization of Tedizolid phosphate nanocrystals for topical ocular application: Improved Solubilization and in vitro drug release. Pharmaceutics **14**, 1328 (2022). https://doi.org/10.3390/pharmaceutics14071328

119. M.S. Khan, P.R. Ravi, D.S. Dhavan, Design, optimization, in vitro and in vivo evaluation of triamcinolone acetonide nanocrystals loaded in situ gel for topical ocular delivery. Colloids Surf. B Biointerfaces **231**, 113539 (2023). https://doi.org/10.1016/j.colsurfb.2023.113539
120. M.S. Khan, P.R. Ravi, T.V.R. Mullapudi, Dose identification of triamcinolone acetonide for noninvasive pre-corneal administration in the treatment of posterior uveitis using a rapid, sensitive HPLC method with photodiode-array detector. Biomed. Chromatogr. **36**, e5264 (2022). https://doi.org/10.1002/bmc.5264
121. R.R. Thakur Singh, I. Tekko, K. McAvoy, H. McMillan, D. Jones, R.F. Donnelly, Minimally invasive microneedles for ocular drug delivery. Expert Opin. Drug Deliv. **14**, 525–537 (2017). https://doi.org/10.1080/17425247.2016.1218460
122. Y. Wu, L.K. Vora, D. Mishra, M.F. Adrianto, S. Gade, A.J. Paredes, R.F. Donnelly, T.R.R. Singh, Nanosuspension-loaded dissolving bilayer microneedles for hydrophobic drug delivery to the posterior segment of the eye. Biomater. Adv. **137**, 212767 (2022). https://doi.org/10.1016/j.bioadv.2022.212767
123. W. Fan, H. Han, Y. Chen, X. Zhang, Y. Gao, S. Li, Q. Jin, J. Ji, K. Yao, Antimicrobial nanomedicine for ocular bacterial and fungal infection. Drug Deliv. Transl. Res. **11**, 1352–1375 (2021). https://doi.org/10.1007/s13346-021-00966-x
124. P.A. Thomas, Fungal infections of the cornea. Eye **17**, 852–862 (2003). https://doi.org/10.1038/sj.eye.6700557
125. C.-M. Phan, L. Subbaraman, L. Jones, Contact lenses for antifungal ocular drug delivery: A review. Expert Opin. Drug Deliv. **11**, 537–546 (2014). https://doi.org/10.1517/17425247.2014.882315
126. A.D. Permana, R.N. Utami, P. Layadi, A. Himawan, N. Juniarti, Q.K. Anjani, E. Utomo, S.A. Mardikasari, A. Arjuna, R.F. Donnelly, Thermosensitive and mucoadhesive in situ ocular gel for effective local delivery and antifungal activity of itraconazole nanocrystal in the treatment of fungal keratitis. Int. J. Pharm. **602**, 120623 (2021). https://doi.org/10.1016/j.ijpharm.2021.120623
127. S. Ghafoorianfar, A. Ghorani-Azam, S.A. Mohajeri, D. Farzin, Efficiency of nanoparticles for treatment of ocular infections: Systematic literature review. J. Drug Deliv. Sci. Technol. **57**, 101765 (2020). https://doi.org/10.1016/j.jddst.2020.101765
128. P. Harasymowycz, C. Birt, P. Gooi, L. Heckler, C. Hutnik, D. Jinapriya, L. Shuba, D. Yan, R. Day, Medical Management of Glaucoma in the 21st century from a Canadian perspective. J. Ophthalmol. **2016**, 6509809 (2016). https://doi.org/10.1155/2016/6509809
129. S. Awwad, R.M. Day, P.T. Khaw, S. Brocchini, H.M. Fadda, Sustained release ophthalmic dexamethasone: In vitro in vivo correlations derived from the PK-eye. Int. J. Pharm. **522**, 119–127 (2017). https://doi.org/10.1016/j.ijpharm.2017.02.047
130. N. Nagai, C. Yoshioka, Y. Mano, W. Tnabe, Y. Ito, N. Okamoto, Y. Shimomura, A nanoparticle formulation of disulfiram prolongs corneal residence time of the drug and reduces intraocular pressure. Exp. Eye Res. **132**, 115–123 (2015). https://doi.org/10.1016/j.exer.2015.01.022
131. A. Tuomela, P. Liu, J. Puranen, S. Rönkkö, T. Laaksonen, G. Kalesnykas, O. Oksala, J. Ilkka, J. Laru, K. Järvinen, J. Hirvonen, L. Peltonen, Brinzolamide nanocrystal formulations for ophthalmic delivery: Reduction of elevated intraocular pressure in vivo. Int. J. Pharm. **467**, 34–41 (2014). https://doi.org/10.1016/j.ijpharm.2014.03.048
132. C.A. Umscheid, D.J. Margolis, C.E. Grossman, Key concepts of clinical trials: A narrative review. Postgrad. Med. **123**, 194–204 (2011). https://doi.org/10.3810/pgm.2011.09.2475
133. L.-C. Liu, Y.-H. Chen, D.-W. Lu, Overview of recent advances in nano-based ocular drug delivery. Int. J. Mol. Sci. **24**, 15352 (2023). https://doi.org/10.3390/ijms242015352
134. S.S. Modi, R.P. Lehmann, T.R. Walters, R. Fong, W.C. Christie, L. Roel, D. Nethery, D. Sager, A. Tsorbatzoglou, B. Philipson, C.E. Traverso, H. Reiser, Once-daily nepafenac ophthalmic suspension 0.3% to prevent and treat ocular inflammation and pain after cataract surgery: Phase 3 study. J Cataract Refract Surg **40**, 203–211 (2014). https://doi.org/10.1016/j.jcrs.2013.07.042

135. T. Kim, K. Sall, E. Holland, R.K. Brazzell, S. Coultas, P.K. Gupta, Safety and efficacy of twice daily administration of KPI-121 1% for ocular inflammation and pain following cataract surgery. Clin. Ophthalmol. **13**, 69–86 (2018). https://doi.org/10.2147/OPTH.S185800
136. L. Schopf, E. Enlow, A. Popov, J. Bourassa, H. Chen, Ocular pharmacokinetics of a novel Loteprednol Etabonate 0.4% ophthalmic formulation. Ophthalmol. Ther. **3**, 63–72 (2014). https://doi.org/10.1007/s40123-014-0021-z
137. K. Beckman, J.A. Katz, P.A. Majmudar, A.T. Rostov, KPI-121 0.25%: A new option for the treatment of dry eye disease, (2021). https://touchophthalmology.com/ocular-surface-disease/journal-articles/kpi-121-0-25-a-new-option-for-the-treatment-of-dry-eye-disease/. Accessed 9 Feb 2025
138. L.R. Schopf, A.M. Popov, E.M. Enlow, J.L. Bourassa, W.Z. Ong, P. Nowak, H. Chen, Topical ocular drug delivery to the Back of the eye by mucus-penetrating particles. Transl. Vis. Sci. Technol. **4**, 11 (2015). https://doi.org/10.1167/tvst.4.3.11
139. S.D. Azhar, N. Shahid, A. Sadiq, A.W. Khan, M.S. Dar, T.K.F. Ahmed, Clobetasol propionate for post-cataract surgery pain and inflammation. Ann. Med. Surg. **86**(11), 6395–6398 (2024). https://doi.org/10.1097/ms9.0000000000002619
140. G.N. Foulks, T.R. Walters, D.L. Wirta, J.V. Greiner, J.L. Boyer, T. Navratil, Prospective Study of NCX 4251 (fluticasone propionate nanocrystals ophthalmic suspension, 0.1%) for Treatment of Blepharitis. ASCRS – 2020 ASCRS Annual Meeting. https://ascrs.confex.com/ascrs/20am/meetingapp.cgi/Paper/64670 (2020, May 16)
141. T. Navratil, J. Bukowski, F. Pilotaz, A. Graves, A. Nadkarni, A. Allen, B. Duquesroix, E. Bastia, L. Storoni, A. Kothe, M.V.W. Bergamini, J. Pennell, B.G. Short, T. Walters, R.D. Fechtner, D.L. Budenz, Preclinical evaluation of NCX 4251, a novel steroid therapy for blepharitis, targeted directly to the eyelid margin to reduce the potential for IOP elevations. Invest. Ophthalmol. Vis. Sci. **60**(9), 6755 (2019)

Drug Delivery to the Lungs Using Drug Nanocrystals

Luca Casula, Anna Maria Fadda, Michele Schlich, Chiara Sinico, and Francesco Lai

Abstract Pulmonary drug delivery has gained significant attention as an efficient route for both local and systemic therapy. This chapter provides an overview of the respiratory system, describing its structure and function in drug absorption. The advantages of pulmonary drug delivery, including rapid onset of action, reduced systemic side effects, and noninvasive administration, are discussed. Aerosol administration is a key strategy for pulmonary drug delivery, and this chapter examines the mechanisms of aerosol deposition, factors influencing drug absorption, and methods for characterizing aerosols. Given the challenges associated with traditional drug formulations, nanocrystals have emerged as a promising approach to enhance drug solubility, dissolution rate, and bioavailability in pulmonary applications. Nanocrystals can be formulated as either liquid nanosuspensions or solid-state nanocrystals, each offering unique advantages for inhalation therapy. The chapter discusses the formulation strategies, stability considerations, and translational potential of these nanocrystal-based systems in improving pulmonary drug delivery.

Keywords Pulmonary delivery · Nanosuspension · Inhalation · Nebulization · Inhalable dry powder · Nano into micro

1 Introduction

1.1 Anatomical and Physiological Overview of the Respiratory System

Drug delivery to the respiratory tract by inhalation has been used for millennia for the local treatment of airway diseases such as asthma, bronchitis, chronic obstructive pulmonary disease (COPD), and cystic fibrosis [1–3]. Airways can also be used as a route for delivering drugs with a systemic effect because of the several advantages

L. Casula (✉) · A. M. Fadda · M. Schlich · C. Sinico · F. Lai
Department of Life and Environmental Sciences, University of Cagliari, Cagliari, Italy
e-mail: luca.casula@unica.it

A. J. Paredes (ed.), *Nanocrystals*, https://doi.org/10.1007/978-3-032-10238-6_10

that this route of administration offers in comparison with oral and parenteral drug delivery. In addition, protein and peptide delivery via inhalation has attracted particular attention in the last decades as this route allows the rapid absorption and avoidance of first-pass metabolism [4].

However, the airways have a very complex anatomical and physiological structure, characterized by a series of cellular and extracellular barriers that prevent the alveolar area from being reached by most foreign substances. Once inhaled, a drug encounters a system of branched tubules whose nominal diameter progressively decreases, thus making it increasingly difficult to reach the deeper lungs [4]. According to Weibel's model, the respiratory tract resembles an upside-down tree with a network of 23 divisions, which, starting from the trachea, lead to a progressive reduction in diameter and thickness of the branches [3, 5].

The main function of the airways is to ensure gas exchange between the body and the environment: the organs of the respiratory system take up oxygen from the air and release carbon dioxide produced by cellular metabolism. Therefore, this system contains pathways of air transport (**conductive zone**) and large surface areas where exchanges occur (**respiratory zone**) between the air present in the lungs and the blood of alveolar capillaries. The lungs, located in the pleural cavities, are the main organs of the respiratory system.

The **conductive zone** includes the oral and nasal cavities, pharynx, larynx, trachea, bronchi, bronchioles, and terminal bronchioles, while the **respiratory zone** includes the respiratory bronchioles and alveolar regions (alveolar ducts, alveolar sacs). The main function of the conductive region is the transport of inhaled air to the lungs (and from the lungs to the environment) during each breath and, moreover, the heating and humidification of the inhaled air.

In the conductive zone, the trachea, a hollow tube with an average length of approximately 10–14 cm and a diameter of about 1.3–2.7 cm in adults, bifurcates into two main bronchi, right and left, which enter the lungs through the hilum, and in turn divide into smaller bronchi that penetrate the pulmonary lobes, two from the left and three from the right. The intrapulmonary bronchi continue to branch into thinner bronchi and bronchioles. The trachea's stability and strength are due to C-shaped cartilaginous rings that are surrounded dorsally by smooth muscles that, upon contraction, can significantly vary the lumen of the respiratory tree. The cartilaginous rings give support to the wall, preventing the closure of the airways. The trachea and bronchi offer little resistance to airflow due to their large diameter and rigid structure [5].

The branching pattern continues to originate into terminal bronchioles that divide to form respiratory bronchioles. The respiratory zone includes all the structures starting from the respiratory bronchioles that divide, giving rise to pulmonary alveoli (alveolar ducts and alveolar sacs), where gas exchange with alveolar capillaries occurs. The alveolar ducts are the short tubes linking the respiratory bronchioles to alveolar sacs, which contain an atrium and several alveoli, the terminal portion of the bronchial tree. Gas exchange takes place at the alveoli wall, which is surrounded by a network of blood capillaries provided by the pulmonary artery [5].

The peculiar structure of alveoli provides a large surface area to maximize gas exchange. In the human body, there are about 480 million alveoli with a total surface area of 40–60 m^2, i.e., approximately half of the total area of the airways. The alveoli wall consists of the alveolar epithelium and vascular connective layer, which are separated from each other by an alveolar septum [6]. The alveolar epithelium is placed in direct contact with the basement membrane of the alveolar capillaries. There are three types of cells in the alveolar epithelium: type 1 pneumocytes, type 2 pneumocytes, and alveolar macrophages. Type 1 pneumocytes are the main cells involved in gas exchanges, while type 2 pneumocytes mainly secrete pulmonary surfactants. Surfactants also play a role in the defense mechanisms of the respiratory system as they can prevent adhesion of the inhaled particles to the lung surface, favoring the phagocytic action of macrophages [7, 8]. The numbers of both pneumocytes are similar, but Type 1 cells account for 96% of the total alveolar surface area [9].

The conductive zone is covered with a vibratile epithelium containing basal cells, goblet cells, secretory cells, and ciliated columnar cells that favor a transport mechanism toward the pharynx. A protective layer of mucus, produced by the goblet cells with submucosal glands, covers the epithelia of the conductive zone [6]. The mucus, a complex hydrogel composed mainly of water (~95%) and glycoproteins (mucins, 2.5% w/v), acts as an important physical filtering barrier for inhaled particles, viruses, and macromolecules, which remain entrapped in the mucus. Mucus can also delay diffusion of the smaller molecules [6]. The mucus layer moves upward from the bronchi toward the pharynx due to the work of the millions of cilia of the respiratory mucosa epithelial cells. The beating of these microscopic cilia is triggered in only one direction, determining the movement of the mucus and corpuscular materials toward the pharynx and the mouth, where it will be expectorated or swallowed. This dynamic process, known as mucociliary clearance, is the main defensive mechanism against the inhaled particles with size >10 μm that become trapped in the mucus and are rapidly removed out of the lungs (within 24 h).

1.2 Advantages and Disadvantages of Drug Administration

The lungs represent an attractive route for drug delivery due to the several advantages they offer in comparison with oral and parenteral routes.

First, the lungs have a large and extensively vascularized area available for drug deposition (140 m^2) that allows for rapid absorption and fast onset of action. Second, pulmonary administration allows to avoid first-pass hepatic metabolism, leading to an enhancement of bioavailability. Indeed, although the lungs are a metabolizing organ, the enzymatic activity is much lower than in the gastrointestinal (GI) tract, thus allowing the administration of drugs at lower doses than the oral route, with consequent reduced side effects. The low enzymatic activities in the lung lining fluid are particularly interesting for the administration of biologics and other drugs that suffer the harsh conditions of the GI tract following oral administration. Third, drug particles can cross the respiratory epithelia and be absorbed also through the lymphatic system, reaching the systemic circulation and then different organs [10, 11].

Inhalation is still predominantly used in the treatment of pathologies of the respiratory tract and consists in the direct administration of the drug in aerosol form to the disease site. In this case, therefore, it can be considered a form of drug targeting whose advantages include the use of a smaller drug dose, rapid absorption and onset of action, ensuring reduction of side effects (e.g., corticosteroids) [1–3].

However, pulmonary administration also presents disadvantages due to the difficulty of achieving efficient and regular drug depositions in the bronchial or alveolar regions of the respiratory tree because of the airways' branched structure, together with the defense mechanisms (i.e., the mucociliary clearance, macrophage phagocytosis, proteolytic enzymes, pulmonary surfactants).

Other disadvantages are linked to the difficulties that some patients experience with the administration procedure, which is more complex than the other common administration routes, as it requires the correct use of an inhaler device. This could be the cause of highly variable drug doses deposited in the lungs. Pulmonary drug delivery is also highly affected by poor patient adherence to therapy [1].

2 Aerosol Administration

The term "aerosol" is used to describe a system of liquid or solid particles finely dispersed in a gas [12]. In order to achieve the desired pharmacological effect to manage a specific condition, the medicinal product needs to be converted into an aerosol and efficiently delivered to the respiratory system. In fact, a poor inhalation technique or improper use of the inhaler provided for the therapy can decrease the drug delivery and therefore impact the disease control [13, 14]. Pharmaceutical inhalers can be divided into three main categories: nebulizers, pressurized metered-dose inhalers (pMDIs), and dry powder inhalers (DPIs) (Fig. 1).

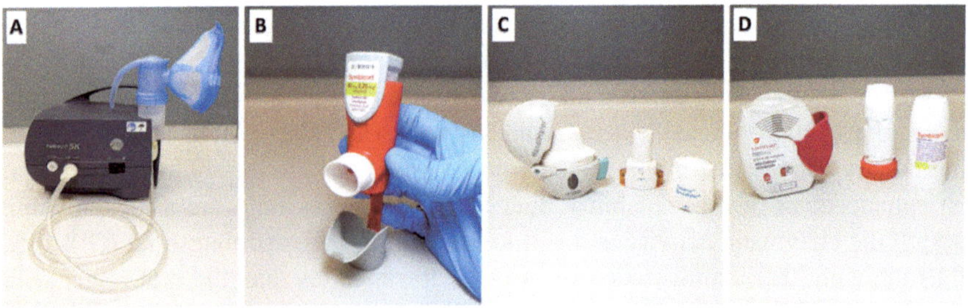

Fig. 1 Examples of commercial (**a**) nebulizer, (**b**) pressurized metered-dose inhaler (pMDI), (**c**) monodose, and (**d**) multidose dry powder inhalers (DPIs)

Nebulizers (Fig. 1a) are able to transform a liquid formulation (solution or suspension) into an inhalable aerosol by different mechanisms, such as the use of a compressed gas (jet nebulizers), ultrasonic vibrations (ultrasonic nebulizer), or a vibrating mesh (mesh nebulizer) [15]. A nebulizer is composed of several parts, including a reservoir for the drug product, a compressor, a baffle, a face mask, and a mouthpiece. These devices are versatile and can be used by patients of all ages and disease severities, including patients with severe airways obstruction or poor coordination [16]. They can deliver a large number of active substances—such as antibiotics, enzymes, or mucolytic drugs—which are not yet deliverable with DPIs or pMDIs and, in some instances, multiple drugs can be combined and administered together using a nebulizer, although this can extend the treatment duration [17]. Additionally, nebulizers do not rely on propellants that may harm the environment and require minimal clinical training to operate. The visible mist generated during use often reassures patients or their caregivers that the medication is being successfully administered [18].

pMDIs (Fig. 1b) are the inhalation devices most frequently prescribed to asthma or COPD patients. They are small and portable and can be manufactured as single-dose, or multidose devices with a dose counter. The device can be activated by pressure on the cannister, leading to the atomization of the liquid formulation by the highly volatile propellant and the release of a precise dose within a brief treatment time, regardless of the patient's respiratory effort [19]. However, the use of a spacer might be needed to (i) reduce the velocity of the aerosol released from the actuator, (ii) decrease the aerosol particle size, and (iii) reduce oropharyngeal deposition and increase lung delivery [20]. Soft mist inhalers (SMIs) are propellant-free MDIs that use mechanical forces to deliver single or multiple doses, reducing the drawbacks of classical pMDIs, such as the need for rigorous actuation-inspiration coordination and the high initial aerosol velocity [21].

DPIs (Fig. 1c, d) are the only class of devices containing micronized inhalable powders. The system usually contains the powder formulation, a dose-counting system, and a mouthpiece, and the principal driving force to deliver the formulation is given by patient's inhalation. They are highly user-friendly, thanks to the fact that they are propellant-free, portable, and provide a fast treatment. Nevertheless, compromised inspiratory ability or humidity-related particle aggregation might affect their delivery efficacy [22].

2.1 Deposition Mechanisms

The first process after aerosol inhalation is the particle deposition, which has a crucial role in determining the clinical efficacy of the treatment. For instance, higher drug deposition in the lungs and lower deposition in the oropharynx region is associated with fewer local or systemic side effects. Particle deposition in the respiratory tract is typically regulated by five different mechanisms (Fig. 2): inertial

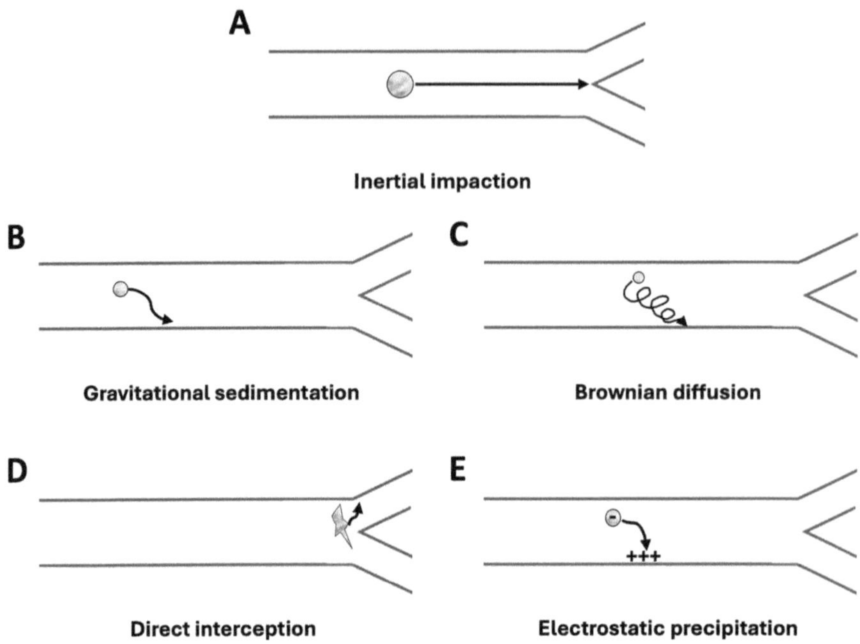

Fig. 2 Deposition mechanism of aerosol particles: (**a**) intertial impaction, (**b**) gravitational sedimentation, (**c**) brownian diffusion, (**d**) direct interception, (**e**) electrostatic precipitation

impaction, (gravitational) sedimentation, Brownian diffusion, interception, and electrostatic precipitation [23].

Inertial impaction is the main deposition mechanism occurring in the higher regions (oropharynx and conducting airways; Fig. 3), characterized by a turbulent flow and higher air velocity (Fig. 2a). Particles larger than 5 μm in diameter, due to their inertia, show a greater resistance to changing their direction when the airflow into the respiratory tract abruptly changes course. As a result, they will deviate from the air streamline, impacting and depositing on the airway walls. Particle properties, such as their diameter, density, and velocity, play a significant role in this type of impaction [24].

Going deeper into the respiratory system, the air velocity decreases, and **sedimentation** by gravitational force (Fig. 2b) represents the driving mechanism for deposition of the particles with a diameter of 0.5–5 μm. Particles of 3–5 μm that have surpassed the impaction show a higher tendency to sediment in the tracheobronchial region, whereas those between 0.5 and 5 μm are expected to sediment or diffuse in the bronchi, bronchioles, and alveolar region. The gravitational force, as well as the density, diameter, and velocity of the particle, influences the probability of deposition by sedimentation [15].

The deposition of particles smaller than 0.5 μm in the lower airways and alveolar region is primarily driven by **diffusion** (Fig. 2c), a process involving the random motion of particles, commonly referred to as Brownian motion. Due to their

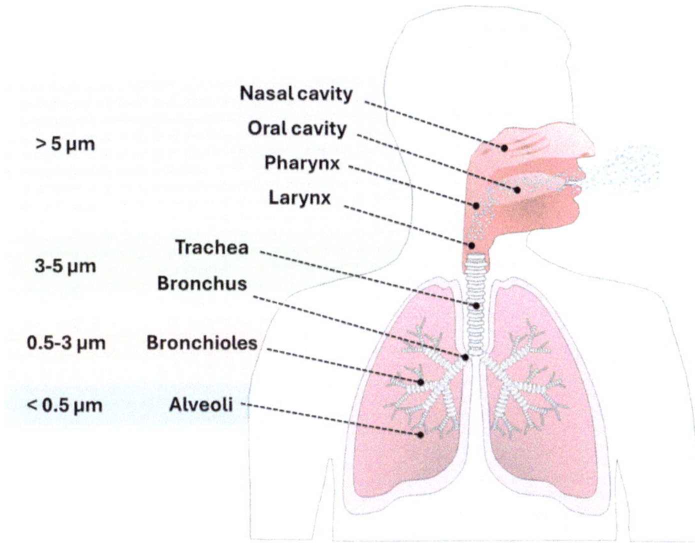

Fig. 3 Distribution of inhaled aerosol particles

extremely small size, most of these particles are exhaled rather than deposited in the respiratory system, resulting in only a small fraction successfully settling within the airways or alveoli. Notably, there is an inverse relationship between the probability of diffusion and particle size, meaning that smaller particles have a higher likelihood of undergoing diffusion. This dynamic plays a crucial role in determining the efficiency of particle deposition in the respiratory system [23].

Direct interception is a phenomenon that usually occurs when a particle moves into an airway smaller in width than the particle diameter itself, as a result of physical contact with the airway surfaces (Fig. 2d). This mechanism is primarily influenced by the geometrical characteristics of the particle. The likelihood of deposition by interception depends on how closely the air streamline passes to the airway surface and the ratio of the particle diameter to the airway width, which is generally small, even in the smallest airways. Compared to other deposition mechanisms, interception is typically less significant and is mostly relevant for fibrous particles.

Deposition of particles through the **electrostatic precipitation** mechanism contributes to less than 10% of the total particle deposition in the respiratory tract. This process is particularly significant for electrically charged particles (Fig. 2e), which can be driven toward the airway walls by the collective influence of similarly charged particles. In the absence of this collective repulsion, a charged particle can still be attracted to a neutral surface through image forces. These image forces arise from the charged particle itself, creating an equal but opposite charge of its own charge. The respiratory tract airway surfaces, often conductive, allow neutralization of charges, but electrically charged particles are still drawn toward the walls through electrostatic attraction. This results in higher deposition rates for charged particles compared to neutral ones [12].

2.2 Factors Affecting Aerosol Deposition

It is well established that the fate of inhaled particles and their deposition in the different regions of the respiratory system depends both on particle properties and physiological factors of the patient.

2.2.1 Particle Properties

Among the first class, the *particle size* certainly represents a key parameter to consider. Given the irregular morphology and heterogeneous density of an aerosol particle, considering the aerodynamic diameter (d_{ae})—rather than the geometric one (d_g)—is preferred. The diameter of a particle is therefore defined as the diameter of a sphere with a density of one and settling at the same speed:

$$d_{ae} = d_g \sqrt{\frac{\rho}{\chi}}$$

where: ρ = particle density; χ = shape correction factor.

However, it is not possible to determine a precise aerodynamic diameter for a pharmaceutical aerosol formulation, since its particle size distribution is heterogeneous. Hence, it is typically described using the mass median aerodynamic diameter (MMAD) and the geometric standard deviation (GSD). MMAD represents the diameter at which half of the aerosol mass consists of particles larger than this size, while the other half is smaller, effectively dividing the mass distribution into two equal parts. Aerosols with an MMAD ranging from 0.5 to 5 μm are generally considered respirable, as smaller particles are exhaled, and larger ones are trapped in the upper airways. GSD, on the other hand, quantifies the variability in particle sizes. Regarding the *shape*, notwithstanding that spherical particles are usually preferred for their easier production, fiber-shaped particles could be manufactured through controlled crystallization to specifically target peripheral regions, as those particles are known to deposit in those areas by interception [15, 25, 26]. As can be seen from the equation, a higher particle *density* increases the aerodynamic diameter, reducing the fraction of particles that reaches deeper regions of the lungs. Therefore, peculiar structures can be used to produce larger but less dense particles, achieving favorable aerodynamic properties, and preventing the formation of aggregates, more frequently observed with smaller, solid, and denser particles [27]. Structural characteristics that can lead to optimal aerodynamic diameters despite the larger size include: (i) hollow particles, with a lower density and larger geometric size due to their void internal structure; (ii) collapsed hollow particles, where the structure has partially collapsed but still retains a reduced density compared to a solid dense particle; (iii) micro-aggregated nanoparticles, which consists of smaller nanoparticles loosely aggregated into a larger structure, mimicking the aerodynamic properties of a single larger particle; and (iv) porous particles, characterized by a network of internal pores that reduce the overall density while increasing the geometric size [26].

2.2.2 Physiological and Pathological Factors

The deposition of respiratory particles is affected by several anatomical and physiological factors. These include the length and diameter of the respiratory airways, alveolar size, and the branching angles of bronchioles. Even among healthy individuals breathing in the same pattern, substantial variation in particle deposition can occur due to individual differences. Factors such as age and changes in the structure and dimensions of the tracheobronchial tree contribute to these differences. In fact, the geometry and conditions of the respiratory airways evolve from infancy to adulthood, influencing particle deposition during childhood. Children have a higher respiratory particle deposition rate compared to adults, typically by a factor of 1.5. Moreover, due to their smaller lung surface area, the particle density deposited per unit of surface area is 4–5 times greater in children than in adults. Gender also plays a role; anatomical differences between men and women lead to variations in deposition. For instance, adult women experience higher deposition rates in the tracheobronchial tree compared to adult men due to smaller thoracic dimensions and lower airway volume. Physiologically, elements such as breathing frequency, inhalation and exhalation patterns, the chosen breathing pathway, airflow dynamics, tidal lung volume (the air exchanged during normal breathing), and functional residual capacity (the gas volume remaining in the lungs post-exhalation) play pivotal roles [28, 29].

Lung diseases can profoundly impact respiratory anatomy and physiology. As a matter of fact, pathological conditions are often associated with airway narrowing, altered lung compliance, and/or mucus production, which can significantly impact the distribution and deposition pattern of inhaled drugs. Specifically, bronchial deposition tends to increase at the expense of pulmonary deposition in conditions like asthma and chronic bronchitis, likely due to reduced airway cross sections in the diseased lungs. In emphysema, deposition is similarly increased in the central regions of the lung, likely because of greater deposition during expiration in airways with flow limitations. Furthermore, deposition patterns in diseased lungs are more uneven than that in healthy lungs, since airway obstruction can lead to poorly ventilated lung regions, preventing sufficient aerosol from reaching and depositing in these areas [28, 30].

2.3 Drug Absorption in the Lungs

Following the deposition within the respiratory system, the effectiveness of pulmonary-administered medications relies heavily on drug absorption within the lungs. The epithelium serves as a barrier to drug absorption, with its thickness varying across the respiratory tract. It is thickest in the trachea (50–60 μm) and progressively thins as it approaches the alveoli (<1 μm) [31]. For a drug particle to be absorbed or metabolized, it must first interact with the mucus layer. However, the chemical nature of the particle can influence its absorption. In particular, small hydrophobic molecules tend to be absorbed rapidly by passive diffusion, with a resulting shorter half-life (seconds to minutes). Nevertheless, when their hydrophobicity increases beyond a certain extent, molecules may become too insoluble for quick absorption, and they may remain in the lungs for hours, days, or weeks. On the other hand, small hydrophilic molecules show a slower absorption through specific transporters or tight junctions, and thus the half-life is <1 h [32]. Since an increase in the molecular weight results in a slower absorption and longer half-life, drug conjugation has been explored as a strategy to obtain controlled release formulations [33, 34].

In addition to the properties of the drug itself, various clearance mechanisms—designed to maintain respiratory health and protect against inhaled foreign particles—significantly influence drug absorption in the lungs. Considerably large particles (≥10 μm) are spontaneously expelled in the upper airways by coughing, sneezing, or swallowing, whereas slightly smaller particles (>6 μm) are generally removed within 24 h of inhalation by mucociliary clearance. Nevertheless, the production of an excessively thick mucus, such as in cystic fibrosis, might affect this mechanism [24, 29].

Particles with reduced size and solubility, able to reach the alveolar regions, might be phagocytized by macrophages before they can be absorbed and then eliminated through the lymphatic system or enzymatic degradation. Lungs are known to contain the same metabolizing enzymes present in the liver, but at a lower concentration. Phase I enzymes, such as oxidases, reductases, and hydrolases, have been found in the lungs and abundantly studied, while there is less information on phase II enzymes. Moreover, the expression of metabolizing enzymes varies significantly in the different lung regions, resulting in variations in metabolism [15, 35].

2.4 Tests and Characterization of Aerosols

Pharmaceutical aerosols must be properly characterized and tested to guarantee their therapeutic efficacy, safety, and quality. Their evaluation focuses on two key aspects. The first is functionality, which ensures patient compliance, reduces waste, and enhances usability. The second aspect is related to the evaluation of the respirable performance of aerosols and their ability to consistently deliver the correct dose at the required rate. A general physical characterization is always required in order to determine the solubility, size, shape, density, charge, and crystallinity of the drug substance and/or excipients used in the formulation. Concerning the performance, particular attention should be given to the determination of the delivered/emitted dose and aerodynamic particle size determination, to predict the deposition of the formulation into the lungs, as well as spray pattern and plume geometry to determine the uniformity and directionality of the aerosol spray [12, 15].

2.4.1 Aerodynamic Assessment of Aerosol Particles

Aerodynamic assessment involves conducting experiments and performing calculations to determine the size distribution of aerosol particles, based on their aerodynamic properties, and evaluate the respirability of a product in vitro. The procedure implies the use of a particular instrument, called the impactor or impinger, that generally includes an induction port, impactor stages, and a vacuum pump. The induction port, typically an aluminum or stainless steel tube with a 90° bend, simulates the patient's mouth and throat while directing the airflow to align with the connected impactor stages. A vacuum pump is used to generate an airflow through the impactor stages, replicating the inhalation act. Before starting the test, the impactor is assembled and the volumetric flow rate at the induction port is accurately set. This flow remains constant throughout the test. The flow rate can vary when testing nebulizers, according to the type of impactor used, whereas it is kept constant at 28–30 L/min when testing pMDIs. With regard to DPIs, it can be between 28 and 100 L/min, and it needs to be accurately chosen in accordance with the DPI resistance. Since the driving force activating the DPIs is the inhalation act of the patient, the flow rate and the duration of airflow through the device are used to simulate the strength and duration of a patient's inspiration. Therefore, the air volume drawn through an inhaler is set at 4 L, reflecting the typical inspiratory capacity of an average adult male weighing around 70 kg. A pressure drop of 4 kPa across the DPI is used to approximate the pressure drop generated by patients when using the device. The duration of the test is then calculated according to the chosen flow rate. The aerosol then flows through the different stages of the impactor. Each stage is equipped with a calibrated mesh that defines a "cutoff size," collecting particles larger than the cutoff while allowing smaller particles to pass through. As a result, the particles are categorized into groups based on their aerodynamic diameter. The amount of sample collected at each stage is then retrieved with a suitable solvent, and quantified using an appropriate analytical method, in order to calculate the proportion of total mass within each aerodynamic size range [36, 37].

2.4.2 Cascade Impactors

The oldest and first cascade impactor described in the European Pharmacopeia as Apparatus A is the Glass Impinger, which can only measure the respirable fraction (Fig. 4a). This easy-to-use apparatus consists of two round-glass flasks vertically connected (representing stage 1) and a final conical flask connected to the last round-glass flask by a curve tube with a 45° angle (stage 2). The second round-glass flask is filled with 7 mL of solvent, while the final conical flask is filled with 30 mL. The inhalation device is connected to the first upper round-glass flask and the test is operated at 60 L/min. Particles smaller than 6.4 μm are expected to reach stage 2, representing the respirable fraction of the delivered dose. This early-stage apparatus was then implemented to also measure the aerodynamic size distribution, obtaining the multistage liquid impinger (MSLI), included both in the European and United States Pharmacopeias (Ph.Eur. and USP) (Fig. 4b). The system is divided into four stages, and a fifth stage comprising a filter to protect the vacuum pump from undeposited particles, and is suitable for flow rates in the range 30–100 L/min. The cutoff size range of each stage can be calculated according to the selected flow rate. The collection solvent is usually put in each stage to avoid particles rebound. The Andersen cascade impactor (ACI) consists of eight vertical stages with a specific number of nozzles, arranged in sequence to progressively collect smaller particles as the aerosol passes through them (Fig. 4c). The inhalation device is connected to the induction port (simulating the throat), and a pre-separator is used in the case of DPIs to block large non-inhalable particles. The cutoff diameter of the different stages can be calculated according to the set flow rate over the range 28.3–100 L/min, but two modified versions are described in the USP when working with 60 or 90 L/min, where some of the initial and final stages are removed and replaced with different plates to improve the discrimination between some cutoff points [37–40].

In the mid-1990s, discussions among the pharmaceutical industry scientists about the limitations of existing impactors led to the creation of the next- generation impactor (NGI) (Fig. 4d). The NGI, the last impactor and the only horizontal one referenced in both the United States and European Pharmacopeias, was developed to improve the recovery process without altering the size distribution of the particles. It includes eight stages, corresponding to seven size-fractionation stages and a micro-orifice collector (MOC), which replaces a final filter. The fixed nozzles ensure optimal performance within the 0.5–5 mm range across the entire flow spectrum and reduce the potential for user error associated with interchangeable nozzles. The NGI has been calibrated using monodisperse particles at flow rates of 15, 30, 60, and 100 L/min, but the cutoff diameter can be specifically calculated when using a different flow rate [41–43]. Similar to the other impactors, when testing DPIs, a pre-separator is indicated to remove large solid aggregates, and pre-coating of the surface of the impaction stages with glycerol, silicone oil, or similar high viscosity liquid deposited from a volatile solvent is required to avoid particle bounce [39, 40, 44].

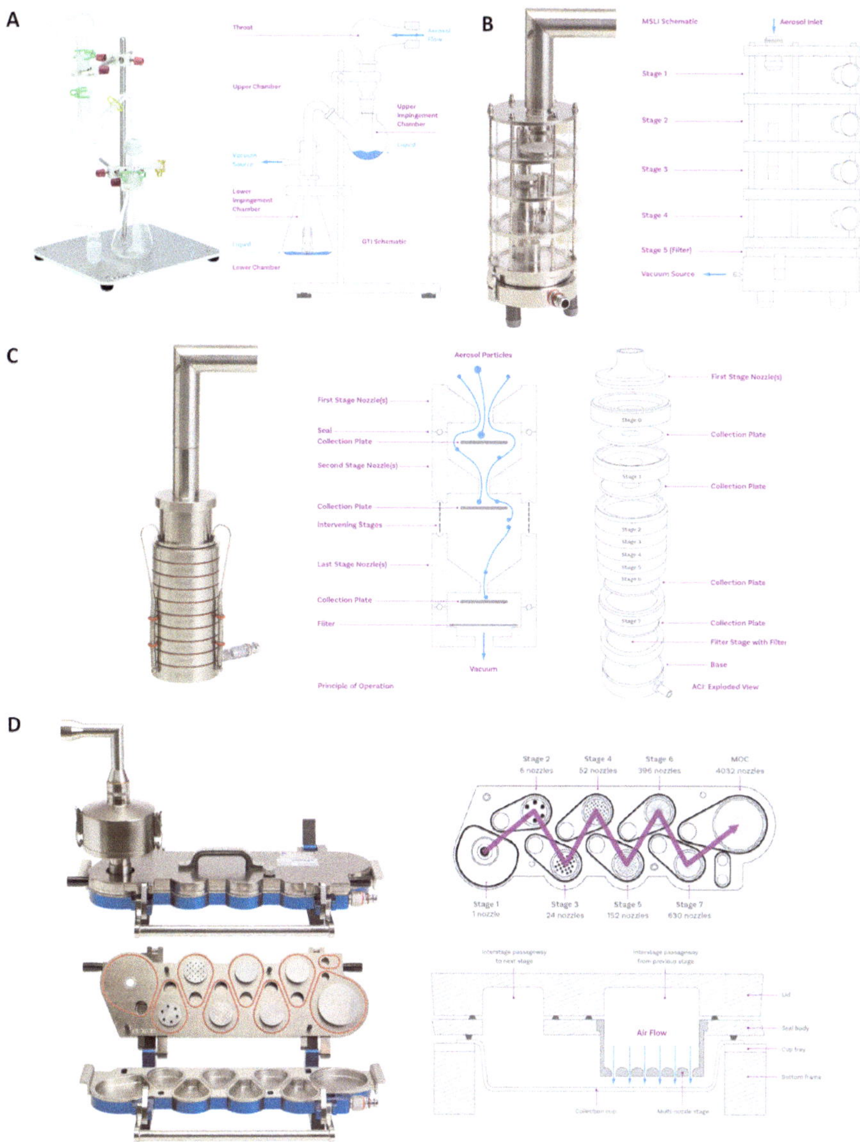

Fig. 4 Design and schematic representation of the working principle of the (**a**) glass impinger, (**b**) multistage liquid impinger (MSLI), (**c**) Andersen cascade impactor (ACI), and (**d**) next-generation impactor (NGI). (Images adapted and reproduced with kind permission of Copley Scientific Ltd)

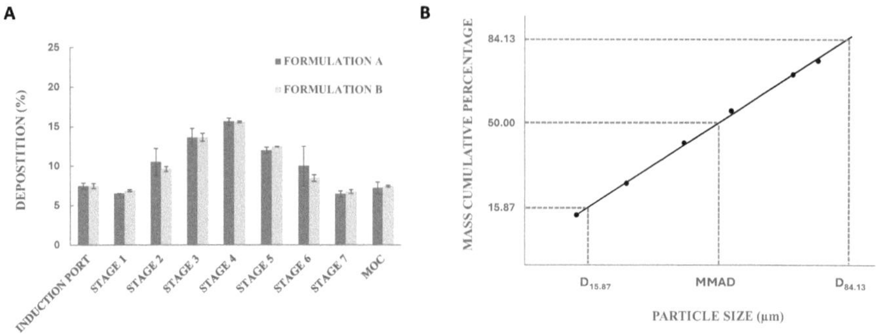

Fig. 5 (**a**) Representative drug deposition (%) on the different stages of NGI of two different formulations. (**b**) Plot of cumulative mass percentage (probability scale) versus aerodynamic diameter (log scale)

2.4.3 Calculation of the Aerodynamic Parameters

The aerodynamic size distribution data obtained from testing the formulations with an impactor are expressed as the mass fractions of active pharmaceutical ingredients (APIs) collected in different parts of the system (Fig. 5a). Each formulation exhibits a characteristic deposition profile, which is a valuable tool for comparing different inhalation products, or a brand-new formulation versus a reference product. By analyzing the amount of drug deposited on various stages, the **fine particle dose** (FPD), **fine particle fraction** (FPF), **mass median aerodynamic diameter** (MMAD), and **geometric standard deviation** (GSD) of the drug particles can be determined.

The data is elaborated by building a collection-efficiency curve by plotting the cumulative percentage undersize sampled at each stage of the impactor on a probability scale against the logarithm of the cutoff diameter of the stages. From this log-probability plot, the MMAD and GSD can be calculated (Fig. 5b). As already mentioned, the MMAD represents the particle diameter at which 50% by weight have a smaller diameter and 50% have a larger diameter, while the GSD represents the distribution of the particle size. Typically, a GSD of 1 indicates a perfectly monodispersed aerosol system, while values higher than 1.22 are considered polydispersed.

The particle diameter when the mass cumulative percentage is 84.13% ($D_{84.13}$) and 15.87% ($D_{15.87}$)—as defined in the Pharmacopeia—should be obtained from the graph and used to calculate the GSD as follows:

$$\mathrm{GSD} = \sqrt{\frac{\mathrm{D84.13}}{\mathrm{D15.87}}}$$

The FPD refers to the mass of drug particles, typically measured in milligrams or micrograms, while FPF represents the percentage of the total drug collected in the impactor or impinger that has a particle size of 5 μm or less. The emitted/delivered dose can be determined by the mass of the deposited drug if the formulation consists solely of the drug or—for formulations containing both the drug and excipients—the weight difference of the inhaler before and after use [25, 37, 39, 40].

3 Nanocrystals Formulation as a Promising Tool for Pulmonary Delivery

As highlighted in previous paragraphs, the efficacy and safety of pulmonary-administered drugs are highly dependent on several physical properties of the formulation, and on how it behaves in the form of an aerosol transported by the respiratory airflow. In the last decades, researchers have explored new formulation strategies to overcome the challenges associated with pulmonary delivery, such as suboptimal lung deposition, impaired absorption, metabolization, and sequestration by macrophages.

Nanocrystals, with their nanoscale dimensions and high surface area-to-volume ratios, represent a potential breakthrough in overcoming these obstacles. Poor water solubility is a major limitation for many pharmaceutical compounds, which hinders their absorption and efficacy. By significantly improving the dissolution and bioavailability of poorly water-soluble drugs, nanocrystals have proven to be a promising technology for pulmonary drug delivery [45, 46]. They increase the drug's surface area exponentially, facilitating faster dissolution in biological fluids [47]. This property is especially critical for pulmonary applications, since the volume of airway fluids is reduced in the lungs, resulting in limited dissolution and thus making poorly soluble drugs challenging to administer effectively. Nanocrystals mitigate this issue by providing a higher concentration of the drug in a soluble form, promoting the absorption into the respiratory epithelium and distribution in the systemic circulation. For instance, studies on drugs like budesonide and formoterol have demonstrated rapid plasma peak concentrations, underscoring the efficiency of nanocrystal-based formulations. These formulations exhibited enhanced dissolution and prolonged lungs retention, contributing to improved bioavailability [48, 49]. Nanocrystals are primarily composed of pure drug particles stabilized by surfactants or polymers, which play a critical role in maintaining their structural integrity. This stabilization prevents particle aggregation and ensures a consistent particle size within the nanoscale range (1–1000 nm) [48]. For pulmonary delivery, stability is paramount, as particle aggregation can lead to uneven drug dispersion and reduced therapeutic efficacy. Stabilizers like Poloxamer 188 and Polysorbate 80 have been widely employed to achieve both electrostatic and steric stabilization of nanocrystals. These stabilizers also contribute to reducing surface tension, further improving the dispersibility of the nanocrystals in respiratory fluids [50, 51]. Furthermore, nanocrystals have demonstrated reduced toxicity compared to other nanoparticle-based systems. Unlike inorganic nanoparticles (e.g., gold or titanium dioxide), drug nanocrystals do not exhibit significant oxidative stress or cytotoxicity in lung tissues, making them safer for chronic pulmonary conditions requiring long-term treatment [52].

As discussed in the previous paragraphs, the deposition of drug particles in the lungs depends on their aerodynamic properties. Traditional formulations often face challenges in reaching the alveolar region due to particle size constraints. Nanocrystals overcome this limitation through their nanoscale dimensions and unique aerodynamic characteristics. Their deposition relies on diffusion rather than impaction or sedimentation, allowing them to penetrate deeper into the alveoli. To optimize aerodynamic properties, nanocrystals are often transformed into micron-sized aggregates using techniques like spray drying. These aggregates dissociate upon contact with mucus, releasing the nanocrystals for efficient absorption. Once released, these drug particles, as small as 100 nm, can diffuse throughout the respiratory tract, with significant deposition observed in both the alveolar and tracheobronchial regions [49].

The lungs possess several physiological defense mechanisms, such as mucociliary clearance, macrophage activity, and the presence of mucus barriers. While these mechanisms protect the respiratory system from harmful agents, they can also impede effective drug delivery. Nanocrystals have shown the ability to partially circumvent these barriers through different mechanisms:

- *Enhanced mucus penetration*: nanocrystals modified with surfactants demonstrate enhanced mucus penetration. By reducing particle size and surface hydrophobicity, these formulations evade entrapment in the mucus gel and reach the epithelium [53]
- *Reduced macrophage uptake*: macrophages in the alveolar region are responsible for engulfing foreign particles, including drug carriers. Nanocrystals, due to their small size and surface properties, are less likely to be recognized and phagocytosed by macrophages. This results in prolonged retention in the lungs and enhanced therapeutic efficacy [49].
- *Controlled-release formulations*: controlled-release nanocrystal formulations ensure sustained drug release, minimizing the frequency of dosing. This is particularly beneficial for chronic conditions like asthma and COPD, where maintaining consistent drug levels is crucial [48].

Nanocrystals offer unparalleled versatility in formulation development, allowing for customization to meet specific therapeutic needs. Techniques like high-pressure homogenization, wet milling, and spray drying enable the production of inhalable powders with optimal properties for pulmonary delivery. Spray drying is a widely used method to obtain dry powders suitable for inhalation [54, 55]. The resulting agglomerates exhibit superior aerodynamic properties and redispersibility, ensuring efficient lung deposition [48].

Excipients like mannitol, lactose, and maltodextrin are commonly used to stabilize nanocrystal formulations during spray drying. These excipients not only improve dispersibility but also enhance the dissolution and aerosolization performance of the final product [54].

Nanocrystals have been explored extensively for treating a wide range of pulmonary diseases [48, 54], including:

- *Asthma and COPD*: enhanced drug deposition and retention effectively addressed inflammation and airflow obstruction;
- *Lung infections*: antibiotic nanocrystals, often in combination, demonstrated high local concentrations and reduced systemic side effects;
- *Lung cancer*: nanocrystal-based formulations of poorly soluble drugs showed promise in delivering therapeutic doses directly to tumor sites.

3.1 Liquid Formulations

As reported in the previous paragraph, the efficacy of a solid-in-liquid aerosolized dispersion to the respiratory tract depends strictly on the particle size distribution of the aerosol droplets, which should have aerodynamic diameter in the range 1–5 μm. Therefore, the dispersed solid particles must have a stable mean diameter much smaller than that of the respirable droplets in which they are incorporated.

In this context, the first study on the potential of nanocrystal suspension as a respirable drug delivery system for poorly water-soluble drugs was proposed by Wiedmann et al. in 1997 [56]. The authors successfully prepared beclomethasone dipropionate (BDP) nanocrystals with a mean diameter of approximately 270 nm, using ball milling of a 5% BDP coarse suspension in an aqueous solution containing 2.5% polyvinyl alcohol as a stabilizer. The main focus of the study was to evaluate the nebulization process of the BDP aqueous nanosuspension. The results indicated that the nebulization process was primarily influenced by the properties of the aqueous medium and was relatively insensitive to the presence of nanocrystals. Additionally, the nebulized dispersions demonstrated a higher fraction of respirable drug compared to the nebulized BDP micronized suspension. These findings suggested that nanocrystal suspensions offer a promising and efficient approach for respiratory drug delivery, particularly for poorly soluble compounds.

Despite the encouraging results, no in vivo data were produced in the following decade. In fact, all the subsequent studies focused on the characterization of the nebulization patterns of nanosuspensions as well as on the optimization of the production process. For instance, a very thorough study carried out by Jacobs and Muller in 2002 [57] focuses on producing a budesonide nanosuspension for pulmonary delivery using high-pressure homogenization, with volumes ranging from 40 to 300 mL. The goal was to create a nebulizable and long-term stable nanosuspension. The particle size was analyzed using laser diffraction and photon correlation spectroscopy, while the zeta potential was also measured. The resulting nanosuspension had a mean particle size of 500–600 nm and showed long-term stability without aggregation or particle growth over 1 year. In vitro aerosolization testing revealed minimal changes in particle size, indicating its suitability for pulmonary delivery. Finally, the scale-up process was successfully carried out.

These results, although limited to a small volume (300 mL), are promising for further development of the formulation. Indeed, a scalable production process for a liquid dosage form for inhalation ensures consistent quality, cost-efficiency, and flexibility to meet increasing demand while maintaining precise dosage control.

With regard to innovative nanosuspension applications for pulmonary drug delivery, a challenging research was published by Casula et al. in 2021 [58], who investigated the ability of electronic nicotine delivery systems (ENDS, e-cigarettes) to deliver drug nanocrystals of poorly water-soluble drugs. The potential success of this approach was based on two main considerations. The first is that nanosuspensions had already shown improved lung bioavailability of poorly soluble drugs as well as superior pharmacokinetic properties when compared to solutions or coarse suspensions of the same drug. The second was that e-cigarettes are devices very easy to use for both former smokers and nonsmokers, thereby being potentially more accepted than other devices such as pMDIs and DPIs. Briefly, a nanocrystal suspension of beclomethasone dipropionate (BDP), a synthetic corticosteroid commonly used by inhalation in the treatment of asthma and chronic obstructive pulmonary disease, was prepared with a wet media milling technique using Poloxamer 188 as the stabilizer. The produced nanocrystals were small in size and uniformly dispersed, exhibiting enhanced solubility compared to the bulk drug. The vaping test results confirmed the presence of drug nanocrystals in the aerosol produced. While it cannot be excluded that new nanoparticles may form during vaporization, the structural similarity observed through orthogonal techniques suggests that the original BDP nanocrystals were carried by the aerosol. Therefore, the ability of ENDS to deliver nanocrystals was proved, highlighting its potential as an alternative medical device for poorly soluble drugs. Alongside, BDP nanocrystals together with curcumin (CUR) nanocrystals were also successfully tested by the same research group in a multicomponent formulation for lung delivery [59]. The formulation exhibited a narrow particle size distribution and no aggregation. The inhalation delivery efficiency was studied with a next-generation impactor (NGI, Apparatus E Ph. Eu), and the nebulization tests showed high ED% and MMAD <5 µm, which are optimal properties for deep lung delivery in the treatment of asthmatic inflammation.

Regarding in vivo administration, Yang et al. in 2008 pioneered the aerosol delivery of fluticasone and budesonide nanosuspensions in a rodent model. To mimic the pulmonary delivery, the animals received a single intratracheal instillation of the selected nanosuspension, showing deep lung deposition, rapid plasma absorption, and fast onset [60]. Later, Chiang et al. demonstrated in 2009 that the aerosol delivery of a fluticasone nanosuspension on mice via intratracheal administration was as efficient as intranasal dosing, and capable of achieving dose-dependent lung deposition [61]. Thereafter, one of the first significant studies on the in vivo use of nanosuspensions was conducted by Costabile and colleagues in 2015 [62]. This research focused on the repositioning of the poorly soluble anthelmintic drug niclosamide (NCL) as an inhaled antibacterial agent targeting *P. aeruginosa*, a common pathogen in cystic fibrosis (CF) patients' airways. The authors successfully produced NCL nanocrystals as a dry powder for reconstitution, in combination with excipients such as appropriate surfactants approved for lung delivery, through a two-step process involving high-pressure homogenization (HPH) and spray drying. Compared to commercial liquid suspensions for inhalation tested under similar conditions, higher fine particle fraction (FPF) and lower mass median aerodynamic diameter (MMAD) values were achieved. The best performance was obtained with the PARI TurboBOY, a conventional nebulizer using a combined compressor/jet technology, and with the Aeroneb Pro, a vibrating mesh nebulizer (VMT). The inhibitory action of NCL nanosuspension was confirmed in vitro against *P. aeruginosa,* without significantly impacting the viability of CF bronchial epithelial cells. In vivo studies in rats showed no acute toxicity at therapeutic doses upon intratracheal instillation. The results support further development of NCL as an antivirulence agent for CF treatment and emphasize the potential cost benefits of repurposing NCL for inhalation therapy. Further advantages on the use of nanosuspensions over microsuspensions were also highlighted by Fu et al. who evaluated fluticasone pulmonary delivery in rats following intratracheal administration [63]. Interestingly, the microsuspension was rapidly cleared by the mucociliary mechanism, leading to a short absorption window. On the other hand, the nanosuspensions reduced mucociliary clearance, enabling dissolution-rate-limited absorption and prolonged pulmonary residence time. This extended pulmonary retention contributed to sustained drug absorption and prolonged anti-inflammatory activity in a lipopolysaccharide-induced lung injury model. Finally, a recent research paper that aimed to create a stable, homogeneous aerosolizable Cannabidiol (CBD) nanosuspension laid the groundwork for developing a more effective CBD inhalation therapy [64]. Using a solvent-antisolvent precipitation method, the CBD was amorphized and reduced to nanoparticles, enhancing its dissolution rate. In vitro tests showed that the nanocrystals dissolved quickly, had favorable aerodynamic properties for lung deposition as demonstrated by nebulization tests with a next-generation impactor (NGI, Apparatus E Ph. Eu), and were biocompatible with human pulmonary cells (Table 1).

3.2 Solid Formulations

Nanocrystalline suspensions can be formulated as solid dosage forms (i.e., nanocrystalline powders) to improve the physical and chemical instabilities associated with liquid-dispersed systems, such as Ostwald ripening, aggregation, etc. The literature contains numerous studies where nanocrystal powders have been obtained as free-flowing powders with excellent aerosolization performance, ensuring homogeneous particle distribution and deposition in the deep lung to achieve therapeutic effects. Various methods are used to produce nanocrystalline powders, with freeze drying and spray drying of nanosuspensions being among the most extensively studied. Several sugars (e.g., lactose, trehalose) and sugar alcohols (e.g., mannitol, xylitol) have been identified as optimal matrix agents that improve powder reconstitution performance by preventing nanocrystal aggregation. Typical stabilizers are

Table 1 Examples of liquid nanosuspensions for pulmonary delivery

Active ingredient	Stabilizer	Method of nanonization	Performance	Ref.
Beclomethasone dipropionate	Polyvinyl alcohol	Wet ball media milling	In vitro nebulization was dominated by the properties of the aqueous medium and showed a greater fraction of respirable fraction compared to the microsuspension	[56]
Budesonide	Lecithin, span85, tyloxapol, cetyl alcohol	High-pressure homogenization	The nanosuspension production was successfully scaled up from 40 to 300 mL and the nanocrystals size remained unvaried after in vitro nebulization	[57]
Beclomethasone dipropionate	Poloxamer 188	Wet ball media milling	The nanosuspension was successfully delivered in vitro using an electronic cigarette, showing its ability to deliver drug nanocrystals through the produced aerosol	[58]
Beclomethasone dipropionate; curcumin	Poloxamer 188	Wet ball media milling	The two drugs were formulated together in a multicomponent nanosuspesion, showing nanocrystal narrow size distribution, no aggregation and optimal nebulization properties following in vitro testing with NGI	[59]
Fluticasone propionate; budesonide	Tween 80	Wet ball media milling	In vivo studies showed deep lung deposition and fast lung absorption following intratracheal administration	[60]
Fluticasone propionate	Tween 80	Wet ball media milling	In vitro nebulization showed a MMAD of 3.7 using a ten-stage cascade impactor; in vivo nebulization in mice was shown to be as efficient as intranasal administration	[61]
Niclosamide	Tween 20, tween 80	High-pressure homogenization	In vitro nebulization showed optimal aerodynamic parameters of both the fresh nanosuspension and the reconstituted mannitol-spray dried one, with inhibition of *P. aeruginosa* and no acute toxicity after in vivo administration in rats	[62]
Fluticasone propionate	Tween 80	Wet ball media milling + high-pressure homogenization	The microsuspension, nanosuspension, and lipid-coated nanosuspension exhibited similar in vitro aerodynamic performances following nebulization with NGI; the nanosuspensions showed reduced mucociliary clearance in vivo, allowing for slower absorption and prolonged pulmonary retention	[63]
Cannabidiol	Tween 80	Solvent/antisolvent nanoprecipitation	The drug was amorphized and reduced to nanocrystals showing improved dissolution rate, in vitro favorable aerodynamic properties with NGI and high biocompatibility with human pulmonary cells	[64]

Poloxamer 188 and 407, Polysorbate 80, D-α-tocopherol polyethylene glycol 1000 succinate (TPGS), cellulosics, polyvinyl alcohol, or sodium dodecyl sulfate. In addition to selecting suitable excipients as well as the drying technique, a successful strategy is to embed nanocrystals into redispersible microparticles, thus combining the favorable features of both the nano- and micro-systems.

The first studies concerning nanocrystal-based dry inhalation powders appeared in the literature in the second decade of the 2000s. In particular, in 2011, S. Onoue et al. successfully developed an inhalable nanocrystalline solid dispersion of tranilast, which provided an interesting alternative to oral therapy with a better safety margin for the treatment of asthma and other airway inflammatory diseases [65].

In the following years, numerous studies focused on optimizing the powder composition. In this regard, for example, M. Malamatari et al. studied respirable powder of different drugs, such as ibuprofen, theophylline, and indomethacin, by combining wet media milling with the spray drying technique using mannitol as a co-milling agent [66–68]. In summary, wet milling of drugs in an organic solvent with mannitol, followed by spray drying, can serve as a formulation strategy for producing respirable particles of water-soluble drugs or those susceptible to crystal transformation in aqueous environments (e.g., hydrate formation). The preservation of the solid phase makes this method preferable to dry milling techniques, which often result in drug amorphization and performance variability.

In this regard, an interesting and highly innovative research was carried out in 2020 by K. Simkova et al. [69]. This study aimed to develop and produce low-density composite particles for pulmonary drug delivery by combining high-energy wet media milling (WMM) for nanoparticle production with spray drying for particle engineering. The approach sought to leverage the benefits of both processes to achieve powders with fast dissolution, high fine particle fraction (FPF) for deep lung deposition, and larger geometric sizes. Budesonide was used as a model drug, with results compared to a commercial product. Mathematical kinetic modeling revealed that formulations involving WMM had superior aerodynamic properties and dissolution rates. Notably, skipping the milling step significantly reduced the FPF, resulting in performance similar to the commercial product. This highlights the importance of both particle size reduction and spray drying for producing efficient inhalable powders. The larger geometric size of spray-dried particles also offers an advantage in avoiding phagocytosis. The choice of additives was found to influence powder properties, with leucine promoting low-density particle formation regardless of WMM usage, while albumin was suggested as an alternative excipient under specific spray drying conditions. Despite morphological differences, all formulations exhibited rapid drug dissolution across different NGI stages, suggesting that the dissolution rate is primarily governed by the properties of drug nanoparticles. The proposed water-based, scalable process was reported to be suitable for high drug load and multidrug formulations, presenting a promising platform technology for inhalable poorly water-soluble products.

Recently, a similar approach was tested by Banat et al. [70] who developed a ketoprofen-containing nanosuspension (~230 nm) via wet media milling, followed by co-spray drying with L-leucine and mannitol. SEM images showed wrinkled, coated, spherical particles (~2 μm) suitable for pulmonary delivery. The mannitol-containing formulation reduced mucin solution viscosity. Key findings included optimal aerodynamic diameter (2.4–4.5 μm), high fine particle fraction (56–71%), rapid dissolution (80% in 5 min), and a fivefold in vitro permeation enhancement. The system also demonstrated significant anti-inflammatory effects with low cytotoxicity in vitro.

Delivering bare nanocrystals to the lungs represents a challenge due to their small size (<1 μm) and tendency to aggregate, which hinders effective deposition. To overcome this issue, Mehanna et al. [71] produced respirable spray-dried rifampicin nanocomposites (SD-RIF NCs) as a potential dry powder for pulmonary tuberculosis treatment. Spray drying was used to produce stable nanocomposites by combining mannitol, maltodextrin, and leucine in a 2:1:1 ratio. The resulting formulation exhibited high powder yield, good flowability, and uniform particle size distribution. The SD-RIF NCs achieved a rapid dissolution rate with 83.78% dissolution efficiency within 10 min and a mean dissolution time of 1.62 min. Aerosol performance tests showed excellent deposition efficiency. Additionally, incorporating rifampicin into nanocomposite particles reduced toxicity to A549 alveolar cells. These findings suggest that the developed system offers a promising approach for efficient pulmonary delivery of anti-tuberculosis drugs.

Still, regarding incorporating nanoparticles into microparticles to enhance aerodynamic properties, Casula et al. recently published a study focused on inhalable mannitol microparticles containing the anti-inflammatory polyphenol curcumin (CUR) nanocrystals as a potential treatment for respiratory diseases. Initially, a curcumin nanosuspension with uniform size distribution was obtained by wet ball media milling [72]. The nanosuspension was then spray dried to form powders where the CUR nanocrystals were embedded in mannitol microparticles. The characterization study showed that the incorporation of CUR nanocrystals influenced the morphological properties of the microparticles, affecting both geometric diameter and flowability. In vitro tests revealed a multiphase dissolution and release pattern, with the mannitol concentration impacting curcumin release. Remarkably, even though changes in aerodynamic parameters due to the presence of CUR nanocrystals occurred after nebulization, suitable MMAD values (<5 μm) for pulmonary delivery were maintained. The promising results discussed above are confirmed by numerous other studies on the development of inhalable powders using nano-in-micro technology for different lung-targeting drugs [51, 73, 74].

Finally, in 2025, Li et al. carried out a very comprehensive study aimed at optimizing nanocrystal-agglomerated particles of posaconazole (PSZ) for dry powder inhalers (DPIs) to improve the solubility, dissolution rates, and pulmonary deposition of this antifungal drug, which faces limited pulmonary bioavailability and significant systemic side effects when administered orally or intravenously [75]. Wet media milling and spray-drying techniques were combined to obtain PSZ nanocrystal-agglomerated dry powders, systematically evaluating the impact of different stabilizers on physicochemical properties. The aerosolization performance was tested with the next-generation impactor, while antifungal efficacy was assessed in vitro and in vivo. The optimized PSZ formulation showed improved solubility and dissolution rates, with a fine particle fraction (FPF) of 78.58 ± 3.21%, ensuring effective lung delivery. In vitro tests confirmed strong antifungal effects and minimal cytotoxicity to lung cells. In vivo studies showed a $C_{max}/AUC_{0 \to \infty}$ ratio in lung tissues 27.32- and 6.76-fold higher than the oral suspension, thus indicating higher local drug concentrations. Overall, this scalable, cost-effective strategy offers high drug loading and promising potential for treating pulmonary fungal infections (Table 2).

4 Conclusions

This chapter explores the promising role of nanocrystals in pulmonary drug delivery, focusing on their formulations as liquid nanosuspensions and dry powders produced via freeze-drying or spray-drying techniques. The ability of nanocrystals to enhance the drug solubility, dissolution rate, and bioavailability has opened new avenues for efficient pulmonary therapies using scarcely soluble drugs. Designing efficient inhalable formulations can also reduce systemic side effects by targeting the drug directly to the respiratory system.

Stabilizers, which play a critical role in maintaining physical stability and preventing aggregation in liquid nanosuspensions, have also been shown to influence aerosol performance and, thus, the deposition patterns and overall efficiency of lung delivery. The choice of stabilizers is therefore crucial for achieving the desired balance between stability and optimal aerodynamic properties for inhalation.

For dry powder formulations, the matrix-forming agents used in drying processes modulate the final powder characteristics, including flowability, particle size distribution, rehydration kinetics, and nanocrystal release. Most importantly, careful design of the drying process and informed selection of excipients is required to achieve optimal aerodynamic properties and effective nebulization. Finally, the inclusion of these excipients can also offer mucoadhesive properties and enhanced lung retention, ensuring consistent drug release once inhaled.

Looking forward, the future of nanocrystal formulations for pulmonary delivery holds great promise. Continued advancements in stabilizer technologies and matrix-forming agents will enhance the precision of these formulations, leading to more efficient drug delivery systems. Research into the long-term stability, safety, and clinical translation of these formulations will be key to their success. With ongoing innovation, nanocrystals have the potential to revolutionize pulmonary drug delivery, offering new, effective therapies for a variety of respiratory diseases.

Table 2 Examples of solid dried nanocrystals for pulmonary delivery

Active ingredient	Method of nanonization	Method of drying	Matrix forming agent	Performance	Refs.
Tranilast	Wet ball media milling	Freeze drying	None—lactose was added post-freeze drying to increase powder dispersibility	The formulation exhibited excellent dispersibility and in vitro lung deposition, with an ED of around 98% and FPF% of about 60%. In vivo inhalation reduced antigen-induced inflammation in rats and was more effective than the physical mixture, likely due to enhanced dissolution	[65]
Indomethacin	Wet ball media milling	Spray drying	Mannitol; L-leucine	Matrix formers enhanced nanocrystals redispersibility upon hydration, further boosting dissolution and significantly improving aerosolization, leading to higher FPF% and smaller MMAD	[66]
Theophylline	Wet ball media milling	Spray drying	Mannitol	Mannitol reduced theophylline crystal size and facilitated microcomposite formation, significantly improving their in vitro aerosol performance compared to the spray-dried suspension of wet-milled theophylline without mannitol	[67]
Ibuprofen	Wet ball media milling	Spray drying	Mannitol; L-leucine	The size of nanocrystals agglomerates was primarily influenced by the leucine-to-drug ratio and the stabilizer type, while redispersibility was significantly affected only by the mannitol-to-drug ratio. The FPF%, assessed using a fast screening impactor, depended on both the leucine and mannitol-to-drug ratios	[68]
Budesonide	Wet ball media milling	Spray drying	Leucine; albumin	The formulation showed improved dissolution and aerodynamic properties when tested in vitro with the NGI. Leucine promoted low-density	[69]

				particle formation, while the use of albumin showed to be related to specific spray drying conditions	
Ketoprofen	Wet ball media milling	Spray drying	Mannitol; L-leucine	In vitro analysis revealed rapid dissolution, optimal MMAD and high FPF% values; the formulation enhanced drug permeation in a cell-based model and exhibited significant anti-inflammatory effects with minimal cytotoxicity	[70]
Rifampicin	Solvent/antisolvent nanoprecipitation + ultrasonication	Spray drying	Maltodextrin; mannitol; L-leucine	The mannitol-based powder exhibited a higher respirable fraction when tested in vitro with a twin stage impactor, compared to the maltodextrin-based formulation; the rifampicin nanocomposite had a significantly higher IC50 value than the free drug, indicating a higher safety profile for lung tissues	[71]
Curcumin	Wet ball media milling	Spray drying	Mannitol	The inclusion of curcumin nanocrystals in mannitol-based microparticles influenced their morphological characteristics, mantaining optimal aerodynamic parameters when tested in vitro with the NGI; multistep dissolution and release studies highlighted the impact of mannitol on drug release behavior	[72]
Resveratrol	High-pressure homogenization	Spray drying	Mannitol	The type and concentration of stabilizers significantly affected the morphology, lung deposition, and in vitro drug release of microparticles. 1% PVA resulted in the highest FPF%, while increasing PVA concentration reduced FPF and altered particle morphology and flowability; in vivo pharmacokinetic studies showed that the presence of *chitosan* enhanced drug retention in the lungs while limiting systemic exposure	[51]

(continued)

Table 2 (continued)

Active ingredient	Method of nanonization	Method of drying	Matrix forming agent	Performance	Refs.
Flavonoid drug	High-pressure homogenization	Spray drying	Maltodextrin; inulin	Water activity had a significant impact on the ability of the formulations to redisperse and their aerodynamic performance; at higher water activity, the maltodextrin-based formulation demonstrated better properties	[73]
Budesonide	Wet ball media milling	Spray drying	Hyaluronic acid	The in vitro dissolution study revealed that budesonide release from the microparticles was not significantly sustained compared to pure nanocrystals; in vivo intratracheal administration to rats showed a budesonide nanocrystals pharmacokinetic profile similar to the intravenous one, while the mucoadhesive microparticles led to a significantly delayed peak concentration (T_{max}) and enhanced bioavailability	[74]
Posaconazole	Wet ball media milling	Spray drying	–	The optimized inhalable powder showed significant improvements in solubility and dissolution rates, with high FPF%. In vivo studies revealed a higher $C_{max}/AUC_{0 \to \infty}$ ratio in lung tissues compared to the oral suspension, indicating higher lung drug concentrations	[75]

References

1. S.P. Newman, Delivering drugs to the lungs: The history of repurposing in the treatment of respiratory diseases. Adv. Drug Deliv. Rev. **133**, 5–18 (2018). https://doi.org/10.1016/j.addr.2018.04.010
2. S.P. Newman, Drug delivery to the lungs: Challenges and opportunities. Ther. Deliv. **8**, 647–661 (2017). https://doi.org/10.4155/tde-2017-0037
3. C. Zhang, D. D'Angelo, F. Buttini, M. Yang, Long-acting inhaled medicines: Present and future. Adv. Drug Deliv. Rev. **204**, 115146 (2024). https://doi.org/10.1016/j.addr.2023.115146
4. K. Berkenfeld, S. Carneiro, C. Corzo, F. Laffleur, S. Salar-Behzadi, B. Winkeljann, G. Esfahani, Formulation strategies, preparation methods, and devices for pulmonary delivery of biologics. Eur. J. Pharm. Biopharm. **204**, 114530 (2024). https://doi.org/10.1016/j.ejpb.2024.114530
5. K. Ahookhosh, O. Pourmehran, H. Aminfar, M. Mohammadpourfard, M.M. Sarafraz, H. Hamishehkar, Development of human respiratory airway models: A review. Eur. J. Pharm. Sci. **145**, 105233 (2020). https://doi.org/10.1016/j.ejps.2020.105233
6. X. Murgia, B. Loretz, O. Hartwig, M. Hittinger, C.M. Lehr, The role of mucus on drug transport and its potential to affect therapeutic outcomes. Adv. Drug Deliv. Rev. **124**, 82–97 (2018). https://doi.org/10.1016/j.addr.2017.10.009
7. A.J. Hickey, Back to the future: Inhaled drug products. J. Pharm. Sci. **102**, 1165–1172 (2013). https://doi.org/10.1002/JPS.23465
8. B. Patel, N. Gupta, F. Ahsan, Barriers that inhaled particles encounter. J. Aerosol Med. Pulm. Drug Deliv. **37**(5), 299–306 (2024)
9. J. Gai, L. Liu, X. Zhang, J. Guan, S. Mao, Impact of the diseased lung microenvironment on the in vivo fate of inhaled particles. Drug Discov. Today **29**, 104019 (2024). https://doi.org/10.1016/j.drudis.2024.104019
10. M. Lippmann, D.B. Yeates, R.E. Albert, Deposition, retention, and clearance of inhaled particles. Occup. Environ. Med. **37**, 337–362 (1980). https://doi.org/10.1136/OEM.37.4.337
11. M. Sanders, Inhalation therapy: an historical review. Prim. Care Respir. J. **16**(2), 71–81 (2007)
12. R.W. Abdo, N. Saadi, N.I. Hijazi, Y.A. Suleiman, *Quality Control and Testing Evaluation of Pharmaceutical Aerosols* (Elsevier Inc., 2019)
13. X. Jia, S. Zhou, D. Luo, X. Zhao, Y. Zhou, Y. Cui, Min effect of pharmacist-led interventions on medication adherence and inhalation technique in adult patients with asthma or COPD: A systematic review and meta-analysis. J. Clin. Pharm. Ther. **45**, 904–917 (2020)
14. J. Mahon, A. Fitzgerald, J. Glanville, R. Dekhuijzen, J. Glatte, S. Glanemann, S. Torvinen, Misuse and/or treatment delivery failure of inhalers among patients with asthma or COPD: A review and recommendations for the conduct of future research. Respir. Med. **129**, 98–116 (2017). https://doi.org/10.1016/j.rmed.2017.05.004
15. N. Rangaraj, S.R. Pailla, S. Sampathi, Insight into pulmonary drug delivery: Mechanism of drug deposition to device characterization and regulatory requirements. Pulm. Pharmacol. Ther. **54**, 1–21 (2019). https://doi.org/10.1016/j.pupt.2018.11.004
16. A.R. Martin, W.H. Finlay, Nebulizers for drug delivery to the lungs. Expert Opin. Drug Deliv. **12**, 889–900 (2015). https://doi.org/10.1517/17425247.2015.995087
17. P.P.H. Le Brun, A.H. De Boer, H.W. Frijlink, H.G.M. Heijerman, A review of the technical aspects of drug nebulization. Pharm. World Sci. **22**, 75–81 (2000). https://doi.org/10.1023/A:1008786600530
18. D.E. Geller, Comparing clinical features of the nebulizer, metered-dose inhaler, and dry powder inhaler. Respir. Care **50** (2005)
19. S.P. Newman, Principles of metered-dose inhaler design. Respir. Care **50**, 1177–1188 (2005)
20. S.P. Newman, Spacer devices for metered dose inhalers. Clin. Pharmacokinet. **43**, 349–360 (2004). https://doi.org/10.2165/00003088-200443060-00001
21. R. Dalby, M. Spallek, T. Voshaar, A review of the development of Respimat® soft mist™ inhaler. Int. J. Pharm. **283**, 1–9 (2004). https://doi.org/10.1016/j.ijpharm.2004.06.018

22. R.W. Dal Negro, Dry powder inhalers and the right things to remember: A concept review. Multidiscip. Respir. Med. **10**, 1–4 (2015). https://doi.org/10.1186/S40248-015-0012-5/TABLES/2
23. M. Ali, Pulmonary drug delivery, in *Handbook of Non-invasive Drug Delivery Systems*, (William Andrew Publishing, 2010), pp. 209–246
24. D.A. Groneberg, C. Witt, U. Wagner, K.F. Chung, A. Fischer, Fundamentals of pulmonary drug delivery. Respir. Med. **97**, 382–387 (2003). https://doi.org/10.1053/rmed.2002.1457
25. B.K. Rubin, Air and soul: The science and application of aerosol therapy. Respir. Care **55**, 911–921 (2010)
26. A.J. Hickey, Controlled delivery of inhaled therapeutic agents. J. Control. Release **190**, 182–188 (2014). https://doi.org/10.1016/j.jconrel.2014.05.058
27. Z. Liang, R. Ni, J. Zhou, S. Mao, Recent advances in controlled pulmonary drug delivery. Drug Discov. Today **20**, 380–389 (2015). https://doi.org/10.1016/j.drudis.2014.09.020
28. W. Hofmann, Modelling inhaled particle deposition in the human lung-a review. J. Aerosol Sci. **42**, 693–724 (2011). https://doi.org/10.1016/j.jaerosci.2011.05.007
29. A. Fernández Tena, P. Casan Clarà, Deposition of inhaled particles in the lungs. Archivos de Bronconeumología (English Edition) **48**, 240–246 (2012). https://doi.org/10.1016/j.arbr.2012.02.006
30. C. Darquenne, Aerosol deposition in health and disease. J. Aerosol Med. Pulm. Drug Deliv. **25**, 140–147 (2012). https://doi.org/10.1089/jamp.2011.0916
31. H. Fischer, J.H. Widdicombe, Mechanisms of acid and base secretion by the airway epithelium. J. Membr. Biol. **211**, 139–150 (2006). https://doi.org/10.1007/s00232-006-0861-0
32. J.S. Patton, C.S. Fishburn, J.G. Weers, The lungs as a portal of entry for systemic drug delivery. Proc. Am. Thorac. Soc. **1**, 338–344 (2004). https://doi.org/10.1513/pats.200409-049TA
33. J.S. Patton, P.R. Byron, Inhaling medicines: Delivering drugs to the body through the lungs. Nat. Rev. Drug Discov. **6**, 67–74 (2007). https://doi.org/10.1038/nrd2153
34. S. Schanker, J.A. Hemberger, Relation between molecular weight and pulmonary absorption rate of lipid- insoluble compounds in neonatal and adult rats. Biochem. Pharmacol. **32**, 2599–2601 (1983)
35. Z. Enlo-Scott, E. Bäckström, I. Mudway, B. Forbes, Drug metabolism in the lungs: Opportunities for optimising inhaled medicines. Expert Opin. Drug Metab. Toxicol. **17**, 611–625 (2021). https://doi.org/10.1080/17425255.2021.1908262
36. F. Buttini, G. Brambilla, D. Copelli, V. Sisti, A.G. Balducci, R. Bettini, I. Pasquali, Effect of flow rate on in vitro aerodynamic performance of NEXThaler® in comparison with Diskus® and Turbohaler® dry powder inhalers. J. Aerosol Med. Pulm. Drug Deliv. **29**, 167 (2016). https://doi.org/10.1089/JAMP.2015.1220
37. F. Buttini, G. Colombo, P.C.L. Kwok, W.T. Wui, Aerodynamic assessment for inhalation products: Fundamentals and current Pharmacopoeial methods, in *Inhalation Drug Delivery: Techniques and Products*, (2013), pp. 91–119
38. V.A. Marple, History of impactors – the first 110 years. Aerosol Sci. Technol. **38**(3), 247–292 (2004)
39. European Directorate for the Quality of Medicine and Healthcare (EDQM) European Pharmacopoeia (Eur. Pharm.), 11th edition. Chapter 2.9.18, Preparations for inhalation: aerodynamic assessment of fine particles 2024
40. United States Pharmacopeial Convention United States Pharmacopeia 35–National Formulary 30, General Chapter <601>. Physical tests and determinations: Aerosols. Nasal sprays, Metered-dose inhalers, and dry powder inhalers. In Pharmacopeial Forum; 2012
41. V.A. Marple, D.L. Roberts, F.J. Romay, N.C. Miller, K.G. Truman, M. Van Oort, B. Olsson, M.J. Holroyd, J.P. Mitchell, D. Hochrainer, Next generation pharmaceutical impactor (a new impactor for pharmaceutical inhaler testing). Part I: Design. J. Aerosol Med. Depos. Clear. Eff Lung **16**, 283–299 (2003). https://doi.org/10.1089/089426803769017659
42. V.A. Marple, B.A. Olson, K. Santhanakrishnan, D.L. Roberts, J.P. Mitchell, B.L. Hudson-Curtis, Next generation pharmaceutical impactor: A new impactor for pharmaceutical inhaler

testing. Part II: Archival calibration. J. Aerosol Med. Depos. Clear. Eff. Lung **16**, 301–324 (2003). https://doi.org/10.1089/jam.2004.17.335

43. V.A. Marple, B.A. Olson, K. Santhanakrishnan, D.L. Roberts, J.P. Mitchell, B.L. Hudson-Curtis, Next generation pharmaceutical impactor: A new impactor for pharmaceutical inhaler testing. Part III. Extension of archival calibration to 15 L/min. J. Aerosol Med. Depos. Clear. Eff Lung **17**, 335–343 (2004). https://doi.org/10.1089/jam.2004.17.335
44. J.P. Mitchell, Practices of coating collection surfaces of cascade impactors: A survey of members of the European Phamaceutical aerosol group (EPAG). Aerosol Soc., 75–78 (2003)
45. J. Cao, Y. Xu, J. Zhang, T. Fang, F. Wu, Y. Zhen, X. Yu, Y. Liu, J. Li, D. Wang, "Nano-in-micro" structured dry powder inhalers for pulmonary delivery: Advances and challenges. J. Drug Deliv. Sci. Technol. **96** (2024). https://doi.org/10.1016/j.jddst.2024.105648
46. F. Lai, M. Schlich, R. Pireddu, F. Corrias, A. Fadda, C. Sinico, Production of nanosuspensions as a tool to improve drug bioavailability: Focus on topical delivery. Curr. Pharm. Des. **21**, 6089–6103 (2015). https://doi.org/10.2174/1381612821666151027152350
47. R.H. Muller, C.M. Keck, Challenges and solutions for the delivery of biotech drugs – a review of drug nanocrystal technology and lipid nanoparticles. J. Biotechnol. **113**, 151–170 (2004). https://doi.org/10.1016/J.JBIOTEC.2004.06.007
48. P. Yue, W. Zhou, G. Huang, F. Lei, Y. Chen, Z. Ma, L. Chen, M. Yang, Nanocrystals based pulmonary inhalation delivery system: Advance and challenge. Drug Deliv. **29**, 637–651 (2022). https://doi.org/10.1080/10717544.2022.2039809
49. J. Zhang, L. Wu, H.K. Chan, W. Watanabe, Formation, characterization, and fate of inhaled drug nanoparticles. Adv. Drug Deliv. Rev. **63**, 441–455 (2011). https://doi.org/10.1016/j.addr.2010.11.002
50. L. Bollenbach, J. Buske, K. Mäder, P. Garidel, Poloxamer 188 as surfactant in biological formulations – an alternative for polysorbate 20/80? Int. J. Pharm. **620**, 121706 (2022). https://doi.org/10.1016/J.IJPHARM.2022.121706
51. Q. Liu, J. Guan, Z. Sun, X. Shen, L. Li, L. Jin, S. Mao, Influence of stabilizer type and concentration on the lung deposition and retention of resveratrol nanosuspension-in-microparticles. Int. J. Pharm. **569**, 118562 (2019). https://doi.org/10.1016/J.IJPHARM.2019.118562
52. W.N. Missaoui, R.D. Arnold, B.S. Cummings, Toxicological status of nanoparticles: What we know and what we don't know. Chem. Biol. Interact. **295**, 1 (2018). https://doi.org/10.1016/J.CBI.2018.07.015
53. G. Costabile, R. Provenzano, A. Azzalin, V.C. Scoffone, L.R. Chiarelli, V. Rondelli, I. Grillo, T. Zinn, A. Lepioshkin, S. Savina, et al., PEGylated mucus-penetrating nanocrystals for lung delivery of a new FtsZ inhibitor against Burkholderia cenocepacia infection. Nanomed. Nanotechnol. Biol. Med **23**, 102113 (2020). https://doi.org/10.1016/J.NANO.2019.102113
54. M. Malamatari, A. Charisi, S. Malamataris, K. Kachrimanis, I. Nikolakakis, Spray drying for the preparation of nanoparticle-based drug formulations as dry powders for inhalation. Process **8** (2020). https://doi.org/10.3390/pr8070788
55. E.F. Craparo, M. Cabibbo, C. Scialabba, L. Casula, F. Lai, G. Cavallaro, Rapamycin-based inhaled therapy for potential treatment of COPD-related inflammation: production and characterization of aerosolizable nano into micro (NiM) particles. Biomater. Sci. **12**, 387–401 (2024). https://doi.org/10.1039/d3bm01210g
56. T.S. Wiedmann, L. DeCastro, R.W. Wood, Nebulization of NanoCrystals(TM): Production of a respirable solid-in-liquid-in-air colloidal dispersion. Pharm. Res. **1** (1997)
57. C. Jacobs, R.H. Müller, Production and characterization of a budesonide nanosuspension for pulmonary administration. Pharm. Res. **19**, 189–194 (2002). https://doi.org/10.1023/A:1014276917363
58. L. Casula, C. Sinico, D. Valenti, E. Pini, R. Pireddu, M. Schlich, F. Lai, A.M. Fadda, A. Maria Fadda, Delivery of beclomethasone dipropionate nanosuspensions with an electronic cigarette. Int. J. Pharm. **596**, 1–9 (2021). https://doi.org/10.1016/j.ijpharm.2021.120293
59. L. Casula, F. Lai, E. Pini, D. Valenti, C. Sinico, M.C. Cardia, S. Marceddu, G. Ailuno, A.M. Fadda, Pulmonary delivery of curcumin and beclomethasone dipropionate in a

multicomponent nanosuspension for the treatment of bronchial asthma. Pharmaceutics **13**, 1300 (2021). https://doi.org/10.3390/pharmaceutics13081300

60. J.Z. Yang, A.L. Young, P.C. Chiang, A. Thurston, D.K. Pretzer, Fluticasone and budesonide nanosuspensions for pulmonary delivery: Preparation, characterization, and pharmacokinetic studies. J. Pharm. Sci. **97**, 4869–4878 (2008). https://doi.org/10.1002/jps.21380
61. P.C. Chiang, J.W. Alsup, Y. Lai, Y. Hu, B.R. Heyde, D. Tung, Evaluation of aerosol delivery of nanosuspension for pre-clinical pulmonary drug delivery. Nanoscale Res. Lett. **4**, 254–261 (2009). https://doi.org/10.1007/s11671-008-9234-1
62. G. Costabile, I. D'Angelo, G. Rampioni, R. Bondì, B. Pompili, F. Ascenzioni, E. Mitidieri, R. D'Emmanuele Di Villa Bianca, R. Sorrentino, A. Miro, et al., Toward repositioning Niclosamide for Antivirulence therapy of Pseudomonas aeruginosa lung infections: Development of inhalable formulations through nanosuspension technology. Mol. Pharm. **12**, 2604–2617 (2015). https://doi.org/10.1021/acs.molpharmaceut.5b00098
63. T.T. Fu, Z.Q. Cong, Y. Zhao, W.Y. Chen, C.Y. Liu, Y. Zheng, F.F. Yang, Y.H. Liao, Fluticasone propionate nanosuspensions for sustained nebulization delivery: An in vitro and in vivo evaluation. Int. J. Pharm. **572**, 118839 (2019). https://doi.org/10.1016/j.ijpharm.2019.118839
64. L. Casula, M. Schlich, M.C. Cardia, E. Lai, S. Marceddu, R. Pireddu, D. Valenti, C. Sinico, F. Lai, E. Pini, Design and bottom-up production of an Aerosolizable Cannabidiol nanosuspension. Mol. Pharm. **22**, 498–508 (2025). https://doi.org/10.1021/acs.molpharmaceut.4c01095
65. S. Onoue, Y. Aoki, Y. Kawabata, T. Matsui, K. Yamamoto, H. Sato, Y. Yamauchi, S. Yamada, Development of inhalable nanocrystalline solid dispersion of Tranilast for airway inflammatory diseases. J. Pharm. Sci. **100**, 622–633 (2011). https://doi.org/10.1002/jps.22299
66. M. Malamatari, S. Somavarapu, M. Bloxham, G. Buckton, Nanoparticle agglomerates of indomethacin: The role of poloxamers and matrix former on their dissolution and aerosolisation efficiency. Int. J. Pharm. **495**, 516–526 (2015). https://doi.org/10.1016/j.ijpharm.2015.09.013
67. M. Malamatari, S. Somavarapu, K. Kachrimanis, M. Bloxham, K.M.G. Taylor, G. Buckton, Preparation of theophylline inhalable microcomposite particles by wet milling and spray drying: The influence of mannitol as a co-milling agent. Int. J. Pharm. **514**, 200–211 (2016). https://doi.org/10.1016/j.ijpharm.2016.06.032
68. M. Malamatari, S. Somavarapu, K. Kachrimanis, G. Buckton, K.M.G. Taylor, Preparation of respirable nanoparticle agglomerates of the low melting and ductile drug ibuprofen: Impact of formulation parameters. Powder Technol. **308**, 123–134 (2017). https://doi.org/10.1016/j.powtec.2016.12.007
69. K. Šimková, B. Joost, G. Imanidis, Production of fast-dissolving low-density powders for improved lung deposition by spray drying of a nanosuspension. Eur. J. Pharm. Biopharm. **146**, 19–31 (2020). https://doi.org/10.1016/j.ejpb.2019.11.003
70. H. Banat, I. Csóka, D. Paróczai, K. Burian, Á. Farkas, R. Ambrus, A novel combined dry powder inhaler comprising Nanosized Ketoprofen-embedded mannitol-coated microparticles for pulmonary inflammations: Development, in vitro–in silico characterization, and cell line evaluation. Pharmaceuticals **17** (2024). https://doi.org/10.3390/ph17010075
71. M.M. Mehanna, S.M. Mohyeldin, N.A. Elgindy, Rifampicin-carbohydrate spray-dried nanocomposite: A futuristic multiparticulate platform for pulmonary delivery. Int. J. Nanomed. **14**, 9089–9112 (2019). https://doi.org/10.2147/IJN.S211182
72. L. Casula, E.F. Craparo, E. Lai, C. Scialabba, D. Valenti, M. Schlich, C. Sinico, G. Cavallaro, F. Lai, Encapsulation of nanocrystals in mannitol-based inhalable microparticles via spray-drying: A promising strategy for lung delivery of curcumin. Pharmaceuticals **17**, 1–16 (2024). https://doi.org/10.3390/ph17121708
73. Y. Luo, Z. Zhang, G. Huang, H. Yu, Y. Ma, Q. Zheng, P. Yue, Roles of maltodextrin and inulin as matrix formers on particle performance of inhalable drug nanocrystal-embedded microparticles. Carbohydr. Polym. **235**, 115937 (2020). https://doi.org/10.1016/J.CARBPOL.2020.115937

74. T. Liu, M. Han, F. Tian, D. Cun, J. Rantanen, M. Yang, Budesonide nanocrystal-loaded hyaluronic acid microparticles for inhalation: In vitro and in vivo evaluation. Carbohydr. Polym. **181**, 1143–1152 (2018). https://doi.org/10.1016/j.carbpol.2017.11.018
75. X. Li, Q. Wang, J. Huang, X. Yue, X. Zhang, X. Fan, Z. Fang, G. Wang, Z. Qiu, D. Luo, et al., Posaconazole nanocrystals dry powder inhalers for the local treatment of invasive pulmonary aspergillosis. Int. J. Pharm. **668**, 124938 (2025). https://doi.org/10.1016/j.ijpharm.2024.124938

Cosmeceutical Nanocrystals

Abraham M. Abraham and Ryan F. Donnelly

Abstract The cosmetic industry occupies a distinctive and influential position in contemporary society, engaging consumers in ways that often surpass those of the pharmaceutical sector. While pharmaceuticals play a critical role in addressing health concerns, they are primarily utilized by individuals who are ill. In contrast, cosmetics are integrated into the daily lives of diverse individuals, transcending age, gender, and cultural boundaries. From skincare to makeup, these products not only enhance physical appearance but also serve as tools for self-expression and confidence enhancement, rendering them essential components of personal care. Despite their widespread use, numerous cosmeceutical active ingredients demonstrate poor water solubility, which can impede their efficacy in final products. Enhancing solubility is crucial to ensure that these ingredients effectively reach their intended sites of action. Various strategies have been employed to improve the solubility of such compounds; however, these methods often present challenges, such as the limited encapsulation efficiency associated with lipid-based nanovesicles. As a result, a matrix-free formulation approach that enhances solubility could offer significant advantages for cosmetics. Nanocrystals represent a promising matrix-free platform that can improve solubility, bioactivity, and the ability to traverse biological barriers effectively. This chapter explores the utilization and application of nanocrystals, as well as plant-derived nanocrystals, within the realm of cosmetics.

Keywords Nanocrystals · Cosmetics · PlantCrystals · Cosmeceutical product · Antioxidants · Poor solubility · Solubility · Skin · High-pressure homogenization · Bead milling · UV protection · Hair product · Cosmetic product

A. M. Abraham (✉) · R. F. Donnelly
School of Pharmacy, Queen's University Belfast, Medical Biology Centre, Belfast, UK
e-mail: abraham.abraham@qub.ac.uk

A. J. Paredes (ed.), *Nanocrystals*, https://doi.org/10.1007/978-3-032-10238-6_11

1 Introduction

The cosmetic industry occupies a unique and influential role in contemporary society, engaging consumers in ways that extend beyond the pharmaceutical sector [1]. While pharmaceuticals are essential for treating various health conditions and managing diseases, their use is primarily restricted to individuals who are unwell or in need of specific medical interventions. In contrast, cosmetics are integrated into the daily routines of countless individuals, transcending age, gender, and cultural barriers. From skincare products to makeup, cosmetics not only enhance physical appearance but also serve as tools for self-expression and confidence enhancement, making them vital components of personal care routines for many people [1, 2]. This widespread acceptance and integration of cosmetics into everyday life underscore the profound impact of the cosmetic industry. It illustrates how beauty and self-care practices resonate with individuals in ways that extend beyond mere health-related products. Consequently, the cosmetic industry fulfils aesthetic desires while playing a crucial role in shaping societal norms and personal identities. In this context, the synergy between advancements in pharmaceuticals and the cosmetic sector has fostered the development of innovative cosmeceutical formulations aimed at improving skin health and overall beauty.

Recent developments in the cosmetics sector have been significantly driven by advancements in nanotechnology [3, 4]. The application of nanotechnology in cosmetics offers numerous benefits, particularly regarding the critical need for effective dermal applications; only dissolved active ingredients can adequately penetrate the skin layers [5, 6]. This is particularly relevant given that many active ingredients in cosmeceuticals originate from natural, insoluble sources, such as plant antioxidants [6, 7]. For these active compounds to be effective, they must be soluble and reach their intended site of action in the skin and maintain sufficient concentrations for extended periods. To address these challenges, various nanotechnology strategies have been developed, including liposomes, ethosomes, and lipid and polymeric nanoparticles [4]. These methods aim to enhance the solubility of cosmeceuticals; however, they often face challenges related to loading efficiency [4]. Additionally, interactions between the lipidic or polymeric matrix (the non-active materials that transport the cosmeceutical) and the vehicle base (such as creams and gels) can lead to the premature release of the cosmeceutical active ingredient before the application to the skin. This premature release can compromise the structural integrity and effectiveness of the nanocarrier [8].

Recent advancements involving nanocrystals present a promising solution to the challenges faced by traditional nanocarriers, particularly in enhancing the solubility of cosmeceuticals [6]. Nanocrystals, which typically measure less than 1000 nm, possess a significantly larger surface area-to-volume ratio compared to their bulk materials [5]. This unique characteristic facilitates improved interaction with solvents, resulting in enhanced dissolution rates and bioavailability of active ingredients. The ability of nanocrystals to increase the saturation solubility of poorly soluble compounds is especially advantageous in cosmetic formulations, where effective delivery of active ingredients is critical [9]. Moreover, nanocrystals can be engineered to enhance their stability, thereby preventing the premature release of active ingredients [10]. By optimizing the formulation process, manufacturers can create nanocrystals that maintain their structural integrity until they reach their target site on the skin [11]. This controlled release mechanism maximizes the efficacy of active ingredients while minimizing potential adverse effects associated with high concentrations [9, 12]. Consequently, consumers can benefit from potent active ingredients without the risk of irritation or adverse reactions.

The implications of nanocrystal technology extend beyond merely improving solubility and stability. The incorporation of nanocrystals into cosmetic formulations can lead to enhanced sensory attributes, such as texture and feel [13]. In an increasingly competitive market, consumers seek products that provide a pleasant application experience, and the use of nanocrystals can contribute to smoother, more luxurious formulations. This sensory enhancement is particularly relevant for product differentiation, which is essential for attracting and retaining consumers in the cosmetic market. Additionally, the versatility of nanocrystals allows for their incorporation into a wide array of cosmetic products, including creams, serums, and lotions [5, 14]. This adaptability makes nanocrystals an appealing option for formulators aiming to create innovative products that cater to diverse consumer preferences. For example, nanocrystals can be utilized in anti-aging formulations to improve the delivery of antioxidants [15] and other active ingredients that promote skin rejuvenation. In sun care products, nanocrystals can enhance the effectiveness of UV (ultraviolet) filters, providing better protection against harmful sun exposure [16, 17].

As the cosmetic industry continues to evolve, the integration of nanocrystals into cosmeceutical formulations is likely to expand, offering innovative solutions for enhancing product performance and addressing consumer demands [18]. The growing awareness of skin health and the desire for effective, safe cosmetic products will drive further research and development in this area. Furthermore, the increasing consumer preference for natural and sustainable ingredients aligns well with the potential of nanocrystals derived from plant sources, paving the way for the development of eco-friendly cosmetic formulations [2, 19]. Thus, the characteristics of nanocrystals in dermal applications and cosmetics offer numerous advantages for improving the effectiveness of active ingredients, particularly antioxidants [6]. Nanocrystals have enhanced solubility, dissolution velocity, adhesiveness, concentration gradient, and bioavailability, positioning them as valuable components in cosmetic formulations [6]. The applications of nanocrystals extend to targeting hair follicles and developing plant-derived nanocrystals, highlighting their versatility and potential within the cosmetic industry [20].

As research continues to advance, the incorporation of nanocrystals into cosmetic formulations is expected to expand further, leading to innovative solutions that enhance product performance and cater to evolving consumer demands. The future of the cosmetic industry appears poised for transformation, with nanocrystal technology paving the way for the creation of effective, safe, and appealing products that resonate with consumers' aspirations for beauty and self-care. This ongoing evolution signifies not only a shift in product formulation but also a broader movement towards embracing the intersection of nanocrystal science and beauty in everyday life.

2 Formulation Principle and Design

Nanocrystals are solid particles with dimensions on the nanoscale, specifically less than 1000 nm, and are dispersed in a surfactant solution [21]. These particles consist of 100% active ingredient without any accompanying matrix [6]. The surfactants utilized in the production of nanocrystals primarily function as stabilizers, reducing interfacial tension and serving as wetting agents [22]. Furthermore, surfactants establish a barrier around the nanosized active ingredient particles, preventing agglomeration and thereby enhancing the stability of the system. Cosmeceutical nanocrystals represent a specialized category of nanocrystals utilized in cosmeceutical applications, predominantly for application on the skin or hair [6]. They exhibit similar characteristics, as well as production and characterization methodologies, to those of drug nanocrystals. In comparison to other nanoformulations, such as liposomes, cosmeceutical nanocrystals can be synthesized using simpler and more cost-effective techniques [23]. For example, Gassmann et al. employed the precipitation method to produce nanosized active ingredients [24]. However, this method has inherent limitations, as it necessitates the use of two solvents: one in which the active ingredient is soluble and another in which it is insoluble, with the latter required to be miscible with the former. To overcome these challenges, a dispersion method was introduced by Muller et al. in 1995 [25]. This approach represents effective strategy to produce nanocrystals and addresses the issue of poor solubility of active ingredients. Various techniques can be utilized to obtain nanocrystals, which can be broadly categorized into top-down and bottom-up approaches, as will be discussed in the following sections. The formulation design of nanocrystals, especially within cosmetic formulations, entails several key elements that affect the stability, effectiveness, and overall quality of the final product. These elements consist of stabilizers, surfactants, cosurfactants, stirring speed, and temperature.

2.1 *Stabilizers*

Stabilizers are vital for preventing nanocrystal aggregation during and after production [22]. They can be ionic or non-ionic polymers that form a protective layer around the nanocrystals, offering electrostatic repulsion and steric stabilization [26, 27]. This stabilization is essential for maintaining the desired particle size and avoiding coalescence, which could reduce efficacy. For example, cellulose nanocrystals often require stabilizers to improve dispersion in aqueous media, thus enhancing their stability and performance in cosmetics. The selection of stabilizers can notably influence the physical characteristics of the nanocrystals, including their size and zeta potential, which indicate stability.

2.2 Surfactants

Surfactants are compounds that lower the surface tension between two immiscible phases, such as oil and water. In nanocrystal production, surfactants stabilize the dispersion of nanocrystals in a liquid medium [27]. They can be anionic, cationic, or non-ionic, and the choice depends on the desired attributes of the final formulation [27]. Surfactants not only reduce interfacial tension but also improve the solubility of poorly soluble active ingredients, enhancing their bioavailability in cosmetic applications. The hydrophilic–lipophilic balance (HLB) of surfactants is crucial, as it determines their effectiveness in stabilizing the nanocrystals. The ratio of the stabilizer used is an important factor for an efficient nanomilling process [11]. For this reason, a surfactant screening is always performed for any nanocrystal formulation.

2.3 Cosurfactants

Cosurfactants are used alongside primary surfactants to improve stability and performance. They can enhance the solubilization of active ingredients and overall stability of the formulation [28]. Cosurfactants may alter the interfacial properties, leading to better dispersion and stability of the nanocrystals. Their use can also help to achieve a uniform particle size distribution, which is essential for the product's performance.

2.4 Stirring Speed

The stirring speed during nanocrystal production significantly influences particle size and uniformity [29]. Higher stirring speeds can produce smaller particle sizes due to increased shear forces that promote the breakup of larger aggregates [30]. However, excessively high speeds may result in smaller, less stable particles that could aggregate over time. Therefore, optimizing stirring speed is crucial for achieving the desired particle size and stability. High-speed stirring is a pre-requirement for some nanocrystals production methods (e.g. producing nanocrystals by high-pressure homogenization).

2.5 *Temperature*

Temperature is another important factor in nanocrystal production [31]. It affects the solubility of active ingredients, the viscosity of the medium, and the kinetics of crystallization [30]. Higher temperatures can enhance the solubility of poorly soluble compounds, facilitating nanocrystal formation. However, temperature must be carefully controlled because excessive heat can degrade sensitive ingredients or alter the properties of stabilizers and surfactants. Thus, maintaining an optimal temperature range during production is vital for achieving stable and effective nanocrystals.

In summary, producing nanocrystals for cosmetic applications necessitates a careful balance of stabilizers, surfactants, cosurfactants, stirring speed, and temperature. Each of these elements plays a crucial role in ensuring the stability, effectiveness, and overall quality of the final product.

3 Cosmeceutical Nanocrystals Production

The production of cosmetic nanocrystals typically employs various processing techniques categorized into top-down and bottom-up approaches [30, 32]. Top-down methods, such as high-pressure homogenization and bead milling [7], are commonly utilized to reduce bulk materials into nanosized particles. These techniques enable precise control over particle size and distribution, which are vital for meeting performance requirements in cosmetic applications. Conversely, bottom-up approaches, including solvent evaporation and precipitation, can be employed to generate nanocrystals from molecular precursors [33, 34]. The choice of processing technique is contingent upon the material characteristics and specific needs of the cosmetic formulation.

3.1 *Top-Down Methods*

These methods involve breaking down bulk materials into nanoscale particles through mechanical processes, such as high-pressure homogenization and bead milling. By combining both methods, “SmartCrystals” can be obtained. All these methods are classified as heat-generating processes. Consequently, for heat-sensitive active cosmeceutical ingredients (e.g. antioxidants and natural products), a cooling system is necessary to maintain the temperature within a specific range and prevent degradation [30]. These methods will be discussed in more detail in the following sections.

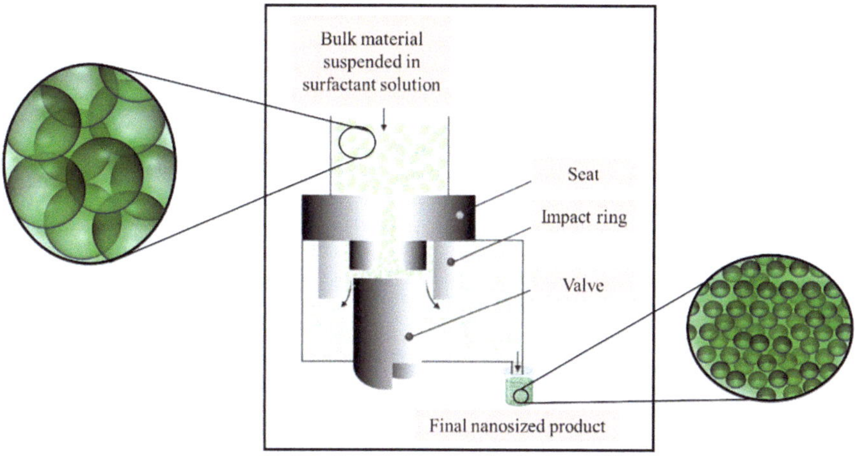

Fig. 1 High-pressure homogenization to produce cosmeceutical nanocrystals. (Modified after [29])

3.1.1 High-Pressure Homogenization

In this method, macroparticles are suspended in a surfactant solution that can be nanosized by forcing the suspension through a small orifice (Fig. 1), which reduces the static pressure below the boiling point of the aqueous suspension containing the macroparticles [18, 29]. When the air pressure is restored and the suspension exits the orifice, boiling of the water occurs, resulting in the formation of air bubbles. These bubbles subsequently shrink, causing the surrounding material containing the drug or cosmetic particles to rush towards the centre, leading to the formation of colloids and a reduction in particle size. This process can be repeated until the desired particle size is achieved [35]. Additionally, there are various commercial high-pressure homogenizers that can operate in continuous mode. High-pressure homogenization (HPH) can be performed in an aqueous medium, referred to as "dissocubes", or in a non-aqueous medium, known as "nanopure". A screening to determine the number of HPH cycles is usually performed for every formulation. High-speed stirring is a pre-requirement for nanocrystal production using HPH.

3.1.2 Bead Milling

This method, developed by Liversidge et al. in the 1990s [33, 36], involves the use of a milling chamber, pearl mills, and high shear media mills. Various types of milling chambers, such as glass vials, can be utilized in conjunction with a milling shaft and recirculation system [37, 38]. The method is simple, involving the addition of cosmetic active ingredients and surfactants to the pearl mills in specific ratios.

Fig. 2 Images of bulk materials suspended in a surfactant (showing large particle sedimentation) before nanosizing in comparison to the homogeneous formulation containing nanosized materials (showing no sedimentation) obtained from the same starting cosmeceutical active ingredient

Temperature control is achieved through a water bath set to a designated degree, which is crucial for certain cosmetic active ingredients sensitive to heat, such as antioxidants. The efficient nanosizing ends up with homogeneous samples and no precipitation (Fig. 2). The nanosization process relies on the principle of impaction, which is facilitated by the application of energy that generates high energy and shear forces [30]. A screening to determine the most efficient nanomilling time is usually performed for every formulation.

3.2 *Bottom-up Methods*

These methods begin at the molecular level and progress through molecular associations to create solid particles. Consequently, this classification is regarded as an assembly method that produces nanocrystals from molecules. This approach encompasses various techniques, such as precipitation, supercritical fluid methods, and spray drying, which will be summarized in the following sections.

3.2.1 Precipitation

This method is a commonly employed technique for generating submicron particles of hydrophobic active pharmaceutical ingredients (APIs) [34]. It involves the use of an organic solvent that is miscible with water, as well as another solution mixed with an antisolvent in which the drug becomes insoluble in the presence of a surfactant. The rapid introduction of the solution into the antisolvent leads to a quick supersaturation of the drug, resulting in the formation of ultrafine nanocrystals. This process consists of two key steps: (1) nucleation and (2) crystal growth [30]. Achieving a high rate of nucleation while maintaining a low rate of crystal growth is essential for producing stable nanocrystals with extremely small size.

3.2.2 Supercritical Fluid Method

This method entails the rapid expansion of a supercritical solution process, specifically utilizing the supercritical antisolvent technique, followed by precipitation *via* a compressed antisolvent process [39]. In this approach, a drug solution is expanded through a nozzle into a supercritical fluid, resulting in the loss of the solvent properties of the fluid and leading to the precipitation of the API as ultrafine particles. To facilitate this process, a chamber containing compressed CO_2 is necessary. As the solvent is removed, the API solution becomes supersaturated, prompting the precipitation of fine nanocrystals. The supercritical antisolvent process employs a supercritical fluid in which the drug has low solubility, along with a solvent that is miscible with the supercritical fluid. The drug solution is injected into the supercritical fluid, where the extraction of the solvent occurs, causing the API solution to reach supersaturation and subsequently precipitate as fine nanocrystals.

3.2.3 Spray Drying Method

This method involves atomizing a liquid solution containing the drug or material of interest into a hot drying chamber, which promotes rapid evaporation of the solvent and leads to the formation of solid nanocrystals. During the spray drying process, the solution is atomized into fine droplets that are then exposed to a controlled temperature environment, facilitating the crystallization of the dissolved substance as the solvent evaporates. The process can be optimized by incorporating excipients that serve as crystallization inducers, enhancing the nucleation and growth of nanocrystals. These excipients can create a physical barrier to crystal growth, allowing for control over the size and morphology of the resulting nanocrystals. Additionally, the conditions under which spray drying is conducted—such as temperature and feed rate—can be adjusted to achieve the desired particle characteristics, including size and surface properties.

There are also other methods to produce nanocrystals like electro spraying and nanojet technology but are not commonly used to obtain cosmeceutical nanocrystals.

4 Characterization of Cosmeceutical Nanocrystals

Characterization of cosmeceutical nanocrystals is crucial for evaluating their physical, chemical, and biological properties to ensure safety and efficacy in cosmetic formulations. Various methods are employed to assess these properties, including particle size and distribution, zeta potential, morphology, resuspendability, crystalline and amorphous forms, saturation solubility, in vitro dissolution studies, biological performance, density, pH values, and colour and odour [30, 40].

4.1 *Particle Size and Distribution*

The particle size and distribution of nanocrystals are critical parameters that affect their stability, bioavailability, and effectiveness in cosmetics. Techniques like laser diffraction and dynamic light scattering (DLS) are commonly used for measuring particle size [37]. Laser diffraction offers a rapid and accurate measurement by analysing the scattering of laser light through a sample. DLS measures the Brownian motion of particles in suspension to calculate their hydrodynamic diameter, providing insights into size distribution and stability. A narrow particle size distribution is often desirable, indicating uniformity and consistency in the formulation.

4.2 *Zeta Potential*

Zeta potential serves as a vital indicator of colloidal system stability, particularly in nanocrystals [14]. It quantifies the electrostatic potential at the slipping plane of a suspended particle, offering insights into the repulsive interactions among particles. A higher absolute zeta potential signifies increased stability, indicating a lower likelihood of particle aggregation. Techniques such as electrophoretic light scattering can measure zeta potential, providing crucial data on the stability of nanocrystals across various formulations. Maintaining an optimal zeta potential is essential for achieving long-term stability in nanocrystal formulations.

4.3 *Morphology*

Various microscopy techniques, including scanning electron microscopy (SEM) and transmission electron microscopy (TEM), are employed to visualize the shape and size of nanocrystals. SEM provides detailed images of surface morphology, and TEM allows for high-resolution imaging of internal structures, offering insights into their surface characteristics. Understanding the morphology of nanocrystals is essential for optimizing their performance in cosmetic formulations [41].

4.4 Resuspendability

Converting nanosuspensions into dry powdered nanocrystals is essential to improve their shelf life and physical stability [7]. Therefore, the capacity of nanocrystals to resuspend is a significant characteristic to be tested, as that impacts their applicability in cosmetic formulations. Resuspendability is evaluated by measuring particle size and distribution of the freshly obtained nanosuspensions and after converting the nanosuspensions into dry nanocrystals (e.g. by freeze drying) [42]. A formulation that readily resuspends without substantial changes in size or distribution is deemed more stable.

4.5 Crystalline and Amorphous Form

The crystalline/amorphous nature of nanocrystals can significantly impact their solubility and stability. Techniques such as X-ray diffraction (XRD) and differential scanning calorimetry (DSC) are commonly utilized to identify the crystalline and amorphous forms [30, 41]. XRD reveals details about the crystalline structure and phase, while DSC measures thermal transitions, such as melting points and glass transition temperatures. Grasping the crystalline nature of nanocrystals is vital for predicting their stability and performance in cosmetic applications.

4.6 Saturation Solubility

Saturation solubility refers to the maximum concentration of a solute that can dissolve in a solvent under specific temperature and pressure conditions [43]. This property is particularly relevant for poorly soluble active ingredients in cosmetic nanocrystal formulations. Saturation solubility can be determined through various methods, including shake-flask technique or equilibrium solubility studies. Higher saturation solubility indicates enhanced bioavailability and efficacy of active ingredients within the formulation.

4.7 In Vitro Dissolution Study

In vitro dissolution studies are conducted to assess the release profile of API from nanocrystals [44]. These studies replicate physiological conditions to evaluate how quickly and effectively active ingredients are released from nanocrystals into a dissolution medium. The findings from these studies yield valuable insights into the bioavailability and therapeutic potential of the formulation. A quicker dissolution rate for cosmeceutical nanocrystals is often desired, as it signifies that active ingredients can be easily absorbed by the skin or other targeted tissues.

4.8 Cosmeceutical Performance

The cosmeceutical performance of cosmeceutical nanocrystals is crucial for their characterization [45], encompassing evaluations of skin penetration and the deposition of active ingredients [6, 37]. In vitro studies utilizing skin models or human skin samples can assess the ability of nanocrystals to effectively penetrate the skin barrier and deliver active ingredients. Techniques such as Franz diffusion cells measure the amount of active ingredient penetrating the skin over time. Additionally, in vivo studies may be conducted to assess the safety and efficacy of nanocrystals on human skin. Other assays are also commonly performed, for example, to determine the antioxidative capacity of cosmetic formulations with antioxidant activity.

4.9 Density

The density of nanocrystals can affect their stability and performance in formulations [30]. Understanding the density of nanocrystals is important for predicting their behaviour in different formulations and ensuring optimal performance in cosmetic applications.

4.10 pH Values

The pH of the nanocrystal formulation is a crucial factor that can affect both stability and product performance. It influences the solubility of active ingredients, the ionization of stabilizers and surfactants, and the overall formulation stability [30]. pH can be determined using a pH meter, and it is vital to maintain the formulation within an appropriate pH range for optimal stability and compatibility with skin. Extreme pH levels may result in the degradation of active ingredients or destabilization of the nanocrystals.

4.11 Colour and Odour

The colour and odour of cosmetic formulations are significant sensory attributes that can affect consumer acceptance [6, 30]. These attributes are particularly crucial for cosmetic products, as they influence the overall user experience. Consistency in colour and odour is vital for maintaining product quality and consumer satisfaction. Furthermore, chemical instabilities can also cause colour and odour changes. This is especially critical in cosmetic products containing antioxidants.

In conclusion, characterizing cosmeceutical nanocrystals involves a thorough assessment of various physical, chemical, and biological properties. By employing multiple characterization methods, formulators can ensure the development of safe and effective cosmetic products that harness the unique advantages of nanocrystals.

5 Nanocrystal Properties in Cosmetics

The exploration of nanocrystals in dermal applications and cosmetics is a complex undertaking, especially regarding their effectiveness in boosting the performance of antioxidants and other APIs. Nanocrystals, which are defined as crystalline entities with dimensions in the nanometre range, possess distinctive physical and chemical characteristics that can be advantageous in cosmetic formulations. The subsequent sections will delve into these attributes in detail, bolstered by relevant scientific principles and findings.

5.1 Enhanced Solubility of Poor Water-Soluble APIs

Many of the cosmeceutical active ingredients are from natural sources (i.e. plants) that have limited water solubility [6, 46, 47]. Therefore, improving their water solubility is essential before turning them into cosmeceutical products. One of the primary benefits of incorporating nanocrystals into cosmetic formulations is their capacity to improve the solubility of APIs that are poorly soluble in water. The solubility of a compound can be explained by the relationship between various factors affecting dissolution rate. By reducing particle sizes to the nanoscale, the surface area significantly increases, resulting in a faster dissolution rate and improved solubility of the API.

The enhanced solubility of poorly soluble APIs via nanocrystals can be effectively understood through the concepts of vapour pressure and particle size [6]. The connection between the vapour pressure of a curved surface and its radius of curvature is particularly significant when comparing nanosized API solubility to that of larger particles. As particle dimensions shrink to the nanoscale, the radius decreases, resulting in higher vapour pressure for the nanocrystals compared to their bulk forms. This increase in vapour pressure corresponds to elevated solubility, as the solubility of a substance is directly linked to its vapour pressure.

Nanocrystals, usually under 1000 nm in size, possess a considerably larger surface area-to-volume ratio than larger particles [15]. This augmented surface area is crucial, as it boosts the interaction between the API and the solvent, thus enhancing the dissolution rate. The correlation between dissolution rate and surface area implies that the increased surface area resulting from nanosizing the API leads to a greater dissolution rate, which is especially advantageous for poorly soluble compounds. Moreover, the increased solubility phenomenon can be explained by the fact that a solid's solubility rises as particle size decreases. As the particle radius diminishes, solubility increases, further supporting the notion that nanocrystals can substantially improve the solubility of poorly soluble APIs. Thus, employing nanocrystals to enhance the solubility of poorly soluble APIs is grounded in the principles of vapour pressure and particle size (based on the Kelvin equation) [48]. The larger surface area-to-volume ratio of nanocrystals clarifies how they improve the dissolution rate and overall bioavailability of poorly soluble compounds.

5.2 Increased Dissolution Velocity and Concentration Gradient

The dissolution velocity of nanocrystals is significantly enhanced due to their larger surface area and reduced particle size [49]. As the radius of particles decreases, the diffusion coefficient increases, resulting in a quicker dissolution rate. This feature is vital in cosmetic applications where rapid absorption of active ingredients is critical. The accelerated dissolution velocity of nanocrystals can be explained through the relationship between dissolution rate and factors such as surface area, concentration gradient, and diffusion properties, as explained by the Noyes–Whitney equation. Nanocrystals, defined as particles with dimensions on the nanoscale, have a much greater surface area-to-volume ratio compared to bulk materials. This increased surface area is essential as it improves the interaction between solid particles and the solvent, promoting a more efficient dissolution process. When a solid dissolves, the molecules at the surface enter the solution, and the rate of this process is directly affected by the available surface area. Therefore, with a larger surface area, more molecules can dissolve simultaneously, resulting in a heightened dissolution velocity. Additionally, reducing particle size to the nanoscale alters the diffusion characteristics of the dissolved molecules. As particles shrink, the distance molecules must travel to escape the solid and enter the solution also shortens. This decreased distance boosts the rate at which molecules diffuses away from the solid surface and into the surrounding solvent. In essence, smaller particles can more easily release their molecules into the solution, contributing to an accelerated dissolution rate.

The concentration gradient also plays a crucial role in the dissolution process. It is defined as the difference in concentration between the saturated solution at the solid's surface and the bulk solution. A steeper concentration gradient leads to a faster dissolution rate, as it drives molecules from areas of higher concentration (near the solid surface) to areas of lower concentration (in the bulk solution). The increased surface area of nanocrystals not only enhances the dissolution rate but also creates a more pronounced concentration gradient, allowing more molecules to enter the solution at once. Thus, the heightened dissolution velocity of nanocrystals is linked to their larger surface area, smaller particle size, and the effects on diffusion and concentration gradients. These traits are especially beneficial in cosmetic applications, where swift absorption of active ingredients is crucial for effectiveness. The ability of nanocrystals to dissolve rapidly and efficiently facilitates improved delivery of active compounds, making them an essential component in cosmetic product formulation.

5.3 *Enhanced Adhesiveness*

The enhanced adhesiveness of nanocrystals can be linked to several important factors that greatly enhance their interaction with surfaces, especially in topical formulation. Their small particle size enables them to infiltrate the microstructures of surfaces like skin or hair follicles, increasing contact area and improving adhesion. Additionally, the high surface area-to-volume ratio of nanocrystals facilitates better interaction and adherence at the site of application (Fig. 3), leading to a more stable and effective reservoir of active ingredients. Specifically, particles ranging from 300 to 700 nm can effectively enter hair follicles, enhancing drug retention and providing a sustained release of active compounds. This property is particularly advantageous in cosmetic products, where lasting adherence and effective delivery of active ingredients are crucial for achieving the desired effects.

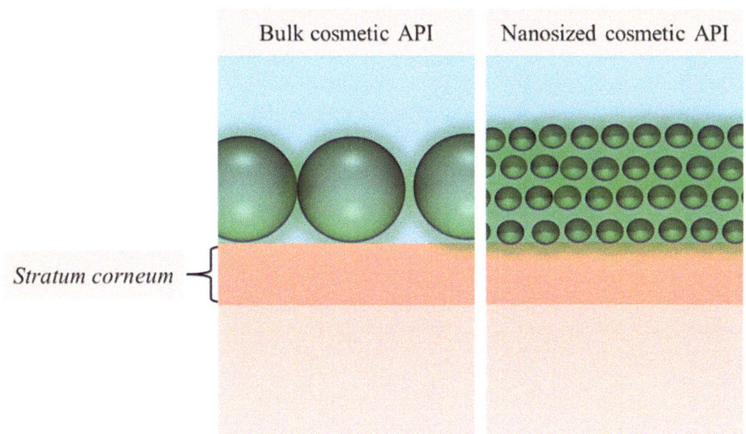

Fig. 3 Comparison of the adhesiveness of large bulk cosmetic active ingredient and cosmeceutical nanocrystals applied to the skin, showing a more efficient skin contact with the active ingredient for the nanocrystals when compared to the large bulk material. (Modified after [9])

5.4 *Enhanced Cosmeceutical Activity*

The use of nanocrystals in cosmetic formulations can enhance their effects due to their distinct physical and chemical characteristics. Nanocrystals feature a high surface area-to-volume ratio, which improves interaction with skin and hair, allowing for better penetration and absorption of active ingredients. This enhanced delivery mechanism is especially advantageous for bioactive compounds, like antioxidants and vitamins, which can reach deeper skin layers, thereby maximizing their benefits [15]. The improvements in solubility, dissolution velocity, and concentration gradient collectively contribute to enhanced bioavailability. This relationship can be expressed in terms of the area under the curve for different formulations, demonstrating that nanocrystal formulations can markedly increase the bioavailability of antioxidants, enhancing their effectiveness in cosmetic applications [7, 47, 50]. Moreover, the optical properties of nanocrystals can improve light scattering and absorption, increasing the visual appeal of cosmetic products. For example, incorporating nanocrystals can also provide protection against harmful UV rays [51]. Furthermore, the stability of cosmetic ingredients is enhanced when formulated as nanocrystals, as they safeguard sensitive compounds from degradation, ensuring the product retains its effectiveness over time. Overall, integrating nanocrystals into cosmetic products not only boosts their performance but also enriches the user experience.

6 Applications of Nanocrystals in Cosmetics

6.1 *Antioxidative Nanocrystals for Skin Application*

The term "anti-aging" is frequently associated with "antioxidant", as many anti-aging formulations aim to mitigate damage caused by free radicals [52]. These free radicals, particularly reactive oxygen species (ROS), play a significant role in skin damage and the aging process by disrupting the skin's natural repair and defence mechanisms. Nanocrystals infused with anti-free radical agents have demonstrated potential in blocking aging pathways by effectively neutralizing free radicals [53]. Quercetin, a flavonoid recognized for its potent antioxidant properties, contains multiple phenolic hydroxyl groups that enhance its ability to protect the skin from oxidative stress [54]. However, its limited water solubility presents challenges for its application in cosmetic formulations [54]. Numerous studies have shown that quercetin nanocrystals can significantly improve their solubility and reduce oxidative stress induced by hydrogen peroxide (H_2O_2) in skin cells [55]. Findings indicate that the viability of cells treated with quercetin nanocrystals exceeded 100%, suggesting a protective effect against oxidative stress [56]. Furthermore, scavenging assays have revealed that the clearance rate of antioxidative nanocrystals on the DPPH (2,2-diphenyl-1-picrylhydrazyl) radical was twice that of conventional micronized particles, highlighting the enhanced antioxidant capabilities of nanocrystal formulations [7]. Another study developed rutin nanocrystals using the high-pressure homogenization (HPH) and bead milling [56], demonstrating considerable progress in enhancing the bioavailability and effectiveness of this powerful antioxidant in cosmetic formulations. Rutin nanocrystals exhibit improved saturation solubility and enhanced skin penetration profiles compared to commercial topical products. This innovative formulation not only facilitates the delivery of rutin to the skin but also preserves its stability over time, rendering it a promising choice for cosmetic products aimed at combating oxidative stress and skin aging.

The application of a cold wet milling system to produce Q10 nanocrystals has demonstrated the potential to enhance the bioavailability of Q10 by up to 13 times compared to traditional formulations [5]. This significant increase in bioavailability positions Q10 nanocrystals as a promising ingredient for enhancing the value of cosmetic products, particularly those aimed at addressing signs of aging and promoting overall skin health. By utilizing these advanced delivery systems, cosmetic formulations can maximize the benefits of Q10, offering consumers effective solutions for maintaining youthful and resilient skin. Consequently, the elimination of excess free radicals is a critical strategy for addressing skin aging. The beauty industry has recognized this necessity, with antioxidant products capturing a significant market share valued in the hundreds of billions. As public awareness of the importance of antioxidants increases, the demand for effective anti-aging solutions is expected to grow, fostering innovation in formulations that leverage the advantages of nanocrystals and other advanced delivery systems.

6.2 Nanocrystals for UV (Ultraviolet) Protection

To improve its effectiveness, rutin has been developed into nanocrystals using a high-pressure homogenization technique. These nanocrystals demonstrate exceptional physical and chemical stability over time, making them ideal for extended use in cosmetic products. Additionally, an alternative method involving anti-solvent nanoprecipitation followed by ultrasonication has led to the creation of rutin nanocrystals stabilized with hydroxypropyl-beta-cyclodextrin [57]. This formulation not only boosts the solubility of rutin but also enhances its ability to penetrate the skin, facilitating better absorption and effectiveness as a protective agent. The protective function of rutin nanocrystals against UV-A radiation can be explained through several essential processes [58]. When UV-A light penetrates the skin, it can cause significant harm, contributing to aging and wrinkle formation. About 50% of UV-A light can reach the deeper skin layer through the outer layer. However, when micron-sized antioxidant crystals are utilized, the release of antioxidants occurs slowly, resulting in a marked decrease in UV-A light penetration into the deeper skin layer. In contrast, when the size of the antioxidant crystals is reduced to the nanoscale, the rapid dissolution of these antioxidants enables immediate and effective neutralization of UV-A radiation. This improved performance is due to the increased permeability, adhesive properties, and sustained effects of the nanocrystals, which work synergically to prevent UV-A light from penetrating deeper and causing damage.

6.3 *Nanocrystals for Hair Follicle Targeting*

The application of nanocrystals for targeting hair follicles represents an innovative approach in cosmetic products aimed at enhancing hair growth and scalp health [59]. Due to their small size and large surface area, nanocrystals have the potential to penetrate hair follicles more effectively than larger particles, facilitating the precise delivery of active ingredients such as growth factors, vitamins, and herbal extracts directly to the follicular region [59]. This targeted delivery is crucial for activating hair follicle function and promoting the transition from the resting phase to the growth phase of the hair cycle. Various studies indicate that specific signalling pathways are instrumental in hair follicle development and growth regulation. By employing a targeted nanocrystal approach, cosmetic products can enhance the bioavailability of substances that influence these pathways, potentially leading to increased hair density and improved overall hair health [60]. Moreover, the sustained release of active ingredients from nanocrystals can provide long-lasting effects, positioning this method as a promising option for tackling hair loss and enhancing scalp health [20]. This cutting-edge strategy not only amplifies the efficacy of hair care products but also addresses the growing consumer demand for targeted and effective solutions in hair restoration and maintenance. While some penetration into the *stratum corneum* may occur, the number of intact particles reaching the viable epidermis (i.e. where the hair follicle is located) is likely to be limited. Therefore, this approach needs further investigation to achieve a better hair follicle targeting using nanocrystals.

7 PlantCrystals: Plant Nanocrystals

7.1 *Plant Nanocrystals Development*

Plants serve as an invaluable reservoir of bioactive compounds that have been utilized for centuries, not only in the realm of traditional medicine but also increasingly within the cosmetics industry [61, 62]. These natural resources are recognized as sustainable sources of phytochemicals that are renowned for their remarkable antioxidant, anti-aging, and antimicrobial properties, rendering them particularly desirable for incorporation into cosmetic formulations [7]. While many of these botanicals are commonly consumed in dietary practices, their beneficial extracts can be procured through various traditional techniques such as infusion, decoction, or maceration [7, 37]. Additionally, advanced methodologies that employ organic solvents are also utilized to enhance extraction efficiency. This is especially critical given that numerous beneficial compounds exhibit low solubility in aqueous environments. However, conventional extraction methods often necessitate substantial quantities of plant material and considerable resources for cultivation, harvesting, and transportation [7]. This situation underscores the pressing need for environmentally sustainable processes that promote eco-friendly production of cosmetic extracts.

In this context, the principles of green chemistry play a pivotal role in the cosmetics sector, advocating for the development of commercially viable products derived from renewable resources through environmentally friendly methodologies [29]. The tenets of green chemistry have spurred innovative extraction techniques aimed at minimizing energy consumption, utilizing renewable natural materials, and reducing waste generated during the extraction process. These modern extraction methods can significantly decrease processing times from several hours to minutes while ensuring consistency and decreasing dependence on organic solvents. There is a burgeoning interest in utilizing plant materials to derive cosmetic extracts with specific targeted properties through these green technologies. However, a significant challenge persists in balancing the allocation of agricultural land for cosmetic ingredients versus the indispensable need for food production. Consequently, there is an urgent demand to explore novel eco-friendly methods that optimize the yields from plant materials specifically for the cosmetics industry. In this regard, the potential of plant nanocrystals emerges as a promising solution to address all the aforementioned challenges.

Plant nanocrystals, or PlantCrystals, can be synthesized using nanosization techniques analogous to those employed in the production of drug nanocrystals, but specifically applied to whole plants or particular plant parts such as seeds, leaves, or bark [7, 29, 37]. This approach effectively facilitates the extraction of valuable phytochemicals for cosmetic applications. Recent advancements in extraction technology have demonstrated that a combination of high-speed stirring and high-pressure homogenization can transform various plant materials into nanosized particles, thereby enhancing their efficacy in cosmetic formulations. These nanosized plant materials (i.e. PlantCrystals) exhibit improved antioxidant, antibacterial, and anti-inflammatory properties, highlighting their potential to optimize the performance of skincare products. By reducing plant materials to small sizes (i.e. in the submicron range), plant cells are disrupted, allowing for the direct release of active cosmetic ingredients. The techniques utilized to generate these nanosized plant materials, referred to as PlantCrystals, closely resemble those involved in the production of drug nanocrystals [7], albeit with a distinct emphasis on cosmetic applications.

Another method for the production of PlantCrystals involves bead milling, a process that was patented in the early 1990s [37]. This technique generates PlantCrystals by employing pearl mills to break down plant cells and facilitate the release of their phytochemicals (Fig. 4). The bead milling process for plant nanocrystals production comprises two distinct stages: the initial phase involves creating a macrobulk suspension by dispersing plant powder in a suitable surfactant

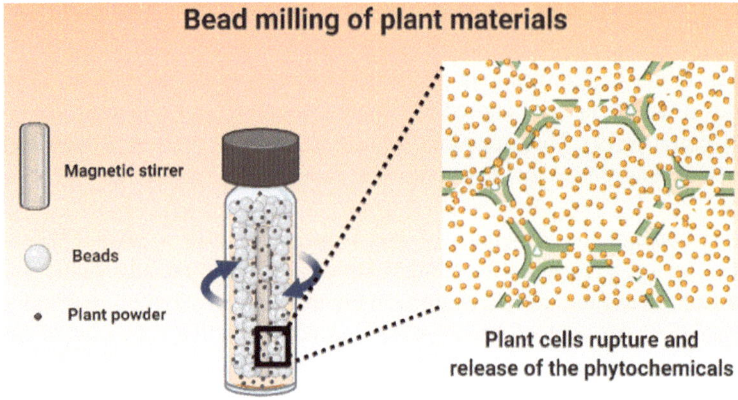

Fig. 4 Schematic representation showing bead milling principle to produce PlantCrystals. (Modified after [37])

solution, followed by the nano-milling phase that employs beads and magnetic stirring rods. The size reduction achieved through bead milling relies on impaction, wherein plant particles are fragmented into nanosized particles through shear forces and the impact of the milling media. While this method is characterized by its simplicity and scalability, it may present certain challenges, such as a protracted milling process that could lead to the oxidation of sensitive phytochemicals and the potential for contamination resulting from the wear of milling beads [37].

The final technique for PlantCrystals production integrates high-pressure homogenization with bead milling [7]. Although both methodologies aim to achieve nanosizing, they serve distinct purposes; for PlantCrystals, the primary focus is on further comminuting larger particles to enhance extraction efficiency and release a greater quantity of active compounds. This synergistic approach begins with high-pressure homogenization to obtain PlantCrystals at a large submicron scale, followed by bead milling to achieve finer particle sizes, thereby optimizing the extract for cosmetic applications.

Thus, PlantCrystals represent a novel class of nano-milled plant materials (i.e. plant nanocrystals) that have been reduced to sizes below one micrometre, offering a groundbreaking avenue for enhancing the bio-efficacy of botanical ingredients in cosmetic formulations. Composed entirely of plant-derived material at submicron dimensions, PlantCrystals are produced through various wet milling processes that are commonly employed for the generation of drug nanocrystals. This innovative technology exemplifies a physical extraction method that enables the complete disruption of plant cells, resulting in a highly efficient extraction process that can significantly elevate the performance of cosmetic formulations. Thus, the exploration and application of PlantCrystals have the potential to revolutionize the cosmetics industry by providing effective, sustainable, and eco-friendly solutions.

8 PlantCrystals for Cosmetics

The exploration of PlantCrystals derived from various botanical sources, including sumac, black tea, argan, cucumber, and many other plants, has shown promising results in enhancing antioxidant activity and improving skin penetration for cosmetic applications [7, 29, 37, 47]. A study on sumac (*Rhus coriaria L.*) demonstrated that processing its fruit into PlantCrystals using high-pressure homogenization (HPH) reduced the particle size to submicron levels, approximately 750 nm, significantly enhancing its antioxidant capacity (AOC). The oxygen radical absorbance capacity (ORAC) assay revealed an increase in AOC by over 500% compared to bulk sumac, emphasizing the effectiveness of HPH in unlocking the antioxidant potential of this plant [29]. This substantial enhancement suggests that sumac PlantCrystals could be a valuable ingredient in formulations aimed at combating oxidative stress in the skin.

Similarly, studies on black tea waste indicated that the antioxidant activity of black tea waste improved markedly through PlantCrystal technology [37]. The resulting "Nano-Wastes" from black tea demonstrated superior antioxidant activity compared to their bulk counterparts, attributed to the higher extraction efficacy seen in total phenolic content (TPC), total flavonoid content (TFC), and total catechin content (TCC) following the nanomilling process. These findings suggest that utilizing plant waste through nanotechnology not only enhances antioxidant properties but also promotes sustainability in cosmetic formulations.

Overall, the studies on PlantCrystals derived from sumac, black tea, argan, cucumber, and many other plants underscore the potential of nano-milling in improving the functional properties of plant compounds by enhancing antioxidant activity and facilitating better skin penetration. Thus, PlantCrystals in the cosmetic industry appears promising, driven by increasing consumer demand for natural and effective skincare solutions. As research continues to unveil the benefits of utilizing nanosized plant materials, the potential applications of PlantCrystals are expanding. The ability to enhance the bioavailability and efficacy of plant-derived antioxidants positions PlantCrystals as a valuable ingredient in formulations aimed at combating oxidative stress and skin aging.

Recent studies have demonstrated that PlantCrystals can significantly improve the antioxidant capacity of various plant extracts by increasing their solubility and facilitating better skin penetration [7, 47]. This is particularly relevant in the context of the growing trend toward green and sustainable cosmetics, as PlantCrystals can be derived from plants and plant waste, thereby promoting environmental sustainability while maximizing the use of valuable phytochemicals. Furthermore, the incorporation of PlantCrystals into topical formulations can enhance the stability of active ingredients, ensuring that they remain effective over time, which is a critical consideration for consumers seeking reliable skincare products. Moreover, the versatility of PlantCrystals facilitates their integration into a wide array of cosmetic formulations, including creams and gels. The incorporation of PlantCrystals in cosmetics not only enhances the solubility of plant materials and their delivery to the skin but may also augment their protective effects against environmental stressors such as UV radiation and pollution. However, this aspect requires further investigation. As the cosmetic industry continues to innovate, the potential for PlantCrystals to function as multifunctional ingredients—providing both aesthetic and therapeutic benefits—will likely accelerate their adoption in mainstream products. In conclusion, the future of PlantCrystals in cosmetics appears promising, with ongoing research and development paving the way for new formulations that leverage the power of nature. As consumer preferences increasingly shift towards natural and effective solutions, the integration of PlantCrystals into cosmetic products will not only meet these demands but also contribute to a more sustainable and health-conscious beauty industry.

9 The Influence of the Vehicle of the Cosmetic Product on the Nanocrystal Penetration

The effectiveness of nanocrystals in cosmetic formulations is greatly influenced by the choice of vehicles, such as creams, gels, or emulsions. Studies have consistently shown that different vehicles can significantly affect the delivery of active ingredients, particularly in terms of how well API formulated as nanocrystals can be better solubilized in the vehicle and thus deliver higher amounts of API through the skin barrier and reach deeper layers [20]. A critical aspect of vehicle selection is its effect on the penetration of particles into skin and hair follicles, where it has been established that while particle size is crucial for penetration, the formulation itself plays a significant role; certain vehicles can enhance or inhibit the capacity of nanocrystals to penetrate the skin effectively, indicating that the design of cosmetic formulations is vital for optimizing the delivery of active ingredients and maximizing their effects. This incorporation might also lead to the dissolution of a greater amount of API in the aqueous component of a topical vehicle that has a role in improving skin permeation and deposition. Further investigations have examined the permeation parameters of various topical formulations, with studies highlighting the effectiveness of nanocrystals in cream formulations, demonstrating their capacity to penetrate the *stratum corneum* through intracellular channels to deliver active ingredients to the dermis [5]. Another study focused on the penetration kinetics of dexamethasone nanocrystals in skin models indicated that the composition of the vehicle significantly influences the bioavailability of the active ingredient [63]. Key formulation characteristics, such as viscosity and polarity, were found to be crucial in determining the amount of API to be dissolved in the vehicle and thus affect the penetration depth and rate of nanocrystals, especially in conditions where the skin barrier function is compromised [64]. This underscores the importance of tailoring the vehicle to meet the specific needs of both the active ingredient and the target skin condition to enhance cosmeceutical outcomes. Studies investigating the influence of excipient selection on the targeting efficiency of nanocrystals to hair follicles reveal that incorporating 2% ethanol into the formulation significantly enhances targeting while concurrently impeding passive diffusion into the *stratum corneum* [20], suggesting that ethanol may improve the retention of nanocrystals in the hair follicle region, enabling more localized delivery of the active ingredient, likely due to ethanol's ability to modify skin barrier properties and facilitate improved penetration of nanocrystals into hair follicles while restricting the diffusion of the active ingredient away from the targeted site. In contrast, the inclusion of glycerol in the formulation adversely affects hair follicle targeting and encourages passive diffusion of nanocrystals [20], indicating that glycerol, typically employed as a humectant in cosmetic formulations, may not serve as an optimal excipient for enhancing targeted delivery of nanocrystals to hair follicles, as it appears to facilitate the escape of the active ingredient from the designated area, potentially diminishing its efficacy in promoting hair growth or addressing scalp conditions.

Interestingly, propylene glycol has emerged as a multifunctional excipient that promotes both hair follicle targeting and passive diffusion [20], suggesting that it can enhance the overall bioavailability of the active ingredient while effectively targeting hair follicles; its capacity to improve skin penetration and solubilization of active compounds renders propylene glycol a valuable component in formulations intended for dermal and hair follicle applications, emphasizing the critical importance of judicious excipient selection in the formulation of nanocrystals for cosmetic applications. The choice of excipients significantly influences the pharmacokinetics of active ingredients, impacting their absorption, distribution, and overall efficacy, and as the cosmetic industry increasingly adopts nanotechnology, a comprehensive understanding of the interactions between nanocrystals and excipients will be essential for developing effective and targeted formulations. Furthermore, these studies highlight the potential of nanocrystals as a delivery system for poorly soluble APIs in cosmetic products, as optimizing formulations with appropriate excipients can enhance the bioefficacy of active ingredients, leading to improved outcomes for consumers, particularly in dermal and hair care products where targeted delivery to skin and hair follicles can foster skin health, hair growth, and promote scalp health.

In summary, excipients play a crucial role in modulating the efficacy of active ingredients, underscoring the need for careful consideration in formulation development, and as research in this area advances, the potential for nanocrystals to transform cosmetic formulations and improve consumer outcomes is significant, with the integration of nanotechnology into the cosmetic industry promising to enhance product performance and meet consumer demand for effective solutions. Ongoing research confirms that the vehicle used in cosmetic formulations is vital for enhancing the performance of nanocrystals, as the choice of formulation—whether cream or gel—along with specific excipients, greatly influences the bioavailability and efficacy of active ingredients, highlighting the paramount importance of vehicle selection in cosmetic formulations. The relationship between the vehicle's properties and the efficacy of nanocrystals is complex yet essential for achieving desired cosmeceutical effects, and future innovations in formulation technology are likely to focus on optimizing these vehicles to enhance active ingredient delivery, thereby improving the effectiveness of cosmetic products and treatments, as the market for skincare and therapeutic products evolves, making it crucial to understand the intricacies of formulation design to meet consumer expectations and advance cosmeceutical care.

10 Conclusion

Nanocrystals are regarded as a promising platform within the cosmetic industry. Their attributes in dermal applications and cosmetic formulations are varied and provide numerous advantages for enhancing the effectiveness of cosmeceutical active ingredients, particularly antioxidants. The benefits of increased solubility, dissolution rate, adhesiveness, concentration gradient, and bioavailability position nanocrystals as valuable components in cosmetic formulations. Their applications extend to targeting hair follicles, as well as the development of plant-derived nanocrystals, highlighting their versatility and potential within the cosmetic sector. As research progresses, the incorporation of nanocrystals into cosmetic formulations and their interactions with delivery vehicles are likely to expand, offering innovative solutions for improving product performance and addressing consumer demands.

References

1. L. Udayanga, N. Subashini, M. Udugama, P. Silva, T. Ranathunge, Knowledge, perceptions, and consumption behaviour of cosmetics among undergraduates of Sri Lanka: a descriptive cross-sectional study. Front. Public Health **11**, 1184398 (2024). https://doi.org/10.3389/fpubh.2023.1184398
2. G.M. Klein, The nature of skincare: Categorising cosmetics with bioactive ingredients in the case of Quenty-cosmetic. Ambix, 1–11 (2025). https://doi.org/10.1080/00026980.2025.2456889
3. H. Xiang, S. Xu, W. Zhang, Y. Li, Y. Zhou, X. Miao, Skin permeation of curcumin nanocrystals: Effect of particle size, delivery vehicles, and permeation enhancer. Colloids Surf. B: Biointerfaces **224**, 113203 (2023). https://doi.org/10.1016/j.colsurfb.2023.113203
4. A.M. Abraham, Q.K. Anjani, M. Adhami, A.R.J. Hutton, E. Larrañeta, R.F. Donnelly, Novel SmartReservoirs for hydrogel-forming microneedles to improve the transdermal delivery of rifampicin. J. Mater. Chem. B **12**, 4375–4388 (2024). https://doi.org/10.1039/D4TB00110A
5. Y. Liu, J. Zhao, J. Chen, X. Miao, Nanocrystals in cosmetics and cosmeceuticals by topical delivery. Colloids Surf. B: Biointerfaces **227**, 113385 (2023). https://doi.org/10.1016/j.colsurfb.2023.113385
6. O. Pelikh, S.F. Hartmann, A.M. Abraham, C.M. Keck, Nanocrystals for dermal application, in *Nanocosmetics*, ed. by J. Cornier, C.M. Keck, M. Van De Voorde, (Springer International Publishing, Cham, 2019), pp. 161–177. https://doi.org/10.1007/978-3-030-16573-4_8
7. A.M. Abraham, R.M. Alnemari, C. Jacob, C.M. Keck, PlantCrystals – nanosized plant material for improved bioefficacy of medical plants. Materials **13**, 4368 (2020). https://doi.org/10.3390/ma13194368
8. B.R. Thompson, B.C. Zarket, E.H. Lauten, S. Amin, S. Muthukrishnan, S.R. Raghavan, Liposomes entrapped in biopolymer hydrogels can spontaneously release into the external solution. Langmuir **36**, 7268–7276 (2020). https://doi.org/10.1021/acs.langmuir.0c00596

9. L. Vidlářová, G.B. Romero, J. Hanuš, F. Štěpánek, R.H. Müller, Nanocrystals for dermal penetration enhancement – effect of concentration and underlying mechanisms using curcumin as model. Eur. J. Pharm. Biopharm. **104**, 216–225 (2016). https://doi.org/10.1016/j.ejpb.2016.05.004
10. M.B. McGuckin, J. Wang, R. Ghanma, N. Qin, S.D. Palma, R.F. Donnelly, A.J. Paredes, Nanocrystals as a master key to deliver hydrophobic drugs via multiple administration routes. J. Control. Release **345**, 334–353 (2022). https://doi.org/10.1016/j.jconrel.2022.03.012
11. A.J. Paredes, N.M. Camacho, L. Schofs, A. Dib, M.D.P. Zarazaga, N. Litterio, D.A. Allemandi, S. Sánchez Bruni, C. Lanusse, S.D. Palma, Ricobendazole nanocrystals obtained by media milling and spray drying: Pharmacokinetic comparison with the micronized form of the drug. Int. J. Pharm. **585**, 119501 (2020). https://doi.org/10.1016/j.ijpharm.2020.119501
12. K. Mitri, R. Shegokar, S. Gohla, C. Anselmi, R.H. Müller, Lutein nanocrystals as antioxidant formulation for oral and dermal delivery. Int. J. Pharm. **420**, 141–146 (2011). https://doi.org/10.1016/j.ijpharm.2011.08.026
13. S. Mohd-Setapar, C. John, H. Mohd-Nasir, M. Azim, A. Ahmad, M. Alshammari, Application of nanotechnology incorporated with natural ingredients in natural cosmetics. Cosmetics **9**, 110 (2022). https://doi.org/10.3390/cosmetics9060110
14. O. Pelikh, P.-L. Stahr, J. Huang, M. Gerst, P. Scholz, H. Dietrich, N. Geisel, C.M. Keck, Nanocrystals for improved dermal drug delivery. Eur. J. Pharm. Biopharm. **128**, 170–178 (2018). https://doi.org/10.1016/j.ejpb.2018.04.020
15. S. Pyo, M. Meinke, C. Keck, R. Müller, Rutin – increased antioxidant activity and skin penetration by nanocrystal technology (smartCrystals). Cosmetics **3**, 9 (2016). https://doi.org/10.3390/cosmetics3010009
16. D.J. Mendoza, M. Maliha, V.S. Raghuwanshi, C. Browne, L.M.M. Mouterde, G.P. Simon, F. Allais, G. Garnier, Diethyl sinapate-grafted cellulose nanocrystals as nature-inspired UV filters in cosmetic formulations. Mater. Today Bio. **12**, 100126 (2021). https://doi.org/10.1016/j.mtbio.2021.100126
17. F. Awan, M.S. Islam, Y. Ma, C. Yang, Z. Shi, R.M. Berry, K.C. Tam, Cellulose nanocrystal–Zno nanohybrids for controlling photocatalytic activity and UV protection in cosmetic formulation. ACS Omega **3**, 12403–12411 (2018). https://doi.org/10.1021/acsomega.8b01881
18. R. Shegokar, What nanocrystals can offer to cosmetic and dermal formulations, in *Nanobiomaterials in Galenic Formulations and Cosmetics*, (Elsevier, 2016), pp. 69–91. https://doi.org/10.1016/B978-0-323-42868-2.00004-8
19. T.O. Ajala, A. Abraham, C.M. Keck, O.A. Odeku, T.O. Elufioye, J.O. Olopade, Shea butter (Vitellaria paradoxa) and Pentaclethra macrophylla oil as lipids in the formulation of nanostructured lipid carriers. Scient. Afr. **13**, e00965 (2021). https://doi.org/10.1016/j.sciaf.2021.e00965
20. O. Pelikh, C.M. Keck, Hair follicle targeting and dermal drug delivery with curcumin drug nanocrystals – essential influence of excipients. Nano **10**, 2323 (2020). https://doi.org/10.3390/nano10112323
21. L. Peltonen, J. Hirvonen, Drug nanocrystals – Versatile option for formulation of poorly soluble materials. Int. J. Pharm. **537**, 73–83 (2018). https://doi.org/10.1016/j.ijpharm.2017.12.005
22. A. Tuomela, J. Hirvonen, L. Peltonen, Stabilizing agents for drug nanocrystals: Effect on bioavailability. Pharmaceutics **8**, 16 (2016). https://doi.org/10.3390/pharmaceutics8020016
23. F. Fontana, P. Figueiredo, P. Zhang, J.T. Hirvonen, D. Liu, H.A. Santos, Production of pure drug nanocrystals and nano co-crystals by confinement methods. Adv. Drug Deliv. Rev. **131**, 3–21 (2018). https://doi.org/10.1016/j.addr.2018.05.002
24. H. Zhan, X. Zhou, Y. Cao, T. Jagtiani, T.-L. Chang, J.F. Liang, Anti-cancer activity of camptothecin nanocrystals decorated by silver nanoparticles. J. Mater. Chem. B **5**, 2692–2701 (2017). https://doi.org/10.1039/C7TB00134G
25. J.P. Möschwitzer, Drug nanocrystals in the commercial pharmaceutical development process. Int. J. Pharm. **453**, 142–156 (2013). https://doi.org/10.1016/j.ijpharm.2012.09.034

26. S. Tanpichai, S. Witayakran, Y. Srimarut, W. Woraprayote, Y. Malila, Porosity, density and mechanical properties of the paper of steam exploded bamboo microfibers controlled by nanofibrillated cellulose. J. Mater. Res. Technol. **8**, 3612–3622 (2019). https://doi.org/10.1016/j.jmrt.2019.05.024
27. H. Cortés, H. Hernández-Parra, S.A. Bernal-Chávez, M.L.D. Prado-Audelo, I.H. Caballero-Florán, F.V. Borbolla-Jiménez, M. González-Torres, J.J. Magaña, G. Leyva-Gómez, Non-ionic surfactants for stabilization of polymeric nanoparticles for biomedical uses. Materials **14**, 3197 (2021). https://doi.org/10.3390/ma14123197
28. B. Morakul, J. Suksiriworapong, J. Leanpolchareanchai, V.B. Junyaprasert, Precipitation-lyophilization-homogenization (PLH) for preparation of clarithromycin nanocrystals: Influencing factors on physicochemical properties and stability. Int. J. Pharm. **457**, 187–196 (2013). https://doi.org/10.1016/j.ijpharm.2013.09.022
29. A.M. Abraham, C. Quintero, L. Carrillo-Hormaza, E. Osorio, C.M. Keck, Production and characterization of sumac PlantCrystals: Influence of high-pressure homogenization on antioxidant activity of sumac (Rhus coriaria L.). Plants **10**, 1051 (2021). https://doi.org/10.3390/plants10061051
30. M.R. Shah et al., *Lipid-Based Nanocarriers for Drug Delivery and Diagnosis* (Elsevier, Oxford, 2017)
31. E.J. Catlin, O.E. Fandiño, L. Lopez-Vidal, M. Sangalli, R.F. Donnelly, S.D. Palma, A.J. Paredes, A novel temperature-controlled media milling device to produce drug nanocrystals at the laboratory scale. Int. J. Pharm. **666**, 124780 (2024). https://doi.org/10.1016/j.ijpharm.2024.124780
32. M. Liu, C. Hong, G. Li, P. Ma, Y. Xie, The generation of myricetin–nicotinamide nanococrystals by top down and bottom up technologies. Nanotechnology **27**, 395601 (2016). https://doi.org/10.1088/0957-4484/27/39/395601
33. L. Gao, D. Zhang, M. Chen, Drug nanocrystals for the formulation of poorly soluble drugs and its application as a potential drug delivery system. J. Nanopart. Res. **10**, 845–862 (2008). https://doi.org/10.1007/s11051-008-9357-4
34. Z. Ma, F. Gu, X. Wang, G. Xiao, Y. Jiang, J. Du, L. He, Y. Yu, X. Song, C. Xu, Preparation of stable liquid disperse dye using the antisolvent precipitation method for enhancing dyeability and eco-friendliness. Dyes Pigments **236**, 112655 (2025). https://doi.org/10.1016/j.dyepig.2025.112655
35. D. Thassu, M. Deleers, Y. Pathak, *Nanoparticulate Drug Delivery Systems* (Informa Healthcare, New York, 2007)
36. G.G. Liversidge, K.C. Cundy, Particle size reduction for improvement of oral bioavailability of hydrophobic drugs: I. Absolute oral bioavailability of nanocrystalline danazol in beagle dogs. Int. J. Pharm. **125**, 91–97 (1995). https://doi.org/10.1016/0378-5173(95)00122-Y
37. A.M. Abraham, R.M. Alnemari, J. Brüßler, C.M. Keck, Improved antioxidant capacity of black tea waste utilizing PlantCrystals. Molecules **26**, 592 (2021). https://doi.org/10.3390/molecules26030592
38. N. Qin, M. Li, L.K. Vora, K. Peng, A.H.B. Sabri, Y. Tao, A.J. Paredes, H.O. McCarthy, R.F. Donnelly, Enhanced long-acting simvastatin delivery via effervescent powder-carrying hollow microneedles and nanocrystal-loaded microneedles. Int. J. Pharm. **665**, 124691 (2024). https://doi.org/10.1016/j.ijpharm.2024.124691
39. L. Padrela, M.A. Rodrigues, A. Duarte, A.M.A. Dias, M.E.M. Braga, H.C. De Sousa, Supercritical carbon dioxide-based technologies for the production of drug nanoparticles/nanocrystals – a comprehensive review. Adv. Drug Deliv. Rev. **131**, 22–78 (2018). https://doi.org/10.1016/j.addr.2018.07.010
40. J. Hecq, M. Deleers, D. Fanara, H. Vranckx, K. Amighi, Preparation and characterization of nanocrystals for solubility and dissolution rate enhancement of nifedipine. Int. J. Pharm. **299**, 167–177 (2005). https://doi.org/10.1016/j.ijpharm.2005.05.014

41. J. Udabe, S. Martin-Saldaña, Y. Tao, M. Picchio, A. Beloqui, A.J. Paredes, M. Calderón, Unveiling the potential of surface polymerized drug nanocrystals in targeted delivery. ACS Appl. Mater. Interfaces **16**, 47124–47136 (2024). https://doi.org/10.1021/acsami.4c07669
42. E. Jakubowska, J. Lulek, The application of freeze-drying as a production method of drug nanocrystals and solid dispersions – a review. J. Drug Deliv. Sci. Technol. **62**, 102357 (2021). https://doi.org/10.1016/j.jddst.2021.102357
43. M. Colombo, S. Staufenbiel, E. Rühl, R. Bodmeier, In situ determination of the saturation solubility of nanocrystals of poorly soluble drugs for dermal application. Int. J. Pharm. **521**, 156–166 (2017). https://doi.org/10.1016/j.ijpharm.2017.02.030
44. Q. Fu, B. Li, D. Zhang, M. Fang, J. Shao, M. Guo, Z. Guo, M. Li, J. Sun, Y. Zhai, Comparative studies of the in vitro dissolution and in vivo pharmacokinetics for different formulation strategies (solid dispersion, micronization, and nanocrystals) for poorly water-soluble drugs: A case study for lacidipine. Colloids Surf. B: Biointerfaces **132**, 171–176 (2015). https://doi.org/10.1016/j.colsurfb.2015.05.010
45. C. Oliveira, C. Coelho, J.A. Teixeira, P. Ferreira-Santos, C.M. Botelho, Nanocarriers as active ingredients enhancers in the cosmetic industry – the european and north america regulation challenges. Molecules **27**, 1669 (2022). https://doi.org/10.3390/molecules27051669
46. M.R. Gigliobianco, C. Casadidio, R. Censi, P. Di Martino, Nanocrystals of poorly soluble drugs: Drug bioavailability and physicochemical sttability. Pharmaceutics **10**, 134 (2018). https://doi.org/10.3390/pharmaceutics10030134
47. A.M. Abraham, S. Wiemann, G. Ambreen, J. Zhou, K. Engelhardt, J. Brüßler, U. Bakowsky, S.-M. Li, R. Mandic, G. Pocsfalvi, C.M. Keck, Cucumber-derived exosome-like vesicles and PlantCrystals for improved dermal drug delivery. Pharmaceutics **14**, 476 (2022). https://doi.org/10.3390/pharmaceutics14030476
48. V.B. Junyaprasert, B. Morakul, Nanocrystals for enhancement of oral bioavailability of poorly water-soluble drugs. Asian J. Pharm. Sci. **10**, 13–23 (2015). https://doi.org/10.1016/j.ajps.2014.08.005
49. D.T. Gadhiya, J.K. Patel, A.A. Bagada, An impact of nanocrystals on dissolution rate of Lercanidipine: Supersaturation and crystallization by addition of solvent to antisolvent. Futur. J. Pharm. Sci. **7**, 128 (2021). https://doi.org/10.1186/s43094-021-00271-x
50. D.A. Yassin, M.J. Nasim, A.M. Abraham, C.M. Keck, C. Jacob, Upcycling culinary organic waste: Production of plant particles from potato and carrot peels to improve antioxidative capacity. CNT **2**, 62–70 (2021). https://doi.org/10.2174/2665978601999200925163905
51. P. Panchal, T.H. Mekonnen, Tailored cellulose nanocrystals as a functional ultraviolet absorbing nanofiller of epoxy polymers. Nanoscale Adv. **1**, 2612–2623 (2019). https://doi.org/10.1039/C9NA00265K
52. G. Bjørklund, M. Shanaida, R. Lysiuk, M. Butnariu, M. Peana, I. Sarac, O. Strus, K. Smetanina, S. Chirumbolo, Natural compounds and products from an anti-aging perspective. Molecules **27**, 7084 (2022). https://doi.org/10.3390/molecules27207084
53. J. Li, W. Ni, M. Aisha, J. Zhang, M. Sun, A rutin nanocrystal gel as an effective dermal delivery system for enhanced anti-photoaging application. Drug Dev. Ind. Pharm. **47**, 429–439 (2021). https://doi.org/10.1080/03639045.2021.1890113
54. N.G. Sahoo, M. Kakran, L.A. Shaal, L. Li, R.H. Müller, M. Pal, L.P. Tan, Preparation and characterization of quercetin nanocrystals. J. Pharm. Sci. **100**, 2379–2390 (2011). https://doi.org/10.1002/jps.22446
55. T.P. Pivetta, L.B. Silva, C.M. Kawakami, M.M. Araújo, M.P.F.M. Del Lama, R.M.Z.G. Naal, S.S. Maria-Engler, L.R. Gaspar, P.D. Marcato, Topical formulation of quercetin encapsulated in natural lipid nanocarriers: Evaluation of biological properties and phototoxic effect. J. Drug Deliv. Sci. Technol. **53**, 101148 (2019). https://doi.org/10.1016/j.jddst.2019.101148
56. J. Wang, X. Xue, X. Miao, Antioxidant effects of quercetin nanocrystals in nanosuspension against hydrogen peroxide-induced oxidative stress in a zebrafish model. Pharmaceuticals **16**, 1209 (2023). https://doi.org/10.3390/ph16091209

57. A.S. Hassan, G.M. Soliman, Rutin nanocrystals with enhanced anti-inflammatory activity: preparation and ex vivo/in vivo evaluation in an inflammatory rat model. Pharmaceutics **14**, 2727 (2022). https://doi.org/10.3390/pharmaceutics14122727
58. M.J. Abla, A.K. Banga, Quantification of skin penetration of antioxidants of varying lipophilicity. Int. J. Cosmetic Sci. **35**, 19–26 (2013). https://doi.org/10.1111/j.1468-2494.2012.00728.x
59. M. Correia, J. Lopes, D. Lopes, A. Melero, P. Makvandi, F. Veiga, J.F.J. Coelho, A.C. Fonseca, A.C. Paiva-Santos, Nanotechnology-based techniques for hair follicle regeneration. Biomaterials **302**, 122348 (2023). https://doi.org/10.1016/j.biomaterials.2023.122348
60. Y. Gu, Q. Bian, Y. Zhou, Q. Huang, J. Gao, Hair follicle-targeting drug delivery strategies for the management of hair follicle-associated disorders. Asian J Pharm Sci **17**, 333–352 (2022). https://doi.org/10.1016/j.ajps.2022.04.003
61. N. Chaachouay, L. Zidane, Plant-derived natural products: A source for drug discovery and development. DDC **3**, 184–207 (2024). https://doi.org/10.3390/ddc3010011
62. K. Mansoor, T. Aburjai, F. Al-Mamoori, M. Schmidt, Plants with cosmetic uses. Phytother. Res. **37**, 5755–5768 (2023). https://doi.org/10.1002/ptr.8019
63. S.B. Lohan, S. Saeidpour, M. Colombo, S. Staufenbiel, M. Unbehauen, A. Wolde-Kidan, R.R. Netz, R. Bodmeier, R. Haag, C. Teutloff, R. Bittl, M.C. Meinke, Nanocrystals for improved drug delivery of dexamethasone in skin investigated by EPR spectroscopy. Pharmaceutics **12**, 400 (2020). https://doi.org/10.3390/pharmaceutics12050400
64. A.S. Alnaim, Nanocrystals in dermal drug delivery: A breakthrough for enhanced skin penetration and targeted skin disorder treatments. Pharmaceutics **16**, 1561 (2024). https://doi.org/10.3390/pharmaceutics16121561

Index

A. J. Paredes (ed.), *Nanocrystals*, https://doi.org/10.1007/978-3-032-10238-6

MIX
Papier aus verantwortungsvollen Quellen
Paper from responsible sources
FSC® C105338

If you have any concerns about our products,
you can contact us on
ProductSafety@springernature.com

In case Publisher is established outside the EU,
the EU authorized representative is:
Springer Nature Customer Service Center GmbH
Europaplatz 3, 69115 Heidelberg, Germany

Printed by Libri Plureos GmbH
in Hamburg, Germany